# Toughened Plastics

ADVANCES IN CHEMISTRY SERIES **233**

# Toughened Plastics I

## Science and Engineering

**C. Keith Riew,** EDITOR
*The BFGoodrich Company*

**Anthony J. Kinloch,** EDITOR
*University of London*

Developed from a symposium sponsored
by the Division of Polymeric Materials:
Science and Engineering
at the 200th National Meeting
of the American Chemical Society,
Washington, D.C.

American Chemical Society, Washington, DC 1993

**Library of Congress Cataloging-in-Publication Data**

Toughened plastics I: science and engineering / C. Keith Riew,
　Anthony J. Kinloch, editors.
　　p.　　cm.—(Advances in chemistry series, ISSN 0065-2393; 233)
　　"Developed from a symposium sponsored by the Division of Polymeric
Materials: Science and Engineering at the 200th National Meeting of
the American Chemical Society, Washington, D.C., August 26-31,
1990."
　　Includes bibliographical references and index.

　　ISBN 0-8412-2500-1
　　1. Plastics—Additives.　2. Elastomers.

　　I. Riew, C. Keith.　II. Kinloch, A. J.　III. American Chemical Society.
Division of Polymeric Materials: Science and Engineering.
　　IV. American Chemical Society. Meeting (200th : 1990 : Washington, D.C.)
　　V. Title: Toughened plastics one.　VI. Series.

QD1.A355 no. 233
[TP1142]
540 s—dc20
[668.4]　　　　　　　　　　　　　　　　　　　　　　　　　　　　　　　92-34578
　　　　　　　　　　　　　　　　　　　　　　　　　　　　　　　　　　　　CIP

The paper used in this publication meets the minimum requirements of American National
Standard for Information Sciences—Permanence of Paper for Printed Library Materials, ANSI
Z39.48-1984.　　　　　　　　　　　　　　　　　　　　　　　　　　　　　　$\infty$

# FOREWORD

The ADVANCES IN CHEMISTRY SERIES was founded in 1949 by the American Chemical Society as an outlet for symposia and collections of data in special areas of topical interest that could not be accommodated in the Society's journals. It provides a medium for symposia that would otherwise be fragmented because their papers would be distributed among several journals or not published at all.

Papers are reviewed critically according to ACS editorial standards and receive the careful attention and processing characteristic of ACS publications. Volumes in the ADVANCES IN CHEMISTRY SERIES maintain the integrity of the symposia on which they are based; however, verbatim reproductions of previously published papers are not accepted. Papers may include reports of research as well as reviews, because symposia may embrace both types of presentation.

# DEDICATION

## A Tribute
## to Professor Willard D. "Bill" Bascom, 1931–1991

The field of rubber-toughened epoxies suffered a great loss on October 8, 1991, with the death of one of its pioneers, Willard D. "Bill" Bascom, Research Professor, Materials Science and Engineering Department, University of Utah, Salt Lake City, Utah.

Dr. Bascom received his B.S. in chemistry from the Worchester Polytechnic Institute in 1953 and his M.S. in physical chemistry from Georgetown University in 1962. In 1970 he received his Ph.D. in physical chemistry from Catholic University Graduate School. He was the head of the Adhesives and Polymer Composite Section, Chemistry Division, Naval Research Laboratory, Washington, D.C.

After making impressive contributions to surface chemistry for 15 years (1956–1981) at the Naval Research Laboratory, he turned his attention to the toughness problem in structural adhesive systems. Despite the importance of the toughening of brittle structural materials, very few researchers had followed up on the initial publications by the colleagues of Frederick J. McGarry at Massachusetts Institute of Technology in the late 1960s.

In the early 1970s Dr. Bascom began a major investigation into the science and engineering of toughening epoxy resins. He was interested in finding failure mechanisms and in particular looked for a relationship between the toughness of bulk resins and bond geometry in adhesive joints. His reports on the dependence of the fracture energy of adhesive joints on temperature and bond thickness gave an impetus to the scientific investigation of fracture behavior in rubber-toughened thermoset resins. His studies on fractography using scanning electron microscopy identified the deformation processes at the crack tip. His observations are the essence of the currently accepted failure mechanisms of epoxy-resin toughening.

In 1976 Dr. Bascom extended his studies to the toughening of epoxy matrix resins for fiber-reinforced composites. Delamination was beginning to be recognized as a crucial cause of composite failure. He established the first clear link between increased matrix-resin toughness and improved interlaminar fracture behavior in composites. This important study led to establishment of new failure mechanisms in composites and to development of new technologies for improved high-performance fiber-reinforced composites.

He continued his study in this field when he became the manager of composite research, Graphite Fibers Department, Hercules, Inc., Magna, Utah, in 1981. This position was followed by a move to the University of Utah in 1986. Dr. Bascom demonstrated the importance of fiber–matrix adhesion in delamination and thus contributed to the improvement of surface treatments for fibers in thermoplastic composites. He established a better understanding of test methods to measure interfacial strength and the composite failure mechanism. In 1989 he received the Adhesives Award sponsored by *Adhesives Age* and was selected by ASTM Committee D-14 on Adhesives for his contributions on the fracture behavior of rubber-toughened epoxy structural adhesives in bulk and in adhesive joints.

To many of us in this field, Professor Bascom was a valued colleague, co-worker, and leader. His work opened up an exciting field of study. His ability to analyze complex problems and find simple solutions brought the science and engineering of toughening plastics to life. Most importantly, however, he was a friend who was always there when we needed him. He will be remembered as a man of keen intelligence and dedication to science. He will be greatly missed by all of us.

DONALD L. HUNSTON
National Institute of Standards and Technology
Polymers Division
Gaithersburg, MD 20899

C. KEITH RIEW
Department of Chemical Engineering
University of Akron
Akron, OH 44325-3906

August 1992

# ABOUT THE EDITORS

C. KEITH RIEW is a Research Professor at the Department of Chemical Engineering, College of Engineering, University of Akron. He has retired from the Corporate Research Division, The BFGoodrich Company, Brecksville, Ohio, 44141, where he was a Research Fellow. He has served as research scientist or manager for New Products Research at the BFGoodrich Company for more than 25 years. He received the M.S. and Ph.D. degrees in organic chemistry from Wayne State University, Detroit, Michigan, and B.S. in chemistry from Seoul National University.

His latest research involves the synthesis, characterization, and engineering of new tougheners for plastics. Other research interests include correlating polymer chemistry, morphology, and structure–property relationship to the failure mechanisms of toughened plastics. He has published over 60 papers and patents on emulsion polymers, hydrophilic polymers, telechelic polymers, and modified or toughened plastics as matrix resins for adhesives and composites.

He organized or co-organized three international symposiums at national meetings of the American Chemical Society, Division of Polymeric Materials: Science and Engineering, Inc., and co-edited or edited the related books: *Rubber-Modified Thermoset Resins* (Advances in Chemistry Series 208) and *Rubber-Toughened Plastics* (Advances in Chemistry Series 222).

ANTHONY J. KINLOCH has a personal chair as Professor of Adhesion in the Department of Mechanical Engineering at Imperial College of Science, Technology and Medicine, which is part of the University of London.

He obtained his Ph.D. in Materials Science in 1972 from Queen Mary College. He leads the Engineering Adhesives and Composites Research Group at Imperial College and has published over 100 patents and papers in the areas of adhesion and adhesives, toughened polymers, and the fracture of polymers and fiber composites. He is co-author of a book on the *Fracture Behavior of Polymers* and author of a book on *Adhesion and Adhesives*.

# CONTENTS

# PREFACE

Toughened Plastics are widely used in many diverse industries and form the basis for engineering plastics, structural adhesives, and matrices for fiber-composite materials. The materials are usually multiphase polymers; the dispersed phase consists of rubbery or thermoplastic domains and the continuous phase is a cross-linked thermosetting or thermoplastic polymer matrix. The basic reason for toughening a plastic that is normally brittle at room temperature is to improve its crack resistance and toughness without significantly decreasing other important properties such as the load-bearing modulus and the mechanical properties at elevated temperatures.

This book is aptly named; its chapters encompass many areas of science and engineering. Various contributions discuss aspects of the toughening of plastics such as the basic chemistry involved, the microstructure of the materials, the resulting mechanical properties, and applications of the toughened plastics. Hence, as would be expected, the authors represent a truly multidisciplinary group. Some of these authors are the foremost authorities on their particular areas within the subject of toughened plastics.

A symposium on "Rubber-Modified Thermoset Resins" was held in 1983 under the auspices of the American Chemical Society in Washington, D.C. The papers were published in 1984 as Volume 208 of the Advances in Chemistry Series. The theme of the 1988 symposium, titled "Rubber-Toughened Plastics," was broadened to cover both thermosets and thermoplastics. The papers from this symposium, held in New Orleans, Louisiana, were published in 1989 as Volume 222 of the Advances in Chemistry Series.

Most of the chapters in this book are based on papers presented at The Third International Symposium on "Toughened Plastics: Science and Engineering" and the workshop on "Toughened Plastics" at the 200th National Meeting, American Chemical Society, Division of Polymeric Materials: Science and Engineering, Inc., Washington, D.C., August 1990.

We are indebted to each contributor to the book and to the referees who provided much valuable advice. We thank the Division of Polymeric Materials: Science and Engineering, Inc. of the American Chemical Society for sponsoring the workshop, the symposium, and this book. The BFGoodrich Company and Imperial College, University of London, also provided administrative assistance. Finally, we thank Mrs. H. Kim Riew for her secretarial contributions to this book and the 1990 workshop in Washington, D.C., Mrs. Gill Kinloch, and our families for their support and patience.

C. KEITH RIEW
Department of Chemical Engineering
The University of Akron
Akron, Ohio 44325–3906

A. J. KINLOCH
Imperial College
University of London
SW7 2BX, United Kingdom

August, 1992

# 1

# Mechanisms of Toughening Thermoset Resins

Y. Huang[1], D. L. Hunston[2], Anthony J. Kinloch[1], and C. Keith Riew[3]*

[1]Imperial College of Science, Technology, and Medicine,
University of London, Department of Mechanical Engineering,
Exhibition Road, London, SW7 2BX, United Kingdom
[2]National Institute of Standards and Technology, Polymers Division,
Gaithersburg, MD 20899
[3]The BFGoodrich Company, Corporate Research Division,
9921 Brecksville Road, Brecksville, OH 44141

*A toughened thermoset generally contains elastic or thermoplastic domains dispersed in discrete form throughout the matrix resin to increase the resistance to crack-growth initiation. Fracture behavior is governed by characteristics of viscoelasticity, shear yielding, and dilatational deformation involving cavitation. In recent years, concepts of theoretical developments of failure mechanisms based on fracture mechanics have been advanced, although controversy over details still exists. The aim of this chapter is to review in detail some of the recent developments in toughening mechanisms and the relationships between microstructure and fracture behavior as illustrated by rubber-toughened epoxy resins. This review considers the behavior of bulk resin, adhesive joints, and matrices of composites. Recent modeling efforts for toughening mechanisms that give a quantitative description of the microstructure–fracture property relationships are also reviewed.*

MATERIALS SCIENTISTS ARE CONCERNED WITH QUANTITATIVE MECHANISMS of toughening and hope to predict the life expectancy of toughened thermosets. This endeavor should enable chemists to develop new polymers or tailor existing polymers to meet new requirements.

*Corresponding author.

0065–2393/93/0233–0001$09.75/0
© 1993 American Chemical Society

Unmodified thermosets are usually single-phase and brittle materials, whereas toughened thermosets are usually multiphase systems. When elastic or thermoplastic domains are correctly dispersed in discrete forms throughout the thermoset matrix, the fracture energy or toughness can be greatly increased. In toughening processes, it is desirable to minimize the sacrifice in thermal and strength properties of the matrix resin. For this reason, the toughener is incorporated by in situ reaction or by adding performed particles as a separate phase. Deformation in the toughened thermoplastic matrix is normally dominated by shear yielding and crazing. Unfortunately, crosslinked materials such as cured epoxy resins have a limited ability to deform in this way, particularly in the triaxial stress field present inside the sample at the crack tip. The addition of an elastomeric or thermoplastic second phase changes this situation and significantly improves the fracture toughness. The fracture behavior of toughened thermosets is dominated by shear yielding and dilatational deformation involving cavitation. Many different mechanisms have been proposed to explain the improved fracture toughness. The purpose of this review is to discuss some of the recent developments in this area.

## Toughening Mechanisms

**General Overview.**  Impact modification of thermosets or thermoplastics has been practiced widely in industry, usually by blending a plastic with an elastomer or a plasticizer (*1*). Such impact modification often sacrifices the inherent and desirable thermal or strength properties of the matrix resin. A typical example is when polystyrene is modified with polybutadiene and the ultimate tensile stress is reduced from ∼ 45 MPa (unmodified) to 15 MPa for the high-impact polystyrene (HIPS) (*2*). To impart ductility or increased impact strength to cured epoxy resin systems, reactive (flexibilizers) or nonreactive (plasticizers) long-chain epoxidized glycols and dimer acids have been added at various levels (*3*). However, most of these modifiers flexibilize the epoxy resins, that is, the heat deflection temperatures and strength properties were reduced with increased strain properties. Because these materials also have relatively high glass-transition temperatures (> 0 °C), toughness improvement at low temperature (< −20 °C) was not realized.

In 1965, McGarry and his colleagues (*4–7*) began toughening thermoset epoxy or polyester resins with difunctional reactive liquid polymers. Their first published results were on toughening liquid epoxy resin (diglycidyl ether of bisphenol A; DGEBA) or unsaturated polyester resin with low-molecular-weight liquid carboxyl-terminated poly(butadiene–acrylonitrile) copolymers (Hycar CTBN; BFGoodrich Company).[1] The reaction of the epoxy groups of the resin with the carboxyl-terminal groups of the CTBN produced alternating block copolymers of DGEBA–CTBN that eventually precipitated as rubbery domains, in situ, during cure (8, 9). The glass-transition temperature of CTBN is below $-30$ °C. Thus, the cured CTBN-toughened epoxy resins also showed improved low-temperature fracture toughness.

McGarry and co-workers compared the chemical bond energy to the fracture energy of CTBN-toughened epoxy resin systems. For example, the work at room temperature required to propagate a crack through brittle cross-linked DGEBA resin is approximately $5–9 \times 10^4$ mJ/m$^2$ of fracture surface produced. The discrepancy between the measured value and calculated values based on carbon–carbon bond breakage (about $5 \times 10^2$ mJ/m$^2$) is caused by localized shear yielding and plastic flow in the material adjacent to the fracture surfaces. This discrepancy can be increased by suspending fine particles of elastomer throughout the cured epoxy matrix. The particles increase the amount and extent of shear yielding and induce other toughening mechanisms that take place in the glassy resin phase, without significant degradation of other mechanical properites of the cross-linked epoxy. The CTBN was an effective toughener in concentrations up to 10 parts per hundred parts by weight of the DGEBA resin. Poor toughening was observed when the average diameter of the particles was $< 1000$ Å, when adhesion between particles and matrix was inadequate, or when the elastomer was miscible and remained in solid solution in the cross-linked epoxy. The particle size was dependent on the initial molecular weight of the elastomer, the nature and concentration of the epoxy cross-linking agent, and the temperature at which the mixture was polymerized. With a fixed content of precipitated rubber domains, as the particle size increased to about 2 to 5 $\mu$m, the crack propagation resistance also increased by five- to tenfold over unmodified epoxy resin.

The major use of rubber-modified (flexibilized or toughened) epoxies in the 1960s was structural adhesives. Researchers were interested in the use of such materials in adhesive joints. Bascom et al. (*10–16*), who were the major contributors to studies of the fracture behavior of CTBN-modified epoxies in adhesive bonds, studied many of the variables that affect adhesive bond fracture to see how the improved fracture of neat (bulk) resin would behave in adhesive joints. The efforts of Bascom et al. included investigations of the basic viscoelastic properties of CTBN-toughened epoxy systems and their fracture behavior in different geometries, that is, bulk, adhesive joints, and matrices of composites. An overview on this subject will be given in the last part of this chapter.

Most of the early publications on toughening technologies or mechanisms drew heavily on the studies of toughened thermoplastics, which were examined extensively (*17–22*). Recent monographs have been dominated by

[1]Certain commercial materials and equipment are identified in this paper in order to specify adequately the experimental procedure. In no case does such identification imply recommendation or endorsement by the authors and their institutions nor does it imply necessarily that they are the best available for the purpose.

developments in the technologies and mechanisms for toughened thermosets (23–30), and these monographs provide a comprehensive description of the science and engineering of toughened plastics.

Early proposals for the toughening mechanisms were based mainly on the phenomena observed in toughened thermoplastics. For example, in 1956, Merz et al. (31, 32) proposed the first mechanism of ultimate failure for heterogeneous mixtures of polystyrene with styrene–butadiene copolymers. The blend was a rubbber-toughened high-impact polystyrene (HIPS) system. Merz et al. postulated that the large strain deformation and the stress-whitening were caused by scattered light from the microcracks and void formation. The basic theory of toughening proposed by Merz et al. was that the rubber particles bridge the opening fracture surfaces at the crack tip, and fracture now requires the tearing of rubber particles.

A similar explanation was later advanced by Kunz-Douglass et al. (33, 34), and Sayer et al. (35), who proposed a tear-energy toughening model for the toughening of thermosets. They suggest that the particles stretch across the crack opening behind the crack tip and hinder the advance of the crack. Based on this idea, they proposed the first quantitative model for toughened thermoset. In this model, the energy absorbed in fracture is the sum of the energy to fracture the matrix and to break the rubber particles. Microcracks due to rubber particles cause tensile yielding and, thus, a large tensile deformation. Voids resulted from opening of the microcracks and permitted large strains.

Kinloch et al. (36) disagreed with the rubber-tear theory because many phenomena, such as stress-whitening, the large amount of plastic deformations, higher fracture toughness at a higher temperature, and so forth, could not be explained. Kinloch et al. suggested that rubber-tear makes a secondary contribution to toughening, but it does not represent the major toughening mechanism. They proposed a mechanism that involves dilatational deformation of the matrix and cavitation of the rubber particles in response to the triaxial stresses near the crack tip, combined with shear yielding between the holes formed by the cavitated rubber particles. The stress-whitening was attributed to light scattering by these holes, and the major energy absorption mechanism was suggested to be the plastic deformation of the matrix. Plastic deformation blunts the crack tip, which reduces the local stress concentration and allows the material to support higher loads before failure occurs. More details of this mechanism will be given later.

The rubber-tear theory also fails to explain recent findings on how some systems may be toughened with nonreactive rigid thermoplastic particles. Indications are that shear yielding and plastic deformation involving cavitation of the matrix surrounding the particles is responsible for the toughening, but not deformation of the particles themselves.

The first report on thermoplastic rigid-particle toughening of epoxy resin was attempted by Raghava (37), who did not obtain large toughness improvement with the hard polyimide particles in the softer epoxy matrix.

Bucknall and Gilbert (38) toughened tetrafunctional epoxy resins using nonreactive thermoplastic poly(ether imide). Toughened epoxy resins were prepared by dissolving a poly(ether imide) (Ultem 1000; General Electric Company) in a tetraglycidyl-4,4'-diaminodiphenyl methane-based resin with 4,4'-diaminodiphenyl sulfone as curing agent at 30 parts per hundred parts of the resin. The poly(ether imide), which had enough incompatibility because of a sufficient surface energy difference from the base epoxy resin, formed a separate phase, with a dynamic loss peak at 200–212 °C. The loss peak of the epoxy occurred at about 265 °C. Three-point bend fracture tests showed a linear increase in the fracture toughness with increasing poly(ether imide) content. The Young's modulus at 23 °C showed a modest reduction from 3.6 to 3.5 GPa when the poly(ether imide) content was increased from 0 to 25%.

A similar study was made by Riew and Smith (39), who toughened polycarbonate with preformed rigid particles using core-shell polymers. They found that thermoplastic polycarbonates deform through shear yielding with or without the presence of the discrete second phase. However, the core of the particles is rubbery and the shell is a rigid plastic phase. Thus, the inner rubbery part may be deformable within the outer plastic shell. SEM micrographs indicated that the toughness was enhanced by simultaneous shear yielding and apparent dilatational deformation involving cavitation. Riew and Smith did not, however, find any evidence of crazing. Cavities are formed in the martrix surrounding the core-shell domains within the plastic zone ahead of the crack tip.

Yee (40) studied the fracture behavior and toughening of nylon 66 matrix with rigid particles of polyphenylene oxide modified with styrene–butadiene–styrene polymer (GTX 910; General Electric Company). This blend is a rigid–rigid polymer alloy. Yee found that the rigid particles must be able to induce localized dilatation in order to increase the toughness of the alloy and that the matrix is capable of some plasticity. The role of cavitation of the toughening particles and the stress concentration effect of a particle near a crack tip also are important in Yee's toughening mechanisms.

Unlike the work of Raghava, the results presented by Bucknall and Gilbert (38), Riew and Smith (39), and Yee (40) showed improvements in toughness with the rigid particles. However, these particles must be capable of plastic deformation, or at least of low modulus, like soft plastics or rigid rubbers, to be effective tougheners.

Particles play an important role in helping the matrix respond to the triaxial stresses present near the crack tip in all but very thin samples. The usual response of an unmodified thermoset is to readily initiate cracks, which cause brittle failure. Thermoplastics, on the other hand, generally craze. Sauer et al. (41) and Hsial and Sauer (42) made a distinction between cracks (microvoids) and crazes. They found that the polystyrene sustained tensile stresses of 20 MN/m$^2$ even when the specimens appeared to be cracked. Examination of the X-ray diffraction of the cracked (crazed) polystyrene showed oriented polymer molecules. Sauer and co-workers concluded that the molecules would absorb energy during the orientation, and the process must be responsible for the observed high strength.

Kambour (18, 43–47) made a major contribution in identifying the crazes in glassy polymers. He observed that the crazing almost always occurred before the fracture of glassy plastics. He characterized the structure and properties of crazes and has published excellent papers on the crazing and fracture behavior of plastics.

Although a craze may be better than a microvoid for supporting a load, a single craze does not provide a major toughening mechanism. Bucknall (48) advanced a multiple crazing theory for toughening in thermoplastics and suggested that the rubber particles both initiate and control the craze growth. This initiation–control mechanism is the most widely accepted explanation for toughening in high-impact polystyrene.

Wang et al. (49–51) and Matsuoka and Daane (52) stated that the toughener particles should do two things to the parent matrix phase: act as stress concentrators (i.e., a large strain will start in the matrix near the interface) and create a multiaxial stress state. Because a multiaxial stress state near the interface further enhances dilatation, the matrix will undergo plastic deformation in the vicinities of the rubber particles.

Wang et al. (49–51) also studied the crazing behavior of polystyrene with spherical rubber or steel inclusions. In rubber-toughened polystyrene the crazing initiated from the interface near the equator of the "rubber" ball at a critical tensile load, and more crazes propagated in the radial direction as the stress increased. The "steel ball," however, initiated crazes on the shoulder at an angle of 37.2° from the equatorial plane and the crazes then propagated in a perpendicular direction to the stress direction. This difference in crazing behavior indicated some differences in deformation of the matrix due to the rigidity of the toughening particles. The stress required to develop interface crazing for rubber is lower than for the steel ball. The crazes in the rubber ball sample were also more stable that those in the steel ball sample, as evidenced by the larger breaking stress vs. craze initiation stress ratio in the rubber ball case.

The preceding examples illustrate the strong influence of the second phase on matrix deformation in thermoplastics. A similar effect is found with thermoset matrices, although the dilatational deformations are different.

Lange et al. (53–55) proposed a different toughening mechanism, that is, the crack-pinning mechanism, which is quite different from the mechanisms of crazing, cavitation, or the shear yielding. They observed increased fracture toughness when aluminum trihydrate was added as a fire-retarding filler to epoxy resins. The toughness improvement was dependent on the volume fraction and particle size of the filler. The increase in fracture energy of a brittle material due to the addition of a brittle and second phase was explained by interactions between the propagating crack and the filler phase. Scanning electron microscopy was used to develop a model that stated that as the crack begins to propagate through the resin, the crack front bows out between the filler particles but remains pinned at the particles. The crack-pinning mechanism operates mainly with inorganic fillers that resist fracture during failure of the epoxy matrix resin. The crack-pinning mechanism is generally less important in ductile matrix materials (56). Although thermosets can be toughened by a crack-pinning mechanism, this is generally less effective than the mechnisms present with soft or ductile particles.

Unmodified epoxy resins cured with alkyl diamines can be made to fail by shear yielding under tensile stress at slow strain rate with no observable stress-whitening. In the case of rubber-toughened epoxies, Rowe and Riew reported shear yielding can occur under tensile stresses in thermoset resins with or without rubber domains (7, 57, 58). Tougheners that are more compatible with epoxy resins, such as CTBN with 27% acrylonitrile content or liquid mercaptan-terminated poly(butadiene–acrylonitrile) copolymers (MTBN), produce particle sizes less than 0.1 $\mu$m. When the rubbery domain sizes are less than 0.1 $\mu$m, shear yielding without dilatational caviatation was observed. In larger rubbery domains with sizes ranging from 1 to 5 $\mu$m, the toughened epoxy resin exhibited extensive cavitation in the rubber particles and surrounding matrix, as evidenced by the large amount of stress-whitening. In general, the rubber would shrink more than the epoxy during cooling after cure, but is prevented from doing so by the chemical bonding between the particle and the matrix. The good adhesion, often due to actual chemical covalent bonding between the rubbery domains and matrix resin, stabilizes the cavity-formation or shear-deformation processes, which in turn are capable of supporting large hydrostatic triaxial stresses ahead of the crack tip in the plastic zone (4, 9). As a result, there is triaxial tension in the rubber. When the sample is loaded, these stresses increase to the point where the particle cavitates.

When the CTBN-toughened epoxy is prepared in the presence of slightly less than a stoichiometric amount of bisphenol A (which like CTBN is a chain extender), a higher molecular weight between cross-links is produced in situ (9) and a bimodal distribution of particle sizes is generated; that is, one family of 1–3-$\mu$m particles and another of less than 0.05 $\mu$m particles (57). For a given amount of added CTBN, this system has a dramatically higher fracture toughness than other formulations (9). A number of explanations have been offered for this. One hypothesis suggests that the small and large particles promote different toughening mechanisms, and the presence of both results in a synergistic effect (9). Both shear yielding and dilatational deformation involving cavitation were observed (58).

A second explanation for increased fracture toughness in bimodal distributions lies in a larger volume fraction of rubbery-phase separated epoxies. Bucknall and Yoshi (59) found that bisphenol A added to epoxy resin in the presence of CTBN increased the volume fraction of rubber particles. The increased volume fraction consequently increased the extent of dilatational deformation that involves cavitation and is the critical parameter for toughening. The measured relationship between volume strain change and longitudinal strain showed the effect of added bisphenol A on the increased dilatational deformation (59–61).

A third explanation for the fracture toughness increase in bimodal distributions points out that the molecular weight between cross-links in this system is higher than in systems without bisphenol A (9). As will be discussed later, it is well known that a lower cross-link density in the matrix of a rubber-toughened system increases the fracture energy (62, 63).

Regardless of the explanation, however, an examination of the fracture surfaces for the epoxies with bimodal rubbery particle distribution shows the same cavitation of rubber particles and matrix shear yielding in the crack-tip region that is observed for other toughened epoxies. The major difference is the size of the region over which these processes occur. As with other toughened thermosets, the size of this region scales with fracture energy. In addition to high fracture energy, the bimodal system also shows very good phase separation and, consequently, very little sacrifice in other thermal and mechanical properties (64).

The morphology of CTBN-toughened epoxies has been further explored by examining the fracture surfaces of the stressed tensile bars. The rubber particles within the shear-yielded area are all elliptic and some particles associated with cavitation show double ellipsoids. These double ellipsoids look like elongated fried eggs with small amounts of egg whites (i.e., cavitated matrix), surrounding the egg yolks (i.e., the rubber particles).

The fracture surface perpendicular to the tensile direction of a tensile bar shows that cavitation around particles eventually progressed to crack formation that caused catastrophic failure. A parallel plane (perpendicular to the stress direction) close to the tensile-bar fracture surface was exposed by rapid hammering on a sharp knife blade to break through the sample after it had been cooled in dry ice. The resulting stress-whitened plane was replicated and observed using transmission electron microscopy to see the frozen-in morphology. This time, the fried egg-like patterns (rubbery domains) were circular and the cavitated matrix (egg-whites region) was twice the size of the rubber particles at the center (egg yolks). The cavities in the egg-whites region have closely associated with the rubber particles (57, 58).

When the specimen is heated above 100 °C, the glass-transition temperature ($T_g$) of the resin, the deformed matrix (i.e., the egg whites) recovered, leaving only the original size rubber particles (i.e., the egg yolks). The stress-whitening disappears also. This thermal response is similar to crazing in thermoplastics, which recovers on heating above $T_g$. Microvoids around the rubbery particles, which were actually microcracks, did not recover and remained as voids. Thus, the stress-whitening is associated with recoverable dilatational deformation just as in the deformation involved in fibril formation of crazes. Additional evidence is obtained by an osmium tetroxide-stained (65) microsection of the stressed and heat-treated material. The particles (stained) are of the same size as the original unstressed sample.

Kinloch et al. (36) suggested that the key factors responsible for toughening are: cavitation of the rubber particles, which permits dilatation of the matrix without the microvoid formation that leads to brittle failure, and shear yielding between the holes formed by particle cavitation.

Pearson and Yee (62, 63) studied the effect of matrix ductility on toughening by varying the molecular weight of epoxy resins. These epoxy resins were cured using diphenyldiamine sulfone and, in some cases, modified with 10 vol % CTBN. Fracture-energy values for the unmodified epoxies showed only a small increase with increasing molecular weight of the epoxy resin. However, the toughness of the CTBN-modified materials was increased dramatically with increased epoxy monomer molecular weight. Tensile dilatometry indicates that the toughening mechanism, when present, is similar to the mechanism found for the piperidine-cured CTBN-modified epoxies studied. Scanning electron and optical microscopy techniques corroborate this finding. The deformation in bulk tensile tests of the unmodified epoxy resins was a shearing process. Significant shear deformation and void formation were observed for some, but not all, of the rubber-modified materials, which is another proof of the importance of initial compatibility and reactivity of CTBN and the epoxy resin. The epoxy molecules should undergo linear chain extension before cross-linking and gelation so as to minimize cross-link density (9).

Pearson and Yee (66) examined the deformation of CTBN-toughened epoxy resins in uniaxial tension and in three-point bending with an edge notch. Scanning electron microscopy of fracture surfaces indicated that cavitation of the rubber particles was a major deformation mechanism. Particle–particle interactions were also found. Optical microscopy of thin sections perpendicular to the fracture surface showed that the cavitated particles generated shear bands. The toughening effect was due to cavitation, which relieved the triaxial tension at the crack tip, and shear-band formation, which created a large plastic zone.

The study by Li et al. (67) on the role of cavitation in the surrounding matrix in the deformation of rubber-toughened epoxy under constrained conditions is very important. Cavitated CTBN particles strongly interact with yielding of epoxy resins in front of the crack tips, thereby blunting the cracks in the tension-bending tests. Uncavitated particles do not have this interaction. Massive shear-yielding mechanisms occurred only when cavitation had occurred. Shear deformation was proposed as the principal mechanism for energy absorption, with plastic dilation absorbing only a small fraction of the total energy.

Yee and Pearson (68) also found that at low strain rates the rubber particles enhanced shear deformation. At sufficiently high strain rates, the rubber particles cavitated and subsequently promoted further shear deformation. No indication of crazing as an important toughening mechanism was found, and there was no significant effect of rubber particle size or type.

**Effect of Rate and Temperature.**  The fracture behavior of toughened thermosets, which is known to vary dramatically with test temperature and loading rate, has been studied extensively by Bitner et al. (64) and Hunston et al. (69, 70). The test temperature and loading rate on fracture warrant detailed discussion because of their importance in deducing the toughening mechanism. In addition to the following review, more details on temperature and loading rate can be obtained in references 64 and 69–73.

An exhaustive examination of a series of four elastomer-modified epoxies plus an unmodified epoxy (Table I) was conducted. The same base resin was used in all the experiments, but three of the toughened systems contained different amounts of an elastomer, whereas the fourth contained bisphenol A (BPA), which altered the morphology to produce a bimodal distribution of elastomer particle sizes. The fracture energies were determined using compact tension specimens at cross-head speeds between 0.05 and 50 mm/min and temperatures between $-60$ and $+60$ °C.

The behavior was found to be viscoelastic in nature, that is, there was an inverse relationship between test temperatures and loading rate. Bitner et al. (64) and Hunston et al. (69), found that the results could be analyzed by time–temperature superposition techniques. The time to failure $(t_f)$ was the parameter chosen to characterize the loading rate. $t_f$ is the time from the initial application of the load until the failure point is reached in the constant cross-head speed experiments. The data for fracture energy were then plotted against $t_f$ for a series of temperatures. Finally, the results for different temperatures were shifted horizontally along the $t_f$ axis until the points superimposed to form a master curve. The data at 20 °C were selected as the reference and were not shifted.

Figure 1 shows results for three of the four toughened epoxies and the unmodified epoxy. The low values of reduced time to failure $(t_f/a_T)$ correspond to low temperatures and high loading rates whereas high values are the

**Table I. Formulation of CTBN-Toughened Epoxy Resins**

| Component[a] | Formulation (parts by weight) | | | | |
|---|---|---|---|---|---|
| | 1 | 2 | 3 | 4 | 5 |
| Epoxy resin (DGEBA) Epoxy equivalent weight = 195 | 100 | 100 | 100 | 100 | 100 |
| Piperidine | 5 | 5 | 5 | 5 | 5 |
| Hycar CTBN | 0 | 5 | 5 | 15 | 18.5 |
| Bisphenol A | 0 | 0 | 24 | 0 | 0 |

[a] Details of composition and cure are given in references 56, 57, and 61.

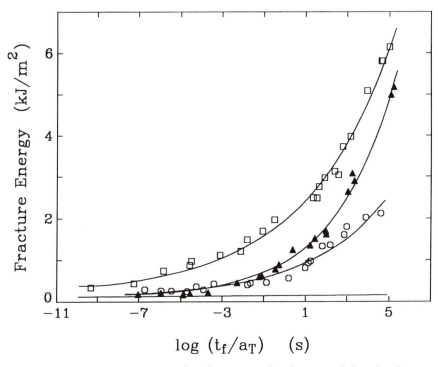

*Figure 1. Fracture energy is plotted against reduced time to failure for three elastomer-toughened materials: formulation 5, □ ; formulation 3, ▲ ; and formulation 2, ○. The straight line at the bottom of the graph represents the unmodified epoxy.*

reverse. Figure 1 presents some interesting information. The unmodified epoxy exhibits very little rate or temperature dependence over the range of test conditions examined. At higher temperatures and slower rates, the fracture energy ($G_{Ic}$) for formulation 3 (Table I) greatly exceeds that for formulation 2. Both systems contain the same concentration of elastomer, 5 phr [parts (by weight of elastomer) per hundred parts (by weight) of epoxy resin] but formulation 3 has the bimodal distribution of particle sizes whereas formulation 2 has the more conventional distribution. This higher $G_{Ic}$ has been noted in the past and is cited as an advantage of a bimodal morphology (8). A reduction in the reduced time to failure by only three decades, however, brings the behavior of the bimodal material to near equivalence with that for the simple 5-phr system. The curves come together at rates that correspond to low-temperature impact. In this region, there appears to be no advantage to the bimodal system. Moreover, if designs are based on measurements made in the range where the bimodal system is superior, the higher rate sensitivity can lead to significant overestimates of toughness.

The shift factors ($a_T$) were obtained empirically as the values needed to superimpose the fracture data. Another approach is to use the shift factors derived from viscoelastic experiments involving simpler stress fields (71, 72). Unfortunately, the linear viscoelastic data can be obtained in a temperature range that overlaps only part of the range tested in the fracture experiments. It is possible, however, to determine the shift factors from yield experiments, and this can be accomplished over the same temperature range as the fracture experiments. These shift factors, determined independently of the fracture test, can then be used successfully to superimpose the toughness data and generate master curves (71, 72). This demonstrates the close relationship between yielding and fracture. Such a relationship is expected based on the proposed mechanism that suggests yielding as the principal mode of energy dissipation.

The data in Figure 1 were analyzed by Bitner et al. (64) and Hunston et al. (69, 70) who suggested that the results could be represented empirically by

$$G_{Ic} = A \left[ \frac{t_f}{a_T} \right]^m + G_{Icb} \tag{1}$$

$$a_T = \exp \left( \frac{\Delta E}{R} \left[ \frac{1}{T} - \frac{1}{T_0} \right] \right) \tag{2}$$

These equations contain five parameters, and although they are strictly empirical, they do have interpretations that can be extremely useful for comparing the behavior of different materials and establishing structure property relationships. For data analysis, the authors chose the reference temperature $T_0$ to be 80 °C below the glass-transition temperature because they felt the fracture behavior was a function of the difference between the test temperature and the glass-transition temperature. The 80 °C was selected to bring the reference temperature within the range where the fracture experiments were conducted. The parameter A provides a measure of the magnitude of the toughening effect, $m$ assesses the loading rate sensitivity, $\Delta E$ measures the temperature dependence, and $G_{Icb}$ is the limiting toughness at low temperatures and high loading rates.

The data presently available are not sufficient to clearly establish relationships between these parameters and the morphology of the materials. Nevertheless, the following speculations have been made based on the limited data available (69, 70, 73). The parameter $\Delta E$ is the same for all the materials tested and is similar in value to that obtained from an equivalent analysis of yield data. Because all of the systems tested had the same matrix resin, the result suggests that $\Delta E$ depends on the matrix phase. The value of $m$ is the same when the distribution of particle sizes is similar but changes for a system with a bimodal distribution. Thus $m$ appears to be dependent on morphology. The parameter A increases with increasing elastomer concentration but also appears to be a function of morphology because it is larger for a bimodal system with the same amount of elastomer.

More experiments are necessary before such correlations can be firmly established. Nevertheless, the acquisition of such data is well worth the effort because such relationships offer a real hope for scientifically optimizing the fracture behavior of these materials.

**Stress Fields around Rubbery Particles**. The toughening mechanisms that operate in rubber-toughened thermosetting polymers obviously involve both the dispersed rubber particles and the polymeric matrix. The first step toward understanding the nature and magnitude of the toughening mechanisms is to establish the stress fields that act in the vicinity of the rubbery particles.

Goodier (74) has derived equations for the stresses in the matrix around an isolated elastic spherical particle embedded in an isotropic elastic matrix, where the matrix is subjected to an applied uniaxial tensile stress remote from the particle. These equations reveal that for a rubbery particle, which typically possesses a considerably lower shear modulus than the matrix, the maximum tensile stress concentration in the matrix due to the presence of the particle occurs at the equator of the particle and has a value of about 1.9. Furthermore, assuming the particle is well bonded to the matrix, the local stress state at this point is one of triaxial tension. The triaxial tension arises essentially because of the volume constraint represented by the bulk modulus of the rubbery particle, which is comparable with that of the matrix. (The low shear modulus of the rubber particle is relative to the thermoset matrix, but its comparable bulk modulus is a consequence of the Poisson's ratio of the rubber being approximately 0.5 whereas that of the matrix is about 0.35–0.4.) In contrast to a void that would produce a stress concentration similar in magnitude, the rubbery particle can fully bear its share of the load across the crack front because of the volume constraint. Indeed, the ability of the rubber particles to be load-bearing as well as stress concentrators, is one reason that rubbery particles may be particularly effective for toughening the glassy polymeric matrix. Nevertheless, the ability of the particle to internally cavitate may be a more important factor.

In rubber-modified polymers, the stress fields of nearby rubbery particles will overlap and, hence, Goodier's solution for an isolated particle is not strictly applicable. Recently, several finite-element analyses (75–78) examined the stress fields in and around rubbery particles or voids in a glassy matrix. The recent analyses claim that when the inclusion is a void, it is equivalent to a cavitated or debonded rubbery particle. In the earlier analyses (75, 76), the multiphase rubber-modified material was modelled as an array of axisymmetric elements. These elastic analyses revealed that the maximum stress concentration inside the matrix occurred adjacent to the equatorial area of the rubbery particle and that this stress concentration increased with the volume fraction ($V_f$) of particles because of the interaction between the stress fields of the particles. For example, for a volume fraction of 0.2, Broutman and Panizza (75) computed the maximum stress concentration is ~ 2.1, whereas for a volume fraction of 0.4, the maximum stress concentration is ~ 2.7. This modest increase in stress concentration does not reflect the relatively steep increase that may be observed in the fracture energy of rubber-modified epoxy polymers when a rubbery volume fraction from ~ 0.05 up to 0.2 is employed (79, 80). In the latest analyses (77, 78), a two-dimensional plane-strain model was employed to simulate the multiphase rubber-modified material. Figure 2 compares the computed von Mises stress concentration factor using this model to that from the conventional axisymmetric model (75, 76). Clearly, the two-dimensional plane-strain model predicts a steeper increase in the stress concentration when the volume fraction of rubbery particles is increased. Thus, the results from the latter model better explain the relatively large increases in the fracture energy that are observed as the volume fraction of the rubbery particles is increased. Furthermore, this two-dimensional plane-strain model may simulate the growth of localized shear yielding.

It is of interest to note that Goodier's analysis also indicates that the stress concentration in the matrix at the particle equator will still be present even at temperatures below the glass-transition temperature of the rubber: Under such circumstances the shear modulus of the rubbery particle still

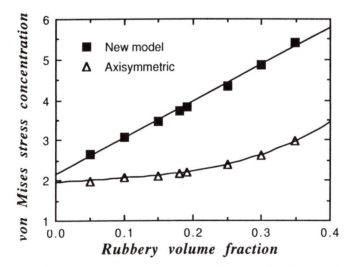

Figure 2. The von Mises stress concentration factor predicted using the two-dimensional plane-strain model and the axisymmetric model.

would be expected to be slightly lower than that of the highly cross-linked thermoset matrix, which is even further below its glass-transition temperature ($T_g$). However, because the shear moduli now will be much closer in value, the extent of the stress concentration will be decreased somewhat. This aspect is important because it readily explains why rubber-modified thermosets are still somewhat tougher than the unmodified polymer at temperatures well below $T_g$ (rubbery phase).

Now the stress field associated with the rubbery particles leads to the initiation of two important deformation processes.

**Matrix Shear-Yielding.**   One process of major importance that occurs is the initiation and growth of multiple, but localized, shear-yield deformations in the matrix (*68, 79, 80*). As a result of the stress concentration in the matrix caused by the presence of the rubbery particles, plastic shear-deformation bands are initiated and, because there are many particles, there is considerably more plastic energy dissipation in the multiphase material than in the unmodified polymer. However, the plastic deformation becomes localized due to (1) the post-yield strain-softening of the epoxy matrix (*81*) and (2) the observation that the shear deformations initiate at one particle but terminate at another particle.

The initiation and growth of shear bands were modeled recently in a rubber-toughened epoxy polymer using the finite-element method (*77, 78*). The epoxy matrix was modeled as an elastic–plastic material possessing the essential features (*81*) of strain-softening and subsequent strain-hardening, whereas the rubbery particles were modeled as elastic materials. Figure 3 shows contours of the equivalent plastic strain in the epoxy matrix during different loading stages. Note that voids, instead of rubbery particles, are shown here because the internal cavitation of the rubbery particles occurred at the yield point of the epoxy matrix in this particular rubber-modified epoxy polymer. Detailed discussions on the importance of the sequence between the internal cavitation of the rubber particles and the shear-yielding of epoxy matrix are given in reference 77. From Figure 3, it is obvious that yielding initiates from the point of maximum stress concentration, that is, the equatorial area of the particle (void). When the load is further increased, a band of shear-yielded material forms at an angle of approximately 45° to the applied stress. Once the band is formed, further plastic-yielding is localized within this band. Thus, the finite element analysis has successfully modeled the localized shear-band plastic-yielding mechanism observed in the fracture of rubber-modified epoxy polymers.

An important point to consider is the previous suggestion that the internal cavitation, or debonding at the particle–matrix interface, can greatly reduce the degree of triaxial stresses acting in the matrix polymer adjacent to the particle, and that this stress reduction then enables the further growth of the shear bands. First, all the stress analyses (*75–78*) have revealed that the

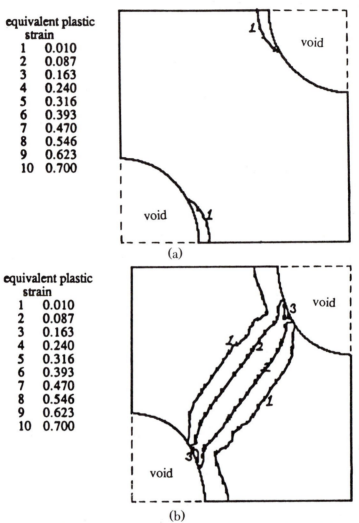

(a)

(b)

*Figure 3. A model of the initiation and growth of the shear-yield bands between rubber particles at different stages of loading. The contours are for a given equivalent plastic strain and the applied strains are (a) 0.025, (b) 0.050, (c) 0.075, and (d) 0.100.*

stresses in the particle are several orders smaller than those in the epoxy matrix if the rubbery particle has a tensile modulus of 1–2 MPa or lower, which is a very reasonable assumption. Further, the stresses in the glassy matrix are of a similar magnitude whether the element of the matrix under consideration is adjacent a rubbery particle or a void, which is equivalent to a cavitated or debonded rubbery particle. Thus, the internal cavitation, or

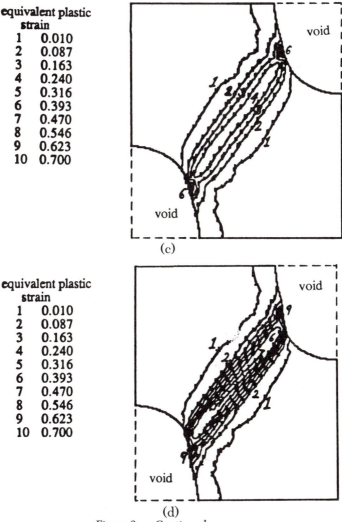

Figure 3. —*Continued*

debonding, of the rubbery particles does not lead to significant relief of the triaxial stresses in the matrix. Second, as experimentally reported by Kinloch and Young (82) and noted by Sjoerdsma (83), the dimensions for a plane-strain to plane-stress transition is on the order of millimeters, as opposed to the order of micrometers that is involved in the matrix ligament between two cavitated (or debonded) rubbery particles. Third, the analysis of Bowden (81) reveals that the initiation and growth of shear bands is only possible under plane-strain constraint. Hence, for all the reasons just discussed, the internal cavitation or debonding of the rubbery particles is not expected to trigger further shear banding.

### Internal Cavitation of the Particle and Subsequent Plastic Void Growth in the Matrix.

Cavitation, or debonding, of the rubbery particle is an important process that enables plastic void growth to occur in the matrix (77, 84). Consideration of this phenomenon necessitates recalling that an overall triaxial (plane strain) stress state usually exists ahead of a crack. This stress state produces dilatation that, in combination with the stresses induced in the particle by cooling after cure, may cause failure and void formation either in the particle or at the particle–matrix interface. Which of these events occurs largely depends upon the degree of interfacial adhesion that is attained at the particle–matrix interface. Use of a rubber with reactive end-groups results in chemical grafting, which gives high adhesion, and internal cavitation of the particle is observed. On the other hand, use of a rubber that has nonreactive end-groups results in no chemical grafting and only relatively weak adhesion, which arises solely from interfacial van der Waals bonds. Debonding of the particle from the matrix is now observed.

Although the growth of voids or microcracks in the rubbery particle or debonding of the particle may dissipate a little energy, a far more important aspect of this process is that it enables plastic void growth in the matrix to occur. Typical micrographs (84) of a rubber-toughened epoxy are shown in Figure 4, where the appearance of the original, undeformed, microstructure (Figure 4a) is compared to the fracture surface of the rubber-modified epoxy (Figure 4b). The fracture test was conducted at 40 °C, when extensive cavitation of the rubbery particles and plastic void growth in the epoxy matrix occurred. These two photographs show that the average size of the voided rubbery particles is far greater than the original rubber particles. (It should be noted that this rubber is chemically grafted to the matrix and, therefore, internal cavitation in the rubbery particle occurs, rather than interface debonding.) Table II shows the measured values of the volume fraction of the rubbery particles (or enlarged voided particles) on the fracture surfaces of a rubber-modified epoxy resin at different test temperatures (70, 77). The original volume fraction of rubbery particles was 0.19. At the two lowest test temperatures, the volume fraction remained at 0.19, which indicates that there was no plastic void growth in the matrix. However, when the test temperature was at or above $-20$ °C, an increase in the size of the (now enlarged) voids is clearly indicated from the data in Table II. This size increase indicates that cavitation of the rubbery particles and plastic void growth in the matrix have occurred under these test conditions (i.e., when the yield stress of the matrix was relatively low).

Note that it is this initiation and voids growth that gives rise to the stress-whitening often observed ahead of the crack tip and on the fracture surfaces.

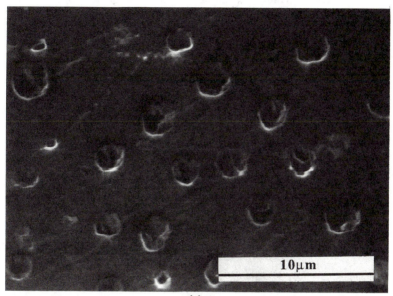

(a)

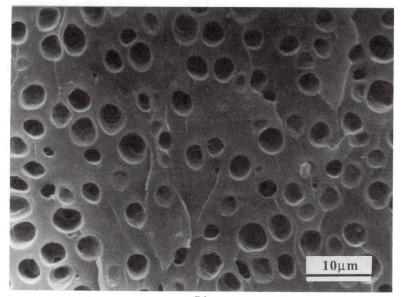

(b)

Figure 4. Microstructures of the rubber-toughened epoxy polymer: (a) original (undeformed) microstructure before fracture test; (b) fracture surface after testing at 40 °C and 2 mm/min. (Reproduced with permission from reference 72. Copyright 1981 Chapman and Hall).

**Table II. Measured Volume Fraction of Rubber-Toughened Epoxy Resin**

| Temperature (°C) | −60 | −40 | −20 | 0 | 23 | 40 |
|---|---|---|---|---|---|---|
| Measured[a] | 0.19 | 0.19 | 0.22 | 0.24 | 0.27 | 0.32 |
| Increase[b] (%) | 0 | 0 | 16 | 26 | 42 | 68 |

[a] The original volume fraction of (undeformed) rubbery particles ($V_{fr}$) was 0.19.
[b] The percentage increase in volume fraction was calculated from the measured volume fraction and the original volume fraction of rubbery particles ($V_{fr}$).

**Criteria for Plastic Void Growth.** A rubber particle with a Poisson's ratio ($v$) approaching 0.5 will possess a relatively high bulk modulus ($K$) that has approximately the same value as the thermosetting resin matrix, as shown by the following equation:

$$K = \frac{E}{3(1 - 2v)} \tag{3}$$

where $E$ is Young's modulus and is, of course, very much lower for the rubber particle than for the matrix. This low Young's modulus demonstrates that a pure rubber particle, like the matrix, is a very rigid elastic body when subjected to triaxial stresses and will be highly resistant to any volumetric deformation. Thus, the relatively "rigid" rubbery phase will not encourage any large-scale plastic dilation in the matrix unless the rubbery particles either internally cavitate or debond from the matrix to form a "void." Indeed, as previously mentioned, the ability of the rubbery particles to internally cavitate and form "voids" in the thermosetting matrix enables plastic void growth to occur in the matrix.

Glassy polymers, such as epoxy polymers, often strain-soften immediately after yield and then strain-harden again (81). During strain-softening, the polymer more readily undergoes plastic deformation. If cavitation or debonding of the rubbery particles has not occurred by the time the matrix starts to strain-harden, then the plastic void growth mechanism will be suppressed due to the rapid strain-hardening of the matrix. Thus, the necessary criteria for the plastic void-growth micromechanism are that the rubbery particles either internally cavitate or debond from the matrix, and largely do so before the onset of strain-hardening of the thermosetting matrix.

**Quantitative Assessment of the Different Mechanisms.** The importance of the various mechanisms to the toughening of epoxy polymers may be demonstrated by using a recently developed mathematical model (77, 78).

The fracture energy of a rubber-modified epoxy ($G_{Ic}$) was first modeled by Kinloch (80):

$$G_{Ic} = G_{Icu} + \Psi \tag{4}$$

where $G_{Ic}$ is the fracture energy of the rubber-toughened epoxy polymer, $G_{Icu}$ is the fracture energy of the unmodified epoxy polymer, and $\Psi$ is the additional energy dissipated per unit area in the rubber-toughened polymer due to the presence of the rubbery particles. The value of $\Psi$ is given by

$$\Psi = \Delta G_r + \Delta G_s + \Delta G_v \tag{5}$$

where $\Delta G_r$ is the contribution from rubbery particles bridging across the crack, behind the crack tip. The terms $\Delta G_s$ and $\Delta G_v$ are the contributions from plastic shear banding in the matrix and plastic void growth in the matrix, respectively; both of these two micromechanisms occur in the plastic zone ahead of the crack tip.

For the rubber-bridging mechanism, Kunz-Douglass et al. (33) proposed that the value of $\Delta G_r$ is given by

$$\Delta G_r = 4\Gamma_t V_{fr} \tag{6}$$

where $\Gamma_t$ is the tearing energy of the rubbery particle and $V_{fr}$ is the volume fraction of rubbery particles.

For the plastic shear-band mechanism then, following an earlier model proposed by Kinloch et al. (79), more recent work (77, 78) has suggested an equation of the form

$$\Delta G_s = 0.5\left(1 + \frac{\mu_m}{\sqrt{3}}\right)^2\left[\left(\frac{4\Pi}{3V_f}\right)^{1/3} - \frac{54}{35}\right]K_{vm}^2 V_f \sigma_{yc}\gamma_f r_{yu} \tag{7}$$

where $V_f$ is either equivalent to $V_{fr}$, which is the original volume fraction of the rubber particles, or to $V_{fv}$, which is the measured volume fraction of the enlarged voided particles. The terms $\sigma_{yc}$ and $\gamma_f$ are the compressive yield stress and the fracture strain of the epoxy matrix, respectively, which may be evaluated using a plane-strain compression test. $K_{vm}$ is the concentration factor of the von Mises stress in the epoxy matrix around the rubbery particles and may be calculated from finite-element analysis (77, 78). The parameter $r_{yu}$ is the radius of the plastic zone for the unmodified epoxy and may be calculated from the mechanical properties of the material (82). The factor $\mu_m$ allows for the pressure dependence of the von Mises yield criterion for glassy polymers (81) as shown by the equation

$$\tau_{vm} = \tau_y - \mu_m P \tag{8}$$

where $\tau_y$ is the yield stress under pure shear, $P$ is the hydrostatic stress, and $\tau_{vm}$ is the von Mises shear stress as defined by

$$(\sigma_1 - \sigma_2)^2 + (\sigma_2 - \sigma_3)^2 + (\sigma_3 - \sigma_1)^2 = 2\sigma_{vm}^2 = 6\tau_{vm}^2 \qquad (9)$$

where $\sigma_1$, $\sigma_2$, and $\sigma_3$ are the principal stresses and $\sigma_{vm}$ is the equivalent von Mises stress. The value of $\mu_m$ has been determined to be between 0.175 and 0.225 for epoxy polymers (6), and was taken as 0.20 in the present study.

For the plastic void growth micromechanism, the strain-energy density may be assessed via

$$U_V(r) = \int_{V_0}^{V_1} P \, d\theta \qquad (10)$$

where $P$ is the local hydrostatic stress and $V_0$ and $V_1$ are the average volume of the voids before and after growth. $d\theta$ is the volumetric strain of the voids during growth and may be expressed as

$$d\theta = V_f \, dV/V \qquad (11)$$

where $V$ is the average volume of the voids during growth. Use of an estimation of the stresses inside the plastic zone (85), assumption that the conditions of linear elastic fracture mechanics (LEFM) are satisfied, and assumption that

$$V_1/V_0 \equiv V_{fv}/V_{fr} \qquad (12)$$

allows $U_v(r)$ to be evaluated and the following expression for $\Delta G_v$ to be derived (77, 78):

$$\Delta G_V = \left(1 - \frac{\mu_m^2}{3}\right)(V_{fv} - V_{fr}) K_{vm}^2 \sigma_{yc} r_{yu} \qquad (13)$$

The measured fracture energy may be calculated by combining equations 6, 7, and 13. The preceding model was applied to a rubber-modified epoxy and was found to be in reasonable agreement with experimental data over a wide range of test temperatures and rates (78), as shown in Table III. Furthermore, the model makes it possible to separate the contributions from the different toughening mechanisms. Table III shows the proportional contributions from the three toughening mechanisms (77, 78), to the total increase in fracture energy of a rubber-modified epoxy. Localized plastic shear-banding is the dominating mechanism under all testing conditions. At high temperatures, the plastic void-growth mechanism, together with the

Table III. Three Proportional Contributions of the Fracture Energy Based on the Toughening Mechanisms

| Fracture Energy | Temperature (°C) | | | | | |
|---|---|---|---|---|---|---|
| | −60 | −40 | −20 | 0 | 23 | 40 |
| Fracture energy of rubber-toughened epoxy $G_{Ic}$ (kJ/m$^2$); from model | 1.30 | 1.49 | 2.05 | 2.72 | 4.79 | 8.25 |
| Fracture energy of rubber-toughened epoxy $G_{Ic}$ (kJ/m$^2$); measured | 1.72 | 1.96 | 2.53 | 3.64 | 5.90 | 7.23 |
| Fracture energy: proportional contribution due to rubber bridging $\Delta G_r/\Psi$ | 0.36 | 0.26 | 0.14 | 0.11 | 0.08 | 0.05 |
| Fracture energy: proportional contribution due to shear yielding $\Delta G_s/\Psi$ | 0.64 | 0.74 | 0.68 | 0.60 | 0.54 | 0.47 |
| Fracture energy: proportional contribution due to cavitation $\Delta G_v/\Psi$ | 0.00 | 0.00 | 0.18 | 0.29 | 0.38 | 0.48 |

Note: $G_{Ic}$ was measured at a displacement rate of 2 mm/min.

shear-banding mechanism, represent the main contributions to the fracture energy increase. The rubber-bridging mechanism plays a minor role. However, note that the contribution from the plastic void-growth mechanism is highly sensitive to the test conditions, and may be completely suppressed at low test temperatures. Under such conditions, the overall increase in fracture toughness is small, that is, only moderate toughening is achieved, and the rubber-bridging mechanism makes a higher proportional contribution to the overall toughness.

These data emphasize the importance of attaining the required microstructure, which consists of a dispersed rubbery phase in the thermosetting matrix, as well as the ability of the thermosetting matrix to be able to undergo plastic deformation. High test temperatures and low test rates, which facilitate plastic yielding in the matrix, lead to higher values of $G_{Ic}$. However, other factors, such as the chemical backbone structure and degree of cross-linking of the matrix, affect the ability of the matrix to plastically deform (63, 86) and so affect the toughness of the rubber-modified polymer.

## Fracture Mechanism: Fracture Behavior in Thin Layers

The foregoing discussion of fracture properties for toughened thermosets has focused on the behavior of bulk samples. In the real world, however, toughened thermosets are generally used in applications such as adhesives and fiber-reinforced composites. Research over the last 15 years has shown that an effective approach to understanding the fracture properties of these complex geometries is to first examine the resin itself and then study how the behavior is modified by features that are specific to an adhesive bond or a composite. There are a number of such features. An example is the presence of an interface. If the bonding at the interface is very poor, an easy path for crack growth is provided. Thus, the fracture resistance of the adhesive or composite can be significantly less than would be expected based on the fracture energy of the resin itself. Two other factors specific to adhesive or composite geometries are (1) residual stresses associated with differences in thermal expansion between the resin and the adherends or fibers and (2) variations in molecular structure or morphology that arise from differences in resin thermal history in bulk samples vs. in adhesives or composites. Although these and other similar factors can have major effects, the research over the last 15 years suggests that in many cases the single most important feature characterizing an adhesive or a composite is the restriction of the resin to a thin layer between adherends or fibers (87).

For brittle resins, this "thin layer" effect is generally small (87). The rationale is that the fracture behavior in such materials depends on events very close to the crack tip, and unless the crack is at the interface or the layer is extremely thin, the crack does not "know" the material is restricted to a thin layer. With tough resins, however, the fracture energy is associated with a large crack-tip deformation zone that is no longer small in size relative to the thickness of the layer. Consequently, there is an interaction that must be considered (87).

**Early Work.**   Although the purpose here is to examine new developments in toughening thermoset resins, a brief review of earlier results is useful because the recent work is built on previous studies. The influence of film thickness was first demonstrated in adhesive joints by Bascom et al. (10), who found that the mode-I fracture energies for bulk samples of brittle adhesives generally agree with the fracture energies obtained for the same materials in adhesive joints. With toughened resins, however, a strong dependence of adhesive fracture behavior on bond thickness (Figure 5) was observed. A very thick bond gave the same results as a bulk sample, but as the bond thickness decreased, the adhesive fracture energy exhibited a peak; that is, it first increased and then decreased. The experiments were performed at different temperatures and the behavior was similar, but the position of the peak shifted (13). Hunston et al. (69, 88) subsequently found a similar shift in peak position when the loading rate was changed.

Bascom et al. (10, 13) offered a partial explanation for this effect by noting that the magnitude of the fracture energy was related to the size of the crack-tip deformation zone. They proposed that as the bond thickness is decreased, the adherends physically constrain the zone size and thereby reduce toughness. The other half of the explanation was offered by Wang (89), who suggested that a second effect was also present. A stress analysis by Wang et al. (90) showed that the crack-tip stress field in the adhesive bond

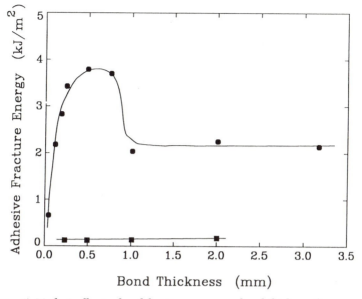

*Figure 5. Mode-I adhesive bond fracture energy vs. bond thickness for an epoxy (■) and a rubber-toughened epoxy (●).*

differs from that in bulk specimens in that the tensile stresses some distance ahead of the crack tip are higher than expected. As the bond thickness is decreased, this effect becomes more pronounced. Based on this result, Wang (89) proposed that these stresses cause the deformation zone to extend farther down the bond line ahead of the crack tip. This increases the zone size (i.e., volume) and, therefore, causes an increase in toughness.

By combining these two effects in the proper way, Hunston et al. (69) and Wang et al. (90) rationalized the peak in adhesive fracture energy. The fracture energy is unaffected for very thick bonds and, therefore, has similar values to those for bulk samples. As the thickness is decreased, the first important effect is associated with the stress field, which causes the zone size to extend down the bond line and increases the fracture energy. When the bond thickness becomes equal to the zone height, the constraining effect of the adherends becomes dominant. Consequently, even though the extension of the zone down the bond line may continue, the decrease in zone height more than compensates, so the zone size and, hence, the toughness, decrease. Using this hypothesis, Bascom and Cottington (13) and Hunston et al. (69) were able to successfully predict the optimum bond thickness (maximum fracture energy) at various temperatures and loading rates based on simple models for the deformation zone size.

Hunston et al. (69) also tested the rationalization for the bond thickness effect by using high speed movies to observe the nature of the crack-tip deformation zone. The stress-whitened region at the crack tip was assumed to be a relative measure of the deformation zone size and shape. The results are consistent with the proposed model. Another study supporting the hypothesis is a stress analysis by Wang. Although the full analysis was never published, the conclusions were reported in several presentations and papers (69, 87, 91–93).

Another area where thin-layer effects are important is composites. The property of interest is interlaminar fracture behavior. The realization that impact damage and delamination are major failure modes (94, 95) motivated the study of interlaminar fracture. Many years ago it was suggested that toughened resin might improve the resistance of a composite to interlaminar fracture. Based on the previously described results for adhesive bonds, some researchers concluded that increasing the toughness of the resin in the composite would have little benefit (96) because the fiber–fiber spacing between plies is very small. It was believed that the constraint resin toughness imposed on the deformation zone size would minimize any benefit that could be achieved.

In the literature, the early data on this situation were conflicting, probably because factors other than resin toughness that influence composite behavior could not be adequately controlled. In the late 1970s, however, Bascom et al. (96) were able to test samples where the increase in resin toughness was so large that it overwhelmed other effects. A large increase in mode-I fracture toughness of the resin produced a modest, but important, increase in mode-I interlaminar fracture toughness of the composite. The improvement was larger than had been expected based on the adhesive bond data. Bascom et al. (96) examined cross sections of the sample perpendicular to the crack plane so they could study the size of the deformation zone. Unlike the situation for bulk resins or adhesive specimens, Bascom et al. were unable to quantify the zone size to an extent that would permit establishment of a direct correlation with toughness, but they assumed that at least a qualitative correlations would exist. This situation suggests that the zone height should be larger than the fiber–fiber spacing with tough resins. In fact, Bascom et al. (96), found that the fibers on either side of the resin layer that contained the crack did not totally constrain the deformation zone. Significant plastic deformation was found in resin that was separated from the crack tip by a number of fibers. Consequently, although the fibers seem to disrupt the deformation zone (to the degree that toughness reflects the zone size), the constraint is not as severe as it is in the adhesive joint.

The work by Bascom et al. (96) was with woven reinforcement, but subsequent studies by a number of authors found similar effects for unidirectional composites. Data summaries in papers by Hunston et al. (97–99) provided a detailed comparison between resins and composites for mode-I fracture behavior; see Figure 6. The comparison results for brittle resins show that it is more difficult to propagate the crack in the composites than in

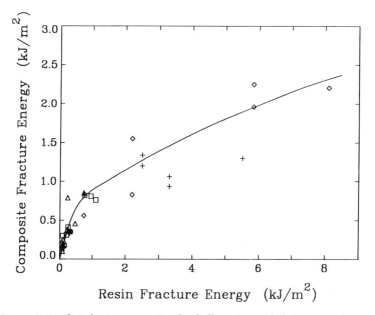

*Figure 6. Mode-I fracture energies for bulk resins and their composites are shown. The curve indicates the general data trend. Samples are: brittle epoxies ( △ ); a novolac epoxy (○); toughened epoxies (◇); a toughened novolac epoxy ( ● ); toughened bismaleimide (×); experimental resins ( □ ); amorphous thermoplastics ( + ); a cross-linkable thermoplastic epoxy ( ▲ ).*

the corresponding bulk sample. This observation was explained by noting that the full resin toughness was translated to the composites and, in addition, the crack growth in the composite involves other energy absorption mechanisms such as breaking of fibers. As the resin toughness is increased, the zone size grows until it reaches a value commensurate with the fiber–fiber spacing between plies, at which point, there is a change in behavior such that every $3\text{-}J/m^2$ increase in resin toughness produces only $\sim 1\ J/m^2$ in interlaminar fracture energy (Figure 6). This outcome is consistent with the earlier speculation that the fibers hinder the toughening by interfering with the deformation zone growth but not totally constraining the zone size.

**Recent Work**. In recent years many of these results have been refined and extended through stress analysis and experimental studies. Although it is not possible to review all of the new work in this area, some of the important papers will be discussed to illustrate the progress. An example in the area of stress analysis for adhesive bonds is the work of Crews et al. (*100*), who predicted the size and shape of the crack-tip deformation zone as a function of bond thickness. Although this analysis is for a brittle resin, the assumed yield stress was significantly less than that typical for a high-performance adhesives. As a result, the predicted deformation zone was much larger than the actual case for a brittle resin. On the other hand, the prediction provides a useful indication of the appearance of the zone in a tough, high-performance resin. The zone size is determined by noting where the stress exceeds the yield value. Because the analysis is elastic rather than elastic–plastic, no provision is made for redistribution of the stress after yielding. Consequently, the results are only qualitative. Nevertheless, the predictions provide a useful comparison for the experimental results discussed herein. Figure 7 shows a plot of the predicted deformation zone size as a function of bond thickness. The similarity with Figure 6 is striking.

In the area of composite stress analysis, Crews et al. (*101*) performed a three-dimensional micromechanics analysis that included the individual fibers. The results support the idea that the deformation zone extends to resin that is separated from the crack tip by a number of fibers. Even with the rather brittle resin used in their analysis, the results show some plastic deformation in resin up to four fibers away from the crack tip. This observation contrasts with suggestions by other authors (*96–98, 102*) that the deformation zone for brittle materials does not extend outside the resin layer surrounding the crack. Moreover, the analysis by Crews and co-workers demonstrates that the deformation zone in the composite is actually larger than that in the bulk resin for brittle materials. This factor may explain why the experimental results for interlaminar fracture energies are greater than the corresponding

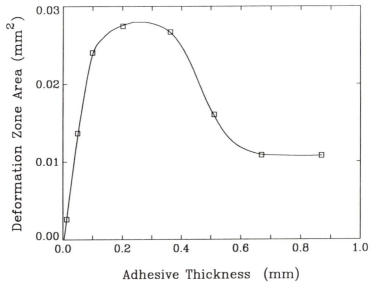

*Figure 7. Predictions of deformation zone size at the critical load as a function of bond thickness.*

resin fracture energies for brittle polymers. Further studies of these contrasting views are needed to clarify the situation.

Although papers such as those by Crews et al. have made an important contribution to understanding failure behavior, they focus on relatively brittle resins. More research, particularly with analyses that go beyond linear elasticity, is necessary because the apparent key to toughened resins is plastic flow or other types of complex deformation. A study by Liechti et al. (*103*) that measured the crack opening displacement behind the crack tip and compared the results with the predictions of a linear elastic analysis, illustrated the need for expanded research. The comparison gave good agreement when the loads were very low, but the differences increased as the loads were raised. As the failure load was approached, the differences were quite significant. Of course, deviations from linear elastic behavior would be expected at the high loads.

Experimental research also has extended the understanding of adhesive and composite fracture. Such studies will be illustrated with examples in three areas: adhesives, composites, and very thin adhesives.

In the first area, new studies have extended the previously mentioned work where movies were used to characterize the crack-tip deformation zone. Recent experiments employed video microscopy (*104, 105*) to provide considerably more detail on the events that occur prior to failure. The data show that the bulk resin, thick adhesive bonds, and thin adhesive bonds not only exhibit different zone sizes and shapes at failure, but also show significant differences in growth patterns for both the deformation zone and any subcritical crack growth that occurs before the onset of rapid crack growth. This data will be very useful for testing future stress analyses and failure models.

A second area of important experimental work is in composites. Bradley (*106*) observed the crack-tip region by conducting an interlaminar fracture test in a scanning electron microscope. The sample can be marked with a grid that permits quantitative measurement of the crack-tip strain field. This grid has been used for mode-I and mode-II experiments and has provided a wealth of extremely useful data that can be compared with results for both bulk resin samples and stress-analysis predictions. The full potential of these data has yet to be realized.

Bradley (*102, 106*) also examined the relationship between resin and composite toughness in mode-I loading. The results are consistent with the earlier work of Hunston et al. (*97–99*), but the resin toughness was extended to higher values and a second change in slope was found. Above a resin fracture energy of 5–6 kJ/m$^2$, a plot such as Figure 8 shows a plateau; that is, further increases in resin toughness have no effect on the interlaminar fracture energy. To test this observation, Bradley (*102*) evaluated polyether-etherketone samples and varied the matrix toughness by changing the loading rate. In this way it was possible to scan through the change in slope using a single material system, so factors other than matrix toughness were held

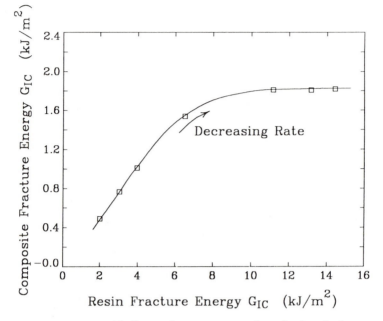

*Figure 8. Composite and bulk resin fracture energies for polyetheretherketone at various loading rates.*

relatively constant. The results shown in Figure 8 clearly support the idea of a plateau.

Bradley (*106*) attributed this change in slope to the presence of a process zone in which the resin is undergoing failure. He indicated that this zone is inside the deformation zone and, like the deformation zone, scales with toughness. Bradley asserted that, unlike the deformation zone, the process zone is truly limited by the fiber–fiber spacing and, therefore, once the process zone fills the resin layer between fibers, no further increase in composite toughness is possible.

The final area of recent experimental work to be discussed here involves very thin adhesive bonds. Thin adhesive bonds represent a bridge between the two previous topics in the sense that the thickness of the resin layer can be varied from that appropriate for adhesives to that characteristic of the resin layer between fibers in the composite. Chai (*107*) performed adhesive experiments at very small bond thicknesses and found that the decreasing fracture energy observed by Bascom for thin bonds (Figure 5) does not decrease monotonically, but exhibits an unexplained plateau at thicknesses in the same general range as the fiber–fiber spacing in composites. A comparison of these plateau values with the interlaminar fracture energies for the corresponding composite gives good agreement despite the limited number of materials examined (*108*).

This agreement is interesting because it suggests that the plastic deformation within the thin layer is adequate to achieve the measured fracture energy in the composite, which is contrary to the previous suggestion that attributes significant amounts of the toughness to plastic deformation of resin that is separated from the crack tip by a number of fibers. Resolution of this interesting conflict will require further study.

Chai (109) also examined failure of thin bonds in mode-II and -III loading. He found that the results for these two types of shear loading were the same within experimental scatter. The fracture energies are much higher than the mode-I value at large and intermediate bond thicknesses. As the thickness is decreased, however, the fracture energy drops dramatically. Chai again compared the results at appropriate bond thicknesses with the mode-II and -III interlaminar fracture energies and obtained a reasonable agreement. Unlike the mode-I case, however, the mode-II and -III adhesive tests did not exhibit a plateau at low thicknesses, which implies that the choice of thickness for use in the comparison makes a large difference. Consequently, there is considerable uncertainty in the correlation. If these results are correct, however, they suggest that mode-II and -III interlaminar fracture energies may be very dependent on the thickness of resin layers between plies. An experimental examination of this possibility would be very interesting.

A final interesting result from the data of Chai (110) is that the adhesive fracture energies for modes I, II, and III appear to be converging when they are extrapolated to very low thicknesses ($\sim 1 \, \mu m$). If this is true, one might speculate that these bond thicknesses are in a range where surface roughness dominates and so the specimen failure looks the same in all cases. Another possibility is that the deformation of the resin is so restricted that only certain common deformations are present. In any case, this convergence is an intriguing question for future study.

## References

1. Drake, R. S.; Siebert, A. R. *SAMPE Q.* **1975**, 6(4), 7.
2. Bucknall, C. B. *Toughened Plastics*; Applied Science Publishers: London, 1977.
3. Lee, H.; Neville, K. *Handbook of Epoxy Resins*; McGraw-Hill: New York, 1967; Chapter 16.
4. McGarry, F. J.; Willner, A. M. *Org. Coat. Plast. Chem.* **1968**, 28(1), 512–525.
5. Sultan, J. N.; McGarry, F. J. *Microstructural Characteristics of Toughened Thermoset Polymers*; Massachusetts Institute of Technology: Cambridge, MA, 1969; Research Report R69–59.
6. Sultan, J. N.; Laible, R. C.; McGarry, F. J. *J. Appl. Polym. Sci.* **1971**, 6, 127.
7. Sultan, J. N.; McGarry, F. J. *Polym. Eng. Sci.* **1973**, 13, 29.
8. Siebert, A. R.; Riew, C. K. *Org. Coat. Plast. Chem.* **1971**, 31, 555.
9. Riew, C. K.; Rowe, E. H.; Siebert, A. R. In *Toughness and Brittleness of Plastics*; Deanin, R. D.; Crugnola, A. M., Eds.; Advances in Chemistry 154; American Chemical Society: Washington, DC, 1976; p 326.

10. Bascom, W. D.; Cottington, R. L.; Jones, R. L.; Peyser, P. *J. Appl. Polym. Sci.* **1975**, *19*, 2545–2562.
11. Bascom, W. D.; Timmons, C. O.; Jones, R. L.; Peyser, P. *J. Mater. Sci.* **1975**, *10*, 1037–1048.
12. Bascom, W. D.; Jones, R. L.; Timmons, C. O. In *Adhesion Science and Technology*; Lee, L. H., Ed.; Plenum: New York, 1975; Vol. 9B, pp. 501–510.
13. Bascom, W. D.; Cottington, R. L.; *J. Adhes.* **1976**, *7*, 333–346.
14. Bascom, W. D.; Cottington, R. L.; Timmons, C. O. *J. Appl. Polym. Sci. Symp.* **1977**, *32*, 165–188.
15. Bascom, W. D.; Hunston, D. L.; Timmons, C. O. *Org. Coat. Plast. Chem.* **1978**, *38*, 179–184.
16. Bascom, W. D.; Oroshnik, J. *J. Mater. Sci.* **1978**, *13*, 1411–1418.
17. Rabionwitz, S.; Beardmore, P. *CRC Rev. Macromol. Sci.* **1972**, *1*, 1.
18. Kambour, R. P. *J. Polym. Sci. Macromol. Rev.* **1973**, *D7*, 1.
19. Kausch, H. H. *Polymer Fracture*; Springer Verlag: Berlin, Germany, 1978; Chapter 9.
20. Kramer, E. J. *Developments in Polymer Fracture*; Andrew, E. H., Ed.; Applied Science Publishers: London, 1979.
21. Kausch, H. H., Ed. *Adv. Polym. Sci.* **1983**, *52/53*.
22. Passaglia, E. *J. Phys. Chem. Solids.* **1987**, *48(11)*, 1075.
23. *Toughness and Brittleness of Plastics*; Deanin, R. D.; Crugnola, A. M., Eds. Advances in Chemistry 154; American Chemical Society: Washington, DC, 1974.
24. Kinloch, A. J.; Young, R. J. *Fracture Behavior of Polymers*; Elsevier Applied Science Publishers: London, 1983.
25. *Proceedings of the International Conference on Toughening of Plastics*; The Plastics and Rubber Institute: London, 1985.
26. Kinloch, A. J. *Adhesion and Adhesives*; Chapman and Hall: New York, 1987.
27. *Toughened Composites*, STP 937; Johnston, N. J., Ed.; American Society for Testing and Materials: Philadelphia, PA, 1987.
28. *Rubber-Modified Thermoset Resins*; Riew, C. K.; Gillham, J. K., Eds.; Advances in Chemistry 208; American Chemical Society: Washington, DC, 1984.
29. *Rubber-Toughened Plastics*; Riew, C. K., Ed.; Advances in Chemistry 222; American Chemical Society: Washington, DC, 1989.
30. *Rubber-Toughened Plastics: Science and Engineering*; Riew, C. K.; Kinloch, A. J., Eds.; Advances in Chemistry 233; American Chemical Society: Washington, DC, 1993, this volume.
31. Merz, E. H.; Claver, G. C.; Baer, M. *J. Polym. Sci.* **1956**, *21*, 482.
32. Merz, E. H.; Claver, G. C.; Baer, M. *J. Polym. Sci.* **1957**, *22*, 325.
33. Kunz-Douglass, S.; Beaumont, P. W. R.; Ashby, M. F. *J. Mater. Sci.* **1980**, *15*, 1109.
34. Kunz-Douglass, S.; Beaumont, P. W. R.; Ashby, M. F. *J. Mater. Sci.* **1980**, *15*, 1109.
35. Sayer, J. A.; Kunz, S. C.; Assink, R. A. *Polym. Prepr.* (*Am. Chem. Soc. Div Polym. Mater. Sci. Eng.*) **1983**, *49*, 442.
36. Kinloch, A. J.; Shaw, S. J.; Tod, D. A.; Hunston, D. L. *Polymer* **1983**, *24(4)*, 1355.
37. Raghava, R. S. *Proc. 29th National Symposium and Exhibition*; Society for the Advancement of Material and Process Engineering: Covina, CA, 1984; pp 1384–1394.
38. Bucknall, C. B.; Gilbert, A. H. *Polymer* **1989**, *30(2)*, 213–217.
39. Riew, C. K.; Smith, R. W. In *Rubber-Toughened Plastics*; Riew, C. K., Ed.; Advances in Chemistry 222; American Chemical Society: Washington, DC, 1989; pp 225.

40. Yee, A. F. *Polym. Mater. Sci. Eng.* **1990**, *63*, 286–290.
41. Sauer, J. A.; Marin, J.; Hsial, C. C. *J. Appl. Phys.* **1949**, *20*, 507.
42. Hsial, C. C.; Sauer, J. A. *J. Appl. Phys.* **1950**, *21*, 1071.
43. Kambour, R. P. *Nature* **1962**, *195*, 1299.
44. Kambour, R. P. *Polymer* **1964**, *5*, 143.
45. Kambour, R. P. *J. Polm. Sci.* **1964**, A-2(*2*), 4159.
46. Kambour, R. P. *J. Polym. Sci.* **1965**, A-3, 1713.
47. Kambour, R. P. *J. Polym. Sci.* **1966**, A-4, 349.
48. Bucknall, C. B. *Toughened Plastics*; Applied Science Publishers: London, 1977; Chapters 6 and 7, p 189.
49. Wang, T. T.; Matsuo, M.; Kwei, T. K. *J. Appl. Phys.* **1971**, *42*, 4188.
50. Wang, T. T.; Matsuo, M.; Kwei, T. K. *Polym. Prepr.* (*Am. Chem. Soc. Div. Polym. Chem.*) **1971**, *12*(*1*), 671.
51. Wang, T. T.; Matsuo, M.; Kwei, T. K. *J. Polym. Sci. Part A-2* **1972**, *10*, 1085.
52. Matsuoka, S.; Daane, J. H. *Polym. Prepr.* (*Am. Chem. Soc. Div. Polym. Chem.*) **1969**, *10*-2, 1198.
53. Lange, F. F.; Radford, K. C. *J. Mater. Sci.* **1970**, *6*, 1197.
54. Lange, F. F. *Philos. Mag.* **1970**, *22*, 983.
55. Lange, F. F. *J. Am. Ceram. Soc.* **1971**, *54*, 614.
56. Owen, A. B. *J. Mater. Sci.* **1979**, *14*, 2521.
57. Rowe, E. H.; Riew, C. K. *Soc. Plast. Eng. Tech. Pap.* **1974**, *20*, 663–665.
58. Rowe, E. H.; Riew, C. K. *Plast. Eng.* **1975**, *31*(*3*), 45.
59. Bucknall, C. B., Yoshii, T. *Br. Polym. J.* **1978**, *10*, 53.
60. Yoshii, T., Ph.D. Thesis, Canfield Institute of Technology, Canfield, England, 1975.
61. Bucknall, C. B. *Toughened Plastics*; Applied Science Publishers: London, 1977; pp 82–87.
62. Pearson, R. A.; Yee, A. F. *Polym. Mater. Sci. Eng.* **1983**, *49*, 316–320.
63. Pearson, R. A.; Yee, A. F. *J. Mater. Sci.* **1989**, *24*(*7*), 2571–2580.
64. Bitner, J. L.; Rushford, J. L.; Rose, W. S.; Hunston, D. L.; Riew, C. K. *J. Adhe.* **1981**, *13*, 2.
65. Riew, C. K.; Smith, R. W. *J. Polym. Sci. Part A-1*, **1971**, *9*, 2737.
66. Pearson, R. A.; Yee, A. F. *J. Mater. Sci.* **1986**, *21*(*7*), 2475–2488.
67. Li, D.; Li, X.; Yee, A. F. *Polym. Mater. Sci. Eng.* **1990**, *63*, 296–300.
68. Yee, A. F.; Pearson, R. A. *J. Mater. Sci.* **1986**, *21*(*7*), 2462–2474.
69. Hunston, D. L.; Kinloch, A. J.; Shaw, S. J.; Wang, S. S. In *Adhesive Joints*; Mittal, K. L., Ed.; Plenum: New York, 1984; p 789.
70. Hunston, D. L.; Bullman, B. W. *Int. J. Adhe. Adhesives* **1985**, *5*(*2*), 69.
71. Hunston, D. L.; Bascom, W. D.; Wells, E. E.; Fahey, J. D.; Bitner, J. L. In *Adhesion and Adsorption of Polymers, Part A*; Lee, L. H., Ed.; Plenum: New York, 1980; p 321.
72. Hunston, D. L.; Carter, W. T.; Rushford, J. L. In *Developments in Adhesives–2*; Kinloch, A. J., Ed.; Applied Science Publishers: London, 1981; Chapter 4.
73. Kinloch, A. J.; Hunston, D. L. *J. Mater. Sci. Lett.* **1987**, *6*, 131.
74. Goodier, J. N. *Trans. ASME* **1933**, *55*, 39.
75. Broutman, L. J.; Panizza, G. *Int. J. Polymeric Mater.* **1971**, *1*, 95.
76. Guild, F. J.; Young, R. J. *J. Mater. Sci.* **1989**, *24*, 2454.
77. Huang, Y. Ph.D. Thesis, University of London, London, 1991.
78. Huang, Y.; Kinloch, A. J. *J. Mater. Sci.* **1992**, *27*, 2763.
79. Kinloch, A. J.; Shaw, S. J.; Tod, D. A.; Hunston, D. L. *Polymer* **1983**, *24*, 1341.
80. Kinloch, A. J. In *Rubber-Toughened Plastics*; Riew, C. K., Ed.; Advances in Chemistry 222; American Chemical Society: Washington, DC, 1989; p 67.

81. Bowden, P. B. In *The Physics of Glassy Polymers*; Haward, R. N., Ed.; Applied Science Publishers: London, 1975; p 279.
82. Kinloch, A. J.; Young, R. J. *Fracture Behavior of Polymers*; Applied Science Publishers: London, 1983; p 303.
83. Sjoerdsma, S. D. *Polym. Commun.* **1989**, *30*, 106.
84. Huang, Y.; Kinloch, A. J. *J. Mater. Sci. Lett.* **1992**, *11*, 484.
85. Knott, J. F. *Fundamentals of Fracture Mechanics*; Butterworths: London, 1979; p 32.
86. Kinloch, A. J.; Finch, C. A.; Hashemi, S. *Polymer* **1987**, *28*, 323.
87. Bascom, W. D.; Hunston, D. L. In *Rubber-Toughened Plastics*; Riew, C. K., Ed.; Advances in Chemistry 222; American Chemical Society: Washington, DC, 1989; Chapter 6 and references therein.
88. Hunston, D. L.; Bitner, J. L.; Rushford, J. L.; Oroshnik, J.; Rose, W. S. *J. Elastomers Plast.* **1980**, *12*, 133.
89. Wang, S. S., personal communication, 1978.
90. Wang, S. S.; Mandell, J. F.; McGarry, F. J. *Int. J. Fracture* **1978**, *14*, 39.
91. Hunston, D. L.; Rushford, J. L.; Wang, S. S.; Kinloch, A. J. *Proceedings of the 37th Annual Conference of the Reinforced Plastics Composite Institute*; The Society of Plastics Industries: New York, 1982; Section 29–C, p 1.
92. Hunston, D. L.; Wang, S. S.; Kinloch, A. J. *Prepr. Am. Chem. Soc. Div. Polym. Mater. Sci. Eng.* **1982**, *47*, 408.
93. Hunston, D. L.; Kinloch, A. J.; Wang, S. S. *19th International SAMPE Technical Conference*, Society for the Advancement of Material and Process Engineering: Covina, CA, 1987; p 142.
94. Williams, J. G.; Rhodes, M. D. *The Effect of Resin on the Impact Damage Tolerance of Graphite–Epoxy Laminates*; NASA Langley Research Center: Hampton, VA, October 1981; Technical Memorandum 83213.
95. Rhodes, M. D.; Williams, J. G. *5th DOD/NASA Conference on Fibr. Composites in Structural Design*; New Orleans, January 1981.
96. Bascom, W. D.; Bitner, J. L.; Moulton, R. J.; Siebert, A. R. *Composites* **1980**, *11*, 9.
97. Hunston, D. L. *Compos. Tech. Rev.* **1984**, *6(4)*, 176.
98. Hunston, D. L.; Moulton, R. J.; Johnston, N. J.; Bascom, W. D. In *Toughened Composites*; Johnston, N. J., Ed.; ASTM STP 937; American Society for Testing and Materials: Philadelphia, PA, 1987; p 74.
99. Hunston, D. L.; Dehl, R. *Technical Paper EM* 87–355; Society of Mechanical Engineers: Dearborn, MI, 1987.
100. Crews, J. H., Jr.; Shivakumar, K. N.; Raju, I. S. *Factors Influencing Elastic Stresses in Double Cantilever Beam Specimens*; NASA Langley Research Center: Hampton, VA; November 1986; Technical Memorandum 89033.
101. Crews, J. H., Jr.; Shivakumar, K. N.; Raju, I. S. *A Fiber–Resin Micromechanics Analysis of the Delamination Front in a DCB Specimen*; NASA Langley Research Center: Hampton, VA; January 1988; Technical Memorandum 100540.
102. Bradley, W. L. In *Thermoplastic Composite Materials*; Carlsson, L. A., Ed.; Composite Materials Series 7; Elsevier: New York, 1991; Chapter 9.
103. Liechti, K. M.; Ginsburg, D.; Hanson, E. C. *J. Adhes.* **1987**, *23*, 123.
104. Hunston, D. L.; Kinloch, A. J.; Wang, S. S. *J. Adhes.* **1989**, *29*, 103–114.
105. Hunston, D. L.; Mizumachi, H.; McDonough, W. *Proceedings of the International Adhesion Conference*; The Plastics and Rubber Institute: London, England, 1990; p 37/1.

106. Bradley, W. L. *Proceedings of the Benibana International Symposium on How to Improve the Toughness of Polymers and Composites*; Yamagata, Japan, October 1990.
107. Chai, H. *Composite Materials Testing and Design*; American Society for Testing and Materials: Philadelphia, PA, 1986: ASTM STP 893, pp 209–231.
108. Chai, H. *Eng. Fracture Mech.* **1986**, *24*, 413–431.
109. Chai, H. *Int. J. Fracture* **1990**, *43*, 117–131.
110. Chai, H. *Int. J. Fracture* **1988**, *137*, 137–159.

RECEIVED for review December 12, 1991. ACCEPTED revised manuscript July 27, 1992.

# TOUGHENED
# THERMOPLASTICS

# Fracture-Toughness Testing of Toughened Polymers

Donald D. Huang

Polymer Products Department, E. I. du Pont de Nemours and Company, Wilmington, DE 19880–0323

*Two versions of the multispecimen J-integral method of ASTM E 813 are under consideration for use with toughened polymers. Although both versions provide reasonable fits to the experimental data, the power-law representation of the resistance curve (E 813–87) is more appropriate for characterization and design. The effects of specimen dimensions and side grooves are investigated, and the relationship between critical J values and fracture-energy initiation values is discussed.*

THE MULTISPECIMEN *J*-INTEGRAL TECHNIQUE HAS BEEN APPLIED to a variety of tough polymers including rubber-toughened polymers (*1–8*) and neat polymers (*9, 10*). This technique is of special interest because it provides a critical value for fracture toughness that can be used for material characterization and design. One advantage of the *J*-integral technique over standard linear elastic fracture mechanics techniques (*6*) is that the minimum specimen size required to obtain valid fracture toughness values is significantly smaller. With many multiphase polymer systems, the morphology in a large fabricated part may not be representative of the morphology found in a small laboratory test specimen because of agglomeration of the toughener phase during fabrication. *J*-integral methodology enables valid fracture-toughness tests to be conducted on test specimens that have morphologies similar to those found in actual parts.

Because a standard for *J*-integral testing of polymers currently does not exist, much of the earlier work was based on either ASTM E 813–81 or E

0065–2393/93/0233–0039$06.25/0

813–87. For example, references 1 and 6–8 followed the E 813–81 method, whereas references 3–5, 9, and 10 have applied the E 813–87 method.

Regardless of which version of E 813 was used in the earlier studies, the experimental data were successfully fitted to functional forms that were proposed for metals. Despite this consistency, the extent to which these techniques can be directly applied to polymers is still unclear.

The primary differences between the two versions lie in the data analysis and selection of the critical plane-strain toughness value ($J_{Ic}$). In E 813–81, the experimental data that comprise the resistance ($J$–$R$) curve are fitted to a bilinear function. The first line, which represents the blunting behavior of the material, is defined as

$$J = 2\sigma_y \Delta a \qquad (1)$$

where $\sigma_y$ is the yield stress that is determined in a separate tensile test and $\Delta a$ is crack growth. The second line is fitted to the experimental data points. The intersection of the two lines determines a critical value of $J$ ($J_c$). The physical interpretation of this critical value is that, with increasing $J$, the initially sharp crack blunts until it reaches $J_c$, from which point the crack grows in a self-similar manner. The total crack growth is composed of a blunting component and a crack-growth component. Below $J_c$, the crack-growth component is zero, whereas above $J_c$, the crack-growth component increases with increasing $J$. Therefore, $J_c$ is the $J$ value at which crack initiation occurs. If the test specimens meet the minimum American Society for Testing and Materials (ASTM) size recommendations, $J_c$ is considered to be a plane-strain value and relabelled $J_{Ic}$.

In E 813–87, the experimentally determined $J$–$R$ curve is fitted with a power law ($J = C_1 \Delta a^{C_2}$, where $C_1$ and $C_2$ are fitting parameters) instead of the bilinear fit. $J_c$ has been redefined to be equal to the $J$ value at which 0.2 mm of crack growth has occurred. Consequently, $J_c$ is determined by the intersection of the power-law fit to the experimental data and a line of slope $2\sigma_y$ that intersects the abscissa at 0.2 mm.

In earlier work (1), a modified version of ASTM E 813–81 was applied to injection-molded plaques of rubber-toughened nylons. The experimental data provided a good fit to the bilinear function. For rubber-toughened amorphous nylon, the fracture-toughness value was valid (according to ASTM recommendations) because it was obtained on specimens that were sufficiently large. However, in the case of rubber-toughened nylon 66, the specimens were 2 mm too thin. Additional tests on side-grooved specimens and on ungrooved specimens tested at slightly lower temperatures produced similar $J_c$ values, thus supporting the notion of a plane-strain value (2). Based on these results, the ASTM recommendations were found to be too conservative for the rubber-toughened nylon 66 system.

In this chapter the data-analysis schemes of both versions of E 813 are investigated to further consider the applicability of the *J*-integral method to toughened polymers. Unique *J*−*R* curves are determined for four toughened polymers. The effects of side grooves, specimen thickness, and depth are described. Finally, the relationships between critical-initiation (*J*) and fracture-energy initiation (*G*) values are examined.

## Experimental Details

**Materials**. The materials used in this study were rubber-toughened nylon 66 (RTN66; Zytel ST801), rubber-toughened amorphous nylon (RTAN; Zytel ST901), acrylonitrile−butadiene−styrene (ABS; Cycolac ABS, grade GSE), and a toughened polyphenylene oxide (TPPO; Noryl EN265). Both rubber-toughened nylons were injection-molded into plaques that were either $100 \times 250 \times 12.7$ mm or $100 \times 250 \times 3.2$ mm. The ABS and TPPO were obtained in both 25- and 50-mm-thick extruded sheets from Westlake Plastics (Lanni, PA). The materials were tested dry as molded at 23 °C and 50% relative humidity. The elastic moduli and yield strengths of the materials are listed in Table I.

**Specimen Geometries**.   Single-edged notched-bend (SENB) specimens were machined from either the plaques or the sheets. The specimens were deeply notched to half of the depth *W*. Unless stated otherwise, *W* was maintained at twice the thickness *B*. The span-to-depth ratio was held at 4 except for the 12.7-mm-thick specimens of toughened nylons where the ratio was 3.5. For the nylons, specimen thicknesses ranged from 12.7 to 3.2 mm. The thinner specimens (down to 6.4 mm) were made by milling equal amounts from the outer surfaces of the 12.7-mm-thick injection-molded plaques. In addition, 3.2-mm-thick specimens were cut from the 3.2-mm-thick plaques. For the ABS and TPPO, 25-mm-thick specimens were made from the 25-mm-thick sheet. Thinner specimens (down to 7.5 mm) were made by sawing the 50-mm sheet through the thickness. The sawn surface was smoothed by milling. The specimens showed no curvature, so it was assumed that residual stresses were minimal. In all cases, the thickness direction of the plaques or sheets was maintained as the thickness direction of the SENB

**Table I. Mechanical Properties**

| Material | E (GPa) | $\sigma_y$ (MPa) |
|---|---|---|
| RTN66 | 2.0 | 50 |
| RTAN | 2.0 | 69 |
| ABS | 2.3 | 48 |
| TPPO | 2.5 | 59 |

specimens. In addition, linear elastic fracture-toughness ($K_c$) tests were performed on 50-mm-thick SENB specimens.

For both RTN66 and RTAN, additional experiments were conducted using SENB specimens that were side-grooved with a blunt cutter (radius of curvature of 250 μm). The total depth of the grooves was 20% of the original thickness. ASTM recommends a total groove depth of 20–25%. Additional information on specimen geometries is given in reference 2.

### Modifications to the Multispecimen *J*-Integral Method.

The *J*-integral method that is under investigation is a multispecimen technique, similar to ASTM E 813. It was originally proposed by Landes and Begley (*11*). The first specimen is completely fractured to determine the ultimate displacement. Subsequent specimens are loaded to different subcritical displacements to obtain different levels of crack growth. From the area under the loading curve of each test, a value of *J* is calculated. Crack growth is marked and measured on the fracture surface. Resistance curves are then constructed. The test is considered valid if the specimen thickness meets the requirement that

$$B > 25(J_c/\sigma_y) \tag{2}$$

The depth *W* should also be greater than twice the minimum *B* determined by eq. 2. Some flexibility is allowed in the depth dimension (as discussed later).

In the current investigation, ASTM E 813 recommendations have not been strictly followed. The modifications to the experimental procedure are discussed in reference 2.

### Data Analysis.

A computer-controlled servohydraulic system was used for all mechanical testing. Software was developed to run the machine, acquire data in the form of load-displacement curves, and numerically integrate the curves to calculate energy values. After the specimen dimensions and ligament length were measured using the travelling microscope, a *J* value for each specimen was calculated.

Because these *J* values were calculated from the total energy measured from the area under the load–load point displacement curve, an indentation energy correction was also made. The correction accounted for local deformation at the loading and support points. A fully supported unnotched specimen of the same thickness and depth as the *J*-test specimen was indented with the load point. This test was conducted at the same rate as the *J* test. Again, a load–load point displacement curve was recorded. The contact stiffness ($S$) was found to be linear up to the maximum load ($P_{max}$) in the individual *J* tests. The energy due to indentation is then

$$U_{in} = 0.75\left( P_{max}^2/S \right) \qquad (3)$$

where $U_{in}$ is the total indentation energy.

For an SENB specimen, the total $J$ is given by $J_T = 2(U_T)/Bb$, where $U_T$ is the total energy, $B$ is the specimen thickness, and $b$ is the ligament. The indentation $J$ ($J_{in}$) can similarly be calculated from $U_{in}$. The real $J$ value for each test specimen is equal to $J_T - J_{in}$. Depending on the size of the specimen and the amount of crack growth, this correction was as high as 14% of $J_T$.

## Results and Discussion

**ASTM E 813–81 Analysis**. In earlier work (*1*), the $J$ results for both RTN66 and RTAN fit the bilinear form of the $J$–$R$ curve. Figures 1 and 2 show $J$–$R$ curves of this form for RTN66 for two different thicknesses. In these tests, the maximum crack growth was limited to 6% of the ligament as recommended in reference 13 to maintain $J$-controlled conditions. This limit differs from the stated recommendation of 1.5 mm in E 813 because the standard metal test specimen has a minimum thickness of 25 mm. In both cases, a straight line was fitted to the crack-growth part of the curve. The data show little scatter. In Figure 1, one datum point is near the blunting line. For

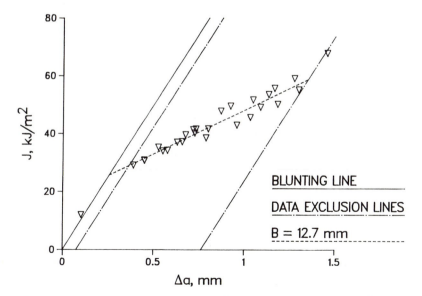

*Figure 1. E 813–81 representation of a* J–R *curve for RTN66;* B = 12.7 mm.

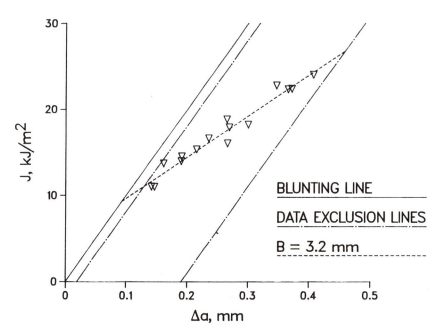

*Figure 2. E 813–81 representation of a* J–R *curve for RTN66;* B = *3.2 mm.*

these two cases, the $J_c$ values are 25.7 and 9.3 kJ/m$^2$ for the 12.7- and 3.2-mm-thick specimens, respectively.

Additional $J_c$ vs. thickness data for RTN66 and RTAN are presented in Figure 3. Although there is scatter in the $J_c$ values, there is a definite trend toward decreasing toughness with decreasing thickness for both materials. Average values are given in Table II, where $B_{min}$ is calculated from the ASTM minimum thickness recommendation ($25J_c/\sigma_y$).

For the RTN66, the $B_{min}$ value calculated from the large specimen tests suggests that valid plane-strain values are obtainable on 12.6-mm-thick or larger specimens. Additional studies on grooved specimens (2) support this recommendation. However, if the $B_{min}$ calculation is valid, larger $J_c$ values (greater than 25 kJ/m$^2$) are expected for the thin specimens because they occurred under mixed-mode conditions. For RTAN, the results are even more confusing. Based on the large specimen data, the ASTM recommendation suggests that specimens as thin as 4.6 mm can provide plane-strain fracture-toughness values. Thus, for this set of tests, the $J_c$ values for test specimens larger than 4.6-mm thickness should have been constant, but, as seen in Figure 3, the $J_c$ values decrease with decreasing thickness.

Because the smaller specimens were made by milling larger pieces of material, one explanation for these results is that the original moldings may have a tough skin with a less tough core. However, the fracture surfaces of all

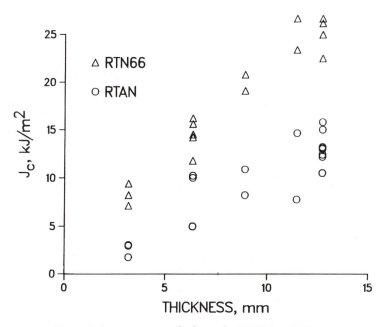

*Figure 3.* $J_c$ *vs. specimen thickness for RTN66 and RTAN.*

**Table II. $J_c$ Results for RTN66 and RTAN (E 813–81)**

| Material | Thickness (mm) | $J_c$ (kJ/m²) | $B_{min}$ (mm) |
|----------|----------------|---------------|----------------|
| RTN66 | 12.7 | 25.1 | 12.6 |
| RTN66 | 11.4 | 25.0 | 12.5 |
| RTN66 | 8.9 | 20.0 | 10.0 |
| RTN66 | 6.4 | 14.5 | 7.3 |
| RTN66 | 3.2 | 8.2 | 4.1 |
| RTAN | 12.7 | 12.6 | 4.6 |
| RTAN | 11.4 | 11.2 | 4.1 |
| RTAN | 8.9 | 9.6 | 3.5 |
| RTAN | 6.4 | 7.5 | 2.7 |
| RTAN | 3.2 | 2.5 | 0.9 |

the different-sized specimens are essentially flat without significant shear lip development (Figure 4). The crack-growth curves are all thumbnail-shaped, but again there is no significant difference between thick and thin specimens. If a tough skin were present on the thick specimens, a greater degree of curvature would be expected. In addition, the elastic moduli obtained from the tests were essentially constant for same-source specimens, which indicates

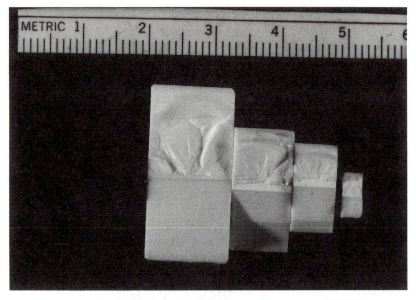

*Figure 4. Fracture surfaces for different-sized specimens of RTAN.*

that there is no strong morphological difference between the different-sized specimens. This hypothesis is supported by image analysis of thin sections from both regions where the morphologies (rubber content, particle-size distribution, and interparticle spacing) were similar. In conclusion, there appears to be no material difference between the core and the skin.

The dependence of $J_c$ on specimen size will be examined in a later section with respect to the E 813–87 analysis.

**ASTM E 813–87 Analysis.** ASTM E 813–87 suggests fitting the crack-growth data to a power law of the form $J = C_1 \Delta a^{C_2}$. Figures 5 and 6 show the power-law fits to the same data that are presented in Figures 1 and 2. Again, the data fit the power-law relationship very well for both cases. However, using this analysis, the $J$–$R$ curves and, therefore, the $J_c$ values for the two different-sized specimens are virtually identical given the experimental scatter in the data. Similar results were obtained by refitting the $J$ data in tests that used intermediate thicknesses. The power-law parameters and the $J_c$ values for the various tests for RTN66 and RTAN are presented in Table III.

Because the specimens were geometrically similar and the failure modes appeared to be the same, all the data from the different-sized $J$ tests were plotted on the same curve. Power-law relationships were fitted to this aggregate data. The results are shown in Figure 7 for RTN66 and RTAN. To

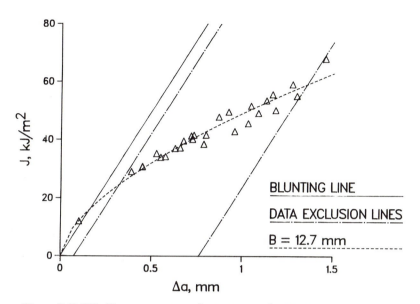

*Figure 5. E 813–87 representation of a* J–R *curve for RTN66; B = 12.7 mm.*

test the generality of the power-law relationship, two additional toughened blends, ABS and TPPO, were tested with specimens that were 7.6, 15.2, and 25.4 mm thick. The results are also given in Figure 7. The power-law representation of the *J–R* curve appears to be appropriate.

The power-law parameters and the $J_c$ values are given in Table IV. The RTN66 and RTAN results have been corroborated by additional *J* tests using 12.7-mm-thick specimens in which the minimum allowable crack-growth range was not limited to 0.6% of the ligament. In this set of experiments, the minimum allowable crack growth was 0.03 mm (0.2% b). Despite the seemingly large differences in the $C_2$ parameter for RTN66, the *J–R* curves for these extended crack-growth tests are similar to the aggregate results (Figure 8). Throughout the remainder of this chapter, the power-law fit to the aggregate *J* data is assumed to be the "reference" resistance curve.

The power-law representations of the *J–R* curves provide an explanation of the unexpected trend of decreasing $J_c$ with decreasing thickness that was found using the E 813–81 technique. E 813–81 represents the *J–R* curve with two straight lines: the blunting line and a linear approximation to part of the power-law curve. Because

$$J = C_1(\Delta a)^{C_2} \qquad (4)$$

$$dJ/da = C_1 C_2 (\Delta a)^{C_2 - 1} \qquad (5)$$

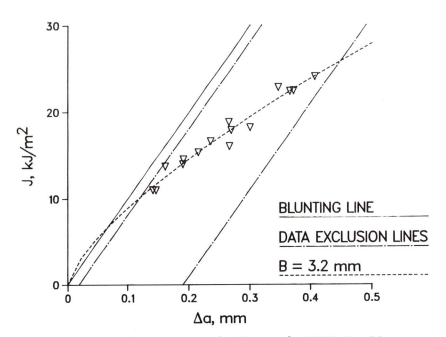

*Figure 6. E 813–87 representation of a* J–R *curve for RTN66;* B = 3.2 *mm.*

**Table III. $J_c$ and Power-Law Parameters**

| Material | Thickness (mm) | $C_1$ | $C_2$ | $J_c$ $(kJ/m^2)$ |
|---|---|---|---|---|
| RTN66 | 12.7 | 49.2 | 0.61 | 33.3 |
| RTN66 | 11.4 | 48.0 | 0.57 | 33.4 |
| RTN66 | 8.9 | 45.1 | 0.58 | 30.1 |
| RTN66 | 6.4 | 42.8 | 0.62 | 26.5 |
| RTN66 | 3.2 | 45.7 | 0.71 | 26.2 |
| RTAN | 12.7 | 35.2 | 0.63 | 17.3 |
| RTAN | 11.4 | 37.7 | 0.69 | 17.3 |
| RTAN | 8.9 | 33.5 | 0.63 | 16.3 |
| RTAN | 6.4 | 32.7 | 0.63 | 15.8 |
| RTAN | 3.2 | 36.4 | 0.74 | 15.2 |

for $C_2 < 1$, $dJ/da$ increases as $\Delta a$ decreases. Because $\Delta a$ is limited by 6% of the ligament to ensure $J$-controlled growth, $\Delta a$ decreases as specimen size decreases. Thus, small specimens will have steeper crack-growth lines. The intersection of these steeper lines with the same theoretical blunting line leads to lower calculated $J_c$ values.

One difference in the current study vs. the E 813 standard is the use of $J$ data obtained at $\Delta a$ values up to 6% of the ligament instead of a fixed value

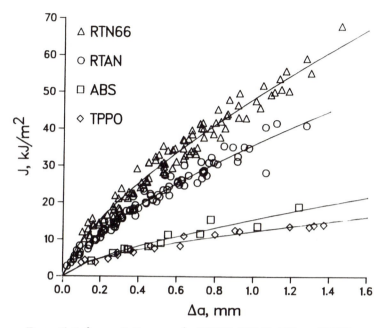

*Figure 7. Reference J–R curves for RTN66, RTAN, ABS, and TPPO.*

**Table IV. Reference Curve Parameters**

| Material | $C_1$ | $C_2$ | $J_c$ $(kJ/m^2)$ |
|----------|-------|-------|------------------|
| RTN66 | 48.2 | 0.70 | 29.2 |
| RTAN | 35.7 | 0.68 | 16.3 |
| ABS | 15.5 | 0.70 | 6.1 |
| TPPO | 12.3 | 0.58 | 5.4 |

(1.5 mm). If the resistance curve were fitted to $J$ data that included maximum $\Delta a$ values up to 1.5 mm instead of 6% of the ligament, the trend of decreasing $J_c$ values with decreasing thickness would result. One example of this trend is given in Figure 9, using 3.2-mm-thick specimens of RTN66, where maximum crack growth was nominally 50% of the ligament $b$. This amount of growth is outside of the 6% $b$ limit for $J$-controlled growth that was suggested by Shih et al. (*13*). Between 0.5 and 2.0 mm, the $J$ data are lower than the data obtained for larger specimens in the same $\Delta a$ range, apparently the effect of being outside of the $J$-controlled region. Thus, the calculated $J_c$ value would also be lower, approximately 18 kJ/m$^2$.

The trend of decreasing $J_c$ with decreasing thickness appears to be an anomaly caused primarily by the approximation of the $J–R$ curve as a bilinear function as recommended in E 813–81.

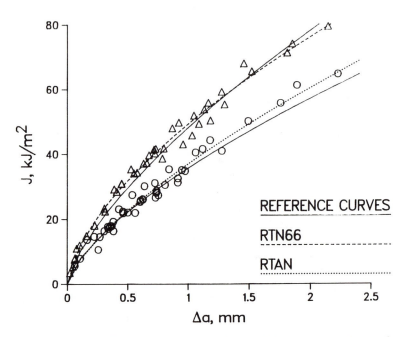

*Figure 8. RTN66 and RTAN J–R curves with extended minimum crack growth.*

It is interesting to note that the $J_c$ values of the nylons are higher than those obtained by E 813–81. This elevation in value results from the new definition of $J_c$ as the energy required for 0.2-mm crack growth. Although this value may be appropriate for design of large (> 25-mm-thick) metal structures, it is unclear whether it is appropriate for plastic components that are usually substantially smaller.

For characterization purposes, the merit of the $J_c$ value is also arguable. The $J_c$ value is obtained from the intersection of the resistance curve and a line that is parallel to the blunting line and has an $x$-axis intercept at 0.2 mm. This definition of $J_c$ is arbitrary. Also, for polymers, questions about the validity of the blunting line and its construction have been raised (7, 12). Use of this construction allows $J_c$ to be used to differentiate between the initial portions of $J–R$ curves if the materials have similar yield strengths. One alternative for characterization purposes is to compare the $J–R$ curves. For the materials tested here, the ranking is RTN66 > RTAN > ABS > TPPO.

**Specimen Size Effects.**   To demonstrate plane-strain conditions, the fracture behavior must be independent of size and geometry. Within experimental scatter and for each material, the $J–R$ curve obtained by fitting the

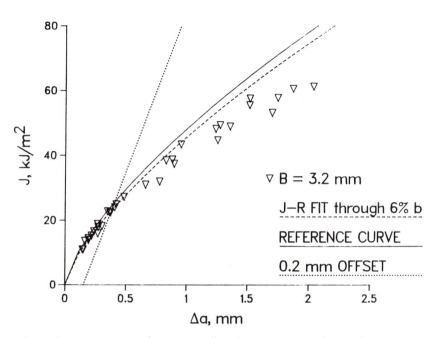

*Figure 9. RTN66 J data (B = 3.2 mm) with maximum crack growth greater than 6% b.*

individual data set for each specimen size was similar to the previously defined reference curve provided the data set was restricted to the *J*-controlled regime. Figure 9 illustrates the similarity of the reference curve with RTN66 *J* data for a specimen thickness of 3.2 mm. The power-law fit to the data that is less than 6% *b* compares favorably with the reference curve. If larger crack-growth data were included in the curve fit, the resulting curve would be flatter and the comparison with the reference curve would be less favorable.

Although a unique *J–R* curve could be generated by a range of specimen sizes, the reference curve met one of the conditions for a plane-strain resistance curve because it described *J–R* behavior that was independent of the in-plane bend bar geometry. (This is especially important because most applications involve thicknesses in this range of sizes.) The E 813–87 minimum thickness recommendations were 15 mm for RTN66, 6 mm for RTAN, 3 mm for ABS, and 2 mm for TPPO. Thus, the recommendation was satisfied for both ABS and TPPO. However, for the toughened nylons, size-independent behavior was found in specimens smaller than recommended by E 813–87. Based on these findings, the E 813–87 thickness recommendation for plane-strain conditions is too conservative.

**Effect of Side Grooves.**   According to the ASTM size recommendation, RTN66 is the only material that should not be in plane strain at the
largest specimen size. However, the recommended specimen sizes were 2
mm too small. For the rubber-toughened nylons, 12.7 mm is the practical
thickness limit for injection-molded plaques. Ideally, to test for geometry
independence and plane-strain conditions, larger specimens should be used
because proportionately more of the crack front is placed under plane-strain
conditions, which leads to lower crack-growth resistance and lower fracture-
toughness values. Another technique that accomplishes this goal is to side-
groove the specimens.

Results for 20% side-grooved specimens are given in Figures 10 and 11
for the rubber-toughened nylons. The $J$–$R$ curves for the side-grooved
specimens are in excellent agreement with curves obtained from the ungrooved specimens. The $J$–$R$ curve for RTAN (Figure 10) is virtually
identical to the reference curve. The $J$–$R$ curve for RTN66 (Figure 11) is
slightly lower at larger crack growths, but is well within the experimental
scatter of data shown in Figure 7. These findings reinforce the notion that the
reference curve is unique.

**Effect of Ligament Size.**   The maximum allowable crack growth
when subsized (less than 25.4-mm-thick) specimens are tested is not clearly
addressed in E 813. Crack growths between 0.15 and 1.5 mm are recom-

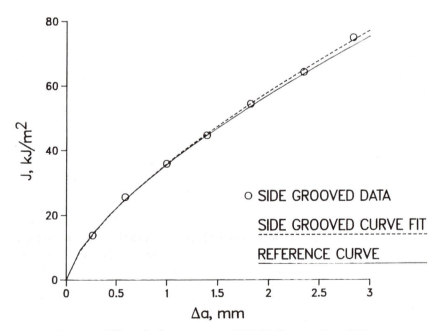

*Figure 10. Effect of side grooving on RTAN* $J$–$R$ *curve;* B = 12.7 *mm.*

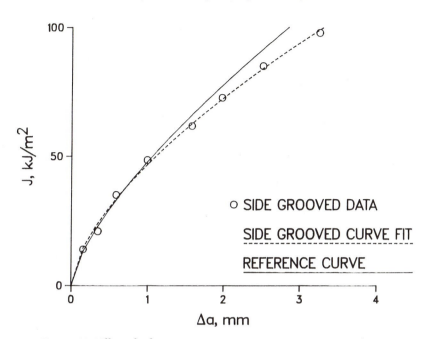

*Figure 11. Effect of side grooving on RTN66 J–R curve; B = 12.7 mm.*

mended for data selection and should be restricted to a maximum of 6% of the ligament to remain under *J*-controlled conditions. For small specimens with depths equal to twice the thickness, satisfaction of both conditions is impossible.

Three different specimen sets of RTN66 and RTAN with different B/W ratios were tested. Figures 12 and 13 show the results for 12.7-mm-thick SENB specimens that had W/B ratios of 1, 2, and 4 for RTN66 and RTAN, respectively. For comparison purposes, the reference curves are also plotted.

In Figure 12, little effect is observed on the *J–R* curve because of changes in the ligament. For the cases W/B = 2 and 4, the data fits the reference curve within experimental scatter. Interestingly, although the *J–R* data for the W/B = 1 set are consistent with the reference curve, the data appear to be on the low end of the scatter of the other sets. At small crack growths (three smallest crack growths corresponding to 6% of the ligament), the data fit fairly well. At intermediate crack growths (middle datum point at 1.06 mm; 9% of the ligament), the *J* value is low but still acceptable. At large crack growths (greater than 17%), the *J* values are significantly lower.

These data are consistent with the findings of McCabe et al. (*14*), who tested specimens of A508 Class 2A tube plate material with short ligaments. In that study, it was suggested that a short ligament is not large enough to sustain crack growths for the full *J–R* curve. Therefore, at small growths, the

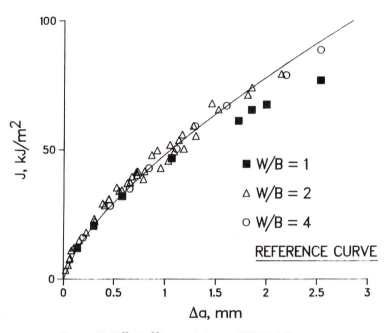

*Figure 12. Effect of ligament size on RTN66 J–R curve.*

data fit well, but at high crack growths, the $J$ values are low. Because of the scatter in the data, more tests are needed to verify this observation.

The $J$–$R$ data for the same specimen geometries for RTAN are given in Figure 13. In this data set, the data for $W/B = 1$ and 2 are reasonably consistent with each other and the reference curve. However, the $J$–$R$ data for the $W/B = 4$ specimen set produce a lower $J$–$R$ curve. Again, because of scatter, more tests are needed.

To explain these results, the minimum depth for plane-strain $J$ tests must be considered. Because there are no general recommendations for polymers, the E 813 standard was used as a first approximation. Use of SENB specimens with $1 < W/B_{min} < 4$, where $B_{min}$ is the minimum plane-strain thickness ($B_{min} = 25(J_{Ic}/\sigma_y)$), is the E 813 recommendation. The $J_c$ and $\sigma_y$ values given in Table III enable the determination of $B_{min} = 14.6$ and 5.9 mm for RTN66 and RTAN, respectively. It is interesting that acceptable results were obtained for RTN66 on specimens with $W/B_{min} = 0.87$ and 1.7, whereas low $J$ results were obtained on specimens with $W/B_{min} = 0.44$. Similarly, acceptable results were obtained for RTAN on specimens with $W/B_{min} = 1.1$ and 2.2, whereas low results were obtained with $W/B_{min} = 4.2$.

Based on these $W/B_{min}$ ratios, it appears that $W/B_{min}$ ratios of slightly less than 1 to slightly less than 4 may be appropriate for polymers if the full $J$–$R$ curve is needed. If only the beginning of the $J$–$R$ curve is needed, lower

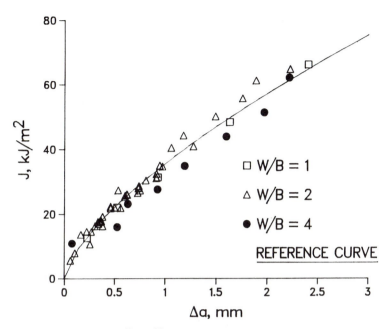

*Figure 13. Effect of ligament size on RTAN J–R curve.*

$W/B_{min}$ ratios may be appropriate. ASTM recommends $W/B_{min} = 2$ as a starting point, which appears to be sensible considering these data. Also, longer ligament depths are preferable to shorter ligament depths within the limits already described. It is premature to suggest that this specific calculation be adopted for $J$ testing of polymers because it is unclear whether the selection of $J_{Ic}$, the E 813 recommended $W/B_{min}$ ratios, or the $B_{min}$ calculation are valid for these materials. However, this calculation does suggest that ligament size limitations should be a function of a fundamental size parameter. Confirmation of this recommendation requires much more data on a variety of polymers.

**Initiation $J_c$–$G_c$ Relationships.** Begley and Landes (*15*) have shown that the $J_{Ic}$ value determined in the $J$-integral test and the $G_{Ic}$ measured in a standard fracture-toughness ($K$) test are equal if both tests are conducted under plane-strain conditions. This relationship was demonstrated on two different steels using the E 813–81 methodology.

   $K$ tests were conducted on 50-mm-thick SENB specimens made from the extruded sheets of ABS and TPPO. The $K$ tests were based on E 399 procedures with important differences as noted in the following text. The depths were 100 mm and $S/W$ was maintained at 4. The $K$ tests were conducted at the same rate as the $J$ tests (25 mm/s). The notches were made by drawing a fresh razor blade through a sawn prenotch. The ratios of the nominal notch depth to the specimen depth ($a/W$) were 0.25 and 0.40. The load displacement curves were slightly nonlinear. As an indication of the degree of nonlinearity, the ratios of the maximum load, to the load determined by the intersection of the curve and the secant line representing 95% of the initial stiffness ($P_{max}/P_5$) were less than 1.1 except for one test where the ratio was 1.14.

Because of the uncertainties associated with defining initiation when the load displacement curves are nonlinear, the initiation point was selected at the load that corresponded to a maximum of 2.5% crack growth. A secant construction analogous to that of E 399 was used. Additional details are provided in reference 16.

The average $K_c$ values for ABS were 5.44 MPa · m$^{1/2}$ for $a/W = 0.25$ and 5.79 MPa · m$^{1/2}$ for $a/W = 0.40$. Assuming a Poisson's ratio of 0.40 and a modulus of 2.3 GPa, the average (calculated) $G_c$ values are 10.9 kJ/m$^2$ and 12.1 kJ/m$^2$, respectively. For TPPO, the average $K_c$ values were 5.60 MPa · m$^{1/2}$ ($a/W = 0.25$) and 5.65 MPa · m$^{1/2}$ ($a/W = 0.38$). Again, assuming a Poisson's ratio of 0.40 and a modulus of 2.3 GPa, the corresponding $G_c$ values are 11.4 and 11.7 kJ/m$^2$, respectively. According to the E 399 size criteria, the specimens would be in plane strain. However, because nonstandard methods were used to determine initiation, it is not clear whether the size recommendations are valid for these tests.

The $G_c$ values for both ABS and TPPO are more than twice as large as their respective $J_{Ic}$ values listed in Table I. Because these values are determined by the E 813–87 construction, the values for the E 813–81 construction should be even lower. Thus, $J_{Ic}$ as defined by the current method may have potential as a design criterion because it is a conservative estimate of the $G_c$ value. However, it remains to be determined whether the $J_{Ic}$ value is too restrictive for polymer design.

For these materials, the discrepancy between $G_c$ (for 2.5% crack growth) and $J_{Ic}$ may be because of the definition of $J_{Ic}$ as the $J$ value required to extend the notch by 0.2 mm, the arbitrary selection of initiation in the $K$ test, or both criteria. For a fairer evaluation of these measurements, the amount of crack growth ($\Delta a$) that has occurred in the $K$ tests should be considered. A direct comparison of the resistance ($J$–$R$) curve from the $J$ test and a resistance ($G$–$R$) curve from the $K$ test can be made. The $K$-test resistance curve is constructed by selecting different points on the experimental load-displacement curve. The change in compliance at each point determines $\Delta a$. The corresponding $G$ value is calculated from the $K$ value determined for each load point.

The resistance curves for ABS are presented in Figure 14. Although there is good agreement between the curves at small crack growths, the $G$ values increase at a slower rate than the $J$ values. For example, up to approximately 1 mm, the curves are similar, but at 2.7 mm, the $G$ value is

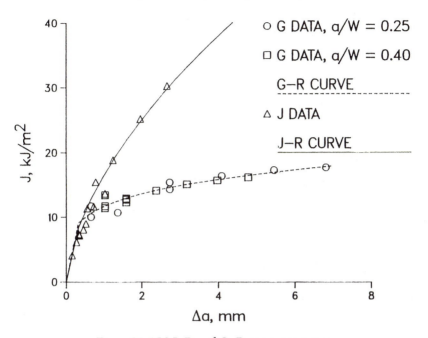

*Figure 14. ABS* J–R *and* G–R *resistance curves.*

approximately 50% of the *J* value. Interestingly, the intersection of these two curves occurs at 9.5 kJ/m². Similar resistance-curve behavior is exhibited by TPPO (*16*). These results suggest that the reference *J–R* curve may be used to calculate critical *G* values if initiation is defined at a specific amount of crack growth and if that amount of crack growth is small.

Although the discrepancies at large crack growths are significant, they may be exaggerated by the different methods used to determine crack growth. The crack growths in the *J* tests are measured from the fracture surfaces, whereas the *G* analyses use the offset secant method to calculate the crack growths. The offset secant method assumes that compliance changes are solely because of crack growth. Therefore, the calculated crack growths are upper limits to the actual crack growth. A more comprehensive study of the compliance–crack growth relationship in these polymeric systems is needed.

## Summary

Two versions of the *J*-integral method of E 813 were investigated to determine their suitability to measure the fracture toughnesses of toughened polymers. Experimentally, the procedures produce results that are consistent in form with *J* results for metals. Either the bilinear approximation of E

813–81 or the power-law approximation of E 813–87 yielded reasonable fits to the initial portions of the resistance curve. For the two cases tested, both methods gave conservative estimates of $G_{Ic}$ (the plane-strain fracture toughness).

Although both forms of E 813 gave reasonable results, the power-law (E 813–87) representation of the $J-R$ curve is more appropriate for toughened polymers because it avoids the issue of defining the initiation point. Instead, the power law defines the critical $J$ value at a specific (though arbitrary) amount of crack growth, thus circumventing the need to distinguish blunting crack growth versus actual crack growth. The power law also provides unique $J-R$ curves as a function of specimen thickness for a range of sizes that are representative of actual applications. Thus, the measured toughness value is independent of specimen size. Relative toughness can be assessed by direct comparison of the $J-R$ curves instead of relying on a single point on the curve.

Some of the recommendations of the $J$ method should be reexamined for use with toughened polymers. In particular, the minimum specimen thickness for plane-strain conditions is conservative and the allowable range of ligament depths requires better definition. Because the $J_{Ic}$ values appear to be conservative estimates of $G_c$, more investigation is needed to determine whether they are too conservative for design.

A promising alternative to a predefined $J_{Ic}$ value for design purposes may be application-dependent critical $J$ values. Using the resistance curve, these critical values may be obtained by determining the $J$ value that corresponds to the maximum allowable crack growth for a given application. In this study, this calculation was most successful when the crack growths were small. More extensive research is needed in this area.

## References

1. Huang, D. D.; Williams, J. G. *J. Mater. Sci.* **1987**, 22, 2503.
2. Huang, D. D. In *Rubber Toughened Plastics*; Riew, C. K., Ed.; Advances in Chemistry Series 222; American Chemical Society: Washington, DC, 1989; p 119.
3. Huang, D. D. In *Seventh International Conference on Deformation, Yield and Fracture of Polymers*; The Plastics and Rubber Institute: London, England, 1988; paper 46.
4. Huang, D. D. *Polymer Preprints* **1988**, 29(2), 159.
5. Huang, D. D. In *Advances in Fracture Research*; Salama, K., et al., Eds.; Pergamon Press: New York, 1989; p 725.
6. Hashemi, S.; Williams, J. G. *Polym. Eng. Sci.* **1986**, 26(11), 760.
7. Narisawa, I.; Takemori, M. T. *Polym. Eng. Sci.* **1989**, 29, 671.
8. Carling, M. J. Ph.D. Thesis, University of London, London, England, 1988.
9. Theuer, T.; Cornec, A.; Krey, J.; Friedrich, K. In *Seventh International Conference on Deformation, Yield and Fracture of Polymers*; The Plastics and Rubber Institute: London, England, 1988; paper 97.
10. Rimnac, C. M.; Wright, T. M.; Klein, R. W. *Polym. Eng. Sci.* **1988**, 28(24), 1586.

11. Landes, J. D.; Begley, J. A. In *Fracture Analysis*, ASTM Standard Technical Publication 560; American Society for Testing and Materials: Philadelphia, PA, 1974; pp 170–186.
12. Hashemi, S.; Williams, J. G. *Polymer* **1986**, 27, 384.
13. Shih, C. F.; Andrews, W. R.; de Lorenzi, H. G.; Vanstone, R. H.; Yukawa, S.; Wilkinson, J. P. D. *Crack Initiation and Growth under Fully Plastic Conditions: A Methodology for Plastic Fracture, Report NP–701–SR*; Electric Power Research Institute: Palo Alto, CA, February 1978; pp 6.1–6.63.
14. McCabe, D. E.; Landes, J. D.; Ernst, H. A. In *Elastic–Plastic Fracture: Second Symposium, Fracture Curves and Engineering Applications*; Shih, C. F.; Gudas, J. P., Eds.; Standard Technical Publication 803; American Society for Testing and Materials: Philadelphia, PA, 1983; Vol. II, p II-562–II-581.
15. Begley, J. A.; Landes, J. D. In *Fracture Toughness*; ASTM Standard Technical Publication 514; American Society for Testing and Materials: Philadelphia, PA, 1972; pp 1–20.
16. Huang, D. D. In *Elastic–Plastic Fracture Test Methods: The User's Experience*; Joyce, J. A., Ed.; ASTM Standard Technical Publication 1114; American Society for Testing and Materials: Philadelphia, PA, 1991; Vol. II, pp 290–305.

RECEIVED for review March 6, 1991. ACCEPTED revised manuscript July 6, 1992.

# Multiple-Phase Toughening-Particle Morphology

## Effects on the Properties of Rubber-Toughened Poly(methyl methacrylate)

**P. A. Lovell, J. McDonald, D. E. J. Saunders, M. N. Sherratt, and R. J. Young**

**Polymer Science and Technology Group, Manchester Materials Science Centre, University of Manchester Institute of Science and Technology, Grosvenor Street, Manchester M1 7HS, United Kingdom**

*Rubber-toughened poly(methyl methacrylate) (RTPMMA) materials were prepared by blending PMMA with multiple-phase toughening particles with radially alternating rubbery and glassy layers. The effects of particle morphology and rubbery-phase volume fraction on the tensile and fracture properties were evaluated. The principal mechanism of deformation operating in these materials, shear yielding with debonding and cavitation of the toughening particles, results in crack-tip blunting. The properties of the RTPMMA materials containing two-layer toughening particles are generally inferior to those containing three- or four-layer toughening particles.*

Eﾠ ARLY ATTEMPTS TO RUBBER-TOUGHEN poly(methyl methacrylate) (PMMA) concentrated on the use of suspension polymerization to produce composite beads consisting of both PMMA and rubbery phases (*1, 2*). Although these materials were commercialized, the morphology (and hence toughness) of artifacts produced from them was very strongly dependent on the molding conditions employed. There was, therefore, a need for better rubber-toughened (RT) PMMA materials.

During the past two decades substantial improvements have been achieved by preparing the toughening particles and the matrix PMMA separately (*3–5*). The toughening particles, prepared by emulsion polymer-

0065–2393/93/0233–0061$06.00/0

ization, typically comprise two to four radially alternating rubbery and glassy layers; the outer layer is always glassy polymer. These particles are cross-linked during their formation so that they retain their morphology and size during blending with PMMA and subsequent molding of the blends.

This route to RTPMMA allows independent control of the properties of the matrix PMMA and the composition, morphology, and size of the dispersed rubbery phase. Although these improved materials were success-fully commercialized by a number of companies (most notably ICI and Rohm and Haas), there are relatively few reports of investigations into their prepara-tion, properties, and deformation behavior. This chapter reports some results from a continuing research program aimed at elucidating the toughening mechanism(s) and optimizing toughening in these materials. The preparation and properties of materials containing two-, three- and four-layer toughening particles are described and discussed.

## Experimental Procedures

**Materials**.   Methyl methacrylate (MMA), ethyl acrylate (EA), $n$-butyl acrylate (BA), styrene (ST), and allyl methacrylate (ALMA) (polymerization-grade monomers, Aldrich, $\geq 98\%$), potassium persulfate (BDH AnalaR, $> 99\%$), and dioctyl sodium sulfosuccinate (Aerosol-OT, Cyanamid) were used as supplied. Distilled water was deionized (Elgacan C114) to a conduc-tivity $< 0.2$ µS/cm before use.

The matrix PMMA used for blending with the toughening particles was poly[(methyl methacrylate)-$co$-($n$-butyl acrylate)] with 8.0 mol % $n$-butyl acrylate repeat units [Diakon LG156 (ICI)]. Analysis of this material by gel permeation chromatography (GPC, polystyrene calibration) in tetrahydrofu-ran showed it to have number-average molar mass ($M_n$) of 47 kg/mol, weight-average molar mass ($M_w$) of 80 kg/mol, and $M_w/M_n$ of 1.7.

**Preparation of Toughening Particles**.   The two-, three- and four-layer (i.e., 2L, 3L, and 4L) toughening particles, shown schematically in Figure 1, were prepared by sequential emulsion polymerizations in which seed particles were first formed and then grown in either two or three stages.

2L          3L          4L                    0·3 µm

*Figure 1. Schematic diagrams of sections through the equators of the 2L, 3L, and 4L toughening particles, showing their sizes and internal structures. Rubbery layers are shown dark and glassy layers appear light.*

The base comonomer formulations used for the formation of the rubbery and glassy layers were BA:ST (78.2:21.8 mol %) and MMA:EA (94.9:5.1 mol %), respectively. ALMA was included at specific constant levels (in the range 0.1 < ALMA < 3.0 mol %) in the comonomer formulations used to form the inner layers (i.e., all of the rubbery layers and the inner glassy layers).

The reactions were performed at 80 °C under a flowing nitrogen atmosphere on a scale of approximately 9 dm$^3$ in a 10-dm$^3$ flanged reaction vessel. For each of the preparations, seed particles of 0.10-$\mu$m diameter were formed first by using Aerosol-OT at 0.6% by weight of the comonomer mixture and potassium persulfate as initiator at an aqueous-phase concentration of 0.5 mmol/dm$^3$. The seed particles then were grown by metered addition of appropriate quantities of the comonomer mixtures together with a synchronized concurrent addition of Aerosol-OT at 0.6 and 0.8% by weight of the comonomer mixture for the formation of the rubbery and glassy layers, respectively. At about 1-h intervals during the growth stages, further additions of potassium persulfate were made to increment its aqueous-phase concentration by approximately 0.4 mmol/dm$^3$ each time. Upon completion of the addition of the comonomer mixture for a given stage, sufficient time was allowed for complete conversion of these monomers before the addition of the comonomer mixture for the next stage was begun.

In the preparation of the 2L particles, the seed particles were grown first to a diameter of 0.19 $\mu$m by formation of rubbery polymer and then to a diameter of 0.23 $\mu$m by formation of the outer glassy layer. In the preparation of the 3L and 4L particles, the respective 0.10-$\mu$m-diameter glassy and rubbery seed particles were grown first to a diameter of 0.20 $\mu$m by formation of glassy polymer, then to a diameter of 0.28 $\mu$m by formation of rubbery polymer, and finally to a diameter of 0.30 $\mu$m by formation of the outer glassy layer.

The formation of these specific particle structures was confirmed by analyzing aliquots removed from the reactions for the percent monomer conversion (gravimetric analysis and gas–liquid chromatography) and for particle size distribution (photon correlation spectroscopy, either Coulter N4SD or Malvern Autosizer IIc).

For each type of toughening particle, the latex obtained from the emulsion polymerization was coagulated by addition to magnesium sulfate solution to yield loose aggregates of the particles. These aggregates were isolated by filtration, washed thoroughly with water, and then dried at 70 °C. In each case the coagulum was analyzed for magnesium by atomic absorption spectroscopy and found to contain less than 60 ppm.

**Preparation, Characterization, and Testing of the Blends.** The dried aggregates of toughening particles were blended with Diakon LG156 at 220 °C by either a single pass through a Werner Pfleiderer 30-mm twin-screw extruder or two passes through a Francis Shaw 40-mm single-screw extruder. This procedure ensured that the aggregates were disrupted to form a uniform dispersion of the primary particles, as confirmed by examining ultramicrotomed sections in a Philips 301 transmission electron microscope (TEM) operated at 100 kV. For each type of particle, four blends containing different weight fractions ($w_p$) of particles were produced. The nomenclature used to define these blends is illustrated by 3L22, which means that the blend contains 3L particles with $w_p = 0.22$.

The matrix PMMA and each of the blends were compression-molded into 3-, 6-, and 12-mm-thick plaques. The 3- and 6-mm-thick plaques were molded at 200 °C, whereas for the 12-mm-thick plaques a mold temperature of 220 °C was needed to obtain fully coherent moldings. Mechanical testing was performed at 20–22 °C by using the following procedures.

For tensile testing, dumbbell-shaped specimens were machined from rectangular strips (145 × 25 mm) cut from the 3-mm-thick plaques. The specimens had a gauge length of 45.0 ± 0.5 mm and width of 10.0 ± 0.5 mm in the constricted region. They were polished with successively finer grades of metallographic polishing paper and finally with Perspex antistatic cleaner. The tensile tests were performed on an Instron model 1122 mechanical testing machine with a cross-head displacement rate of 5 mm/min, which corresponds to a nominal strain rate of $2 \times 10^{-3}$ s$^{-1}$. Initial strains were measured using a 25-mm Instron extensometer. Six specimens of each material were tested, and mean values of the derived tensile properties were obtained.

Low-strain-rate three-point-bend fracture tests also were performed on the Instron model 1122 testing machine by using rectangular bars (length, 105.0 ± 0.5 mm; width, 24.0 ± 0.5 mm) that were cut from the 12-mm-thick plaques and milled to remove edge markings. A chevron notch approximately 10.8 mm long was machined into each specimen in accordance with ASTM E 399–83 and sharpened, just prior to testing, by pressing a fresh razor blade into the notch. The support span was set to four times the width of the specimen. The test was carried out by using a cross-head displacement rate of 10 mm/min, which corresponds to a nominal strain rate of $3 \times 10^{-3}$/s. The load-displacement curve was recorded; the specimen thickness, width, and initial crack length were measured after fracture by using a traveling microscope. The critical value of the stress-intensity factor (or fracture toughness) ($K_{Ic}$) was evaluated in accordance with an extension of ASTM E 399–83 that was described in a published protocol for plastics testing (6). The critical value of the strain-energy release rate (or fracture energy) ($G_{Ic}$) was then calculated from

$$G_{Ic} = \frac{K_{Ic}^2 (1 - \nu^2)}{E} \qquad (1)$$

where $E$ is Young's modulus and $\nu$ is Poisson's ratio. Values of $\nu$ were determined from volume-strain measurements performed on an Instron model 6025 mechanical testing machine with specimens (7) (gauge length, 100.0 ± 0.5 mm; width, 8.0 ± 0.5 mm) cut from the 3-mm-thick plaques and polished as for the tensile-test specimens. Axial strains were measured with a 50-mm Instron extensometer, and lateral strains with two low-mass, nonindenting extensometers (designed and made by ICI). Three specimens were tested to determine $v$ and 10 specimens were tested to determine $K_{Ic}$ and $G_{Ic}$ for each material. Confirmation that $K_{Ic}$ and $G_{Ic}$ were measured under plane-strain conditions was obtained by comparing the specimen dimensions to the usual criteria (6).

Instrumented falling-weight impact tests were carried out with an instrument designed and built by ICI (D. R. Moore, ICI Wilton Centre, Wilton, United Kingdom). Specimens (length, 70.0 ± 0.5 mm; width, 12.5 ± 0.5 mm) were cut from the 6-mm-thick plaques and polished with silicon carbide paper to remove edge markings. For each material, specimens were notched to predetermined depths to produce at least eight samples for each of five different notch-to-width ratios in the range 0.1–0.5. Just before a specimen was tested, the notch was sharpened by pressing a fresh razor blade into it. The specimen was then mounted on supports (50-mm span) and fractured in Charpy mode with a striker impact velocity of approximately 1 m/s. The amplified voltage output from a piezoelectric crystal positioned in the striker was recorded as a function of time and converted to a force-displacement curve from which the load and the energy absorbed at the onset of fracture were obtained. The specimen thickness, width, and initial crack length were measured after fracture by using a traveling microscope. The results for each material were analyzed by using established procedures to determine $G_{Ic}$ (8–10).

Fracture surfaces were examined by mounting samples on a metal stub, sputter-coating with a thin layer of gold (to prevent charging), and analyzing in an International Scientific Instruments 100A scanning electron microscope (SEM) operated at 10 kV.

## Results and Discussion

**Preparation of the RTPMMA Materials.** The design of the toughening particles takes into account a number of general requirements. Because many applications for RTPMMA materials demand a high percent transmission of visible light (i.e., comparable to that of the matrix PMMA), it is essential to prevent light scattering by the dispersed toughening particles. This goal is achieved by selecting the compositions of the rubbery and glassy layers so that their refractive indices are equal to that of the matrix PMMA (11). The particles have an outer glassy layer of just sufficient thickness (10–20 nm) to prevent particle coalescence during their isolation (i.e., so that they are only loosely aggregated after latex coagulation and drying, and can be redispersed to give uniform blends). The inner layers are cross-linked by using ALMA, which also serves to provide chemical graft-linking (3–5) at the interfaces between the layers, including that between the outer glassy layer and the preceding rubbery layer. Thus the particles retain their size and morphology upon blending, and a good particle–matrix interface results from mixing of the matrix PMMA with the outer glassy layer.

Figure 2 shows TEM micrographs of thin films cast from dispersions containing 0.04–0.16 $g/dm^3$ of toughening particles in a 40-$g/dm^3$ solution of matrix PMMA in toluene. The layer structures of the toughening particles are evident without the need for staining agents; the rubbery phases appear darker than the glassy phases, presumably because of the presence of the styrene repeat units. The outer glassy layer of the particles is not visible because its composition is almost identical to that of the matrix PMMA with which it mixes. The micrographs, therefore, give further confirmation that the toughening particles are of the required size and structure.

Some representative TEM micrographs of the RTPMMA materials, shown in Figure 3, illustrate the good dispersions of toughening particles achieved by the blending process.

**Tensile Testing.** The stress–strain curves presented in Figure 4 show that the toughening particles induce large-scale yielding in the matrix PMMA. This yielding leads to much greater fracture strains and greatly increases the energy required to cause fracture. In this respect the 2L particles are significantly less effective than the 3L and 4L particles.

For each type of toughening particle, Young's modulus ($E$) of the RTPMMA material decreases approximately linearly with the volume fraction ($V_p$) of the rubbery phase. For the materials prepared from the 3L and 4L particles, this volume fraction includes the internal glassy layers. The linear decrease can be seen from Figure 5a, which also shows that the decrease of $E$ with $V_p$ is greatest for the 2L particles and marginally greater for the 3L particles as compared to the 4L particles. Larger values of $E$ for the materials prepared from 4L particles (compared to those from 2L particles) have recently been predicted by finite element analysis (F. J. Guild, Queen Mary College, London, United Kingdom, unpublished work). These larger values result from the additional constraints introduced by the presence of internal layers in the 4L particles, in particular the internal glassy layer.

Yield stress ($\sigma_y$) and fracture stress ($\sigma_f$) also decrease with $V_p$ (see Figures 5b and 5c). Again the trend with the 2L particles is significantly different from the essentially identical trends with 3L and 4L particles. In general, for materials that show extensive yielding, the 2L particles give rise to lower values of $\sigma_y$ and $\sigma_f$ for a given value of $V_p$.

Further inspection of Figure 4 reveals that, for each type of RTPMMA material, the fracture strain ($\epsilon_f$) passes through a maximum as $w_p$ increases.

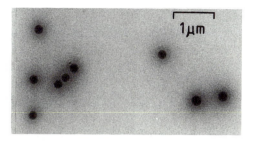

(a)

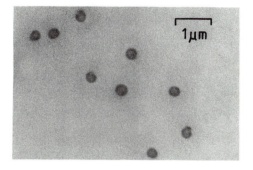

(b)

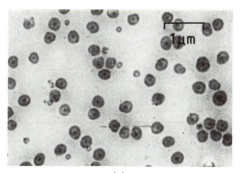

(c)

*Figure 2. TEM micrographs of ultrathin films of matrix PMMA: a, two-layer; b, three-layer; and c, four-layer toughening particles. The rubbery phases appear much darker than the glassy phases. The outer glassy layer of the particles is not visible because its composition is almost identical to that of the matrix PMMA with which it mixes.*

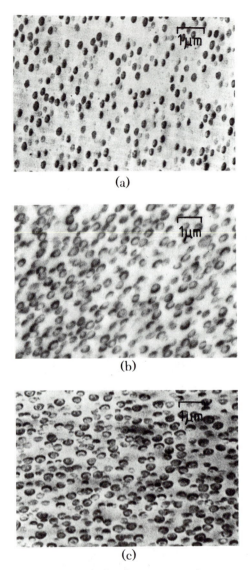

*Figure 3. TEM micrographs of ultrathin sections of representative RTPMMA materials: a, 2L20; b, 3L22; and c, 4L37.*

The many possible reasons for the decrease in $\epsilon_f$ at high $w_p$ include increased interactions among toughening particles, increased stresses in the interstitial matrix PMMA, and increased probabilities for the presence of defects arising from direct contacts among toughening particles.

A characteristic feature in the deformation of the RTPMMA materials (with the exception of 2L11) is the appearance of stress-whitened regions.

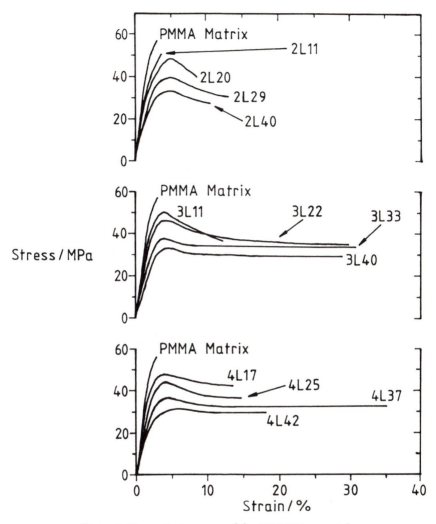

*Figure 4. Stress–strain curves of the RTPMMA materials.*

These regions, first observed just before the yield point at 3–4% strain, coincide with the appearance of shear bands at approximately 45° angles to the tensile axis. The whitening is concentrated in the shear bands and intensifies continuously as the materials are deformed to fracture, at which point they are densely whitened. The curves of volume strain against axial strain (obtained for measurements of $v$) show that the volume strains are negligible, in agreement with the observations of Hooley et al. (*12*) and Bucknall et al. (*13*) on RTPMMA materials similar to those reported here. Thus it may be generally concluded that RTPMMA materials deform principally by shear yielding. The subsequent section describes the results from

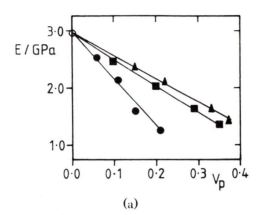

(a)

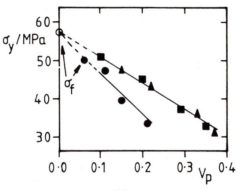

(b)

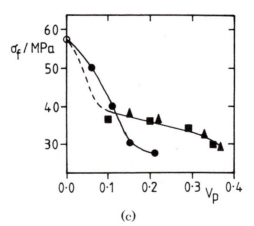

(c)

*Figure 5. Variation of tensile properties with particle structure and rubbery-phase volume fraction* ($V_p$): *a, Young's modulus* (E); *b, tensile yield stress* ($\sigma_y$); *and c, tensile fracture stress* ($\sigma_f$). *Key:* ●, 2L; ■, 3L; *and* ▲, 4L.

SEM analysis of fracture surfaces; evidence presented indicates that the stress-whitening arises from debonding and cavitation of the toughening particles.

**Fracture Testing**. The low-strain-rate values of $K_{Ic}$ for the RTP-MMA materials are significantly higher than that of the matrix PMMA (for which $K_{Ic}$ is 1.22 MPa m$^{1/2}$). This contrast demonstrates the increase in toughness resulting from inclusion of the toughening particles. Maxima observed in the variation of $K_{Ic}$ with $V_p$ (*see* Figure 6a) are probably a consequence of the opposing effects of increasing $V_p$ upon shear yielding and Young's modulus. Separate trends are observed for the 2L materials and for

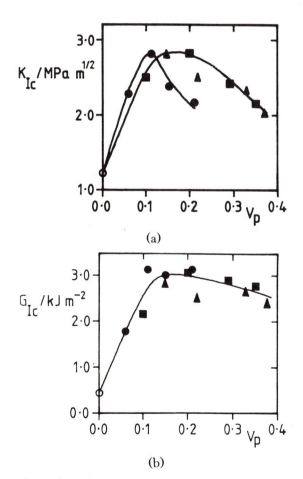

(a)

(b)

*Figure 6. Effects of toughening particle structure and rubbery-phase volume fraction ($V_p$) on low-strain-rate values of a, fracture toughness ($K_{Ic}$) and b, fracture energy ($G_{Ic}$). Key: ●, 2L; ■, 3L; and ▲, 4L.*

the 3L and 4L materials, though the maximum values of $K_{Ic}$ are similar ($\sim 2.8$ MPa m$^{1/2}$) and correspond to an increase in $K_{Ic}$ by a factor of approximately 2.3. The values of $G_{Ic}$ calculated from $K_{Ic}$, $E$, and $\nu$ show no strong dependence upon particle structure and also show a weaker maximum in their variation with $V_p$ (*see* Figure 6b). These trends are principally a consequence of the lower values of $E$ for the 2L materials and of higher values of $V_p$ for the 3L and 4L materials. The toughening effect is more clearly evident in the values of $G_{Ic}$, which for most of the RTPMMA materials are factors of 5.5–7.1 higher than that of the matrix PMMA (for which $G_{Ic}$ is 0.44 kJ/m$^2$).

The critical crack-tip radius ($\rho_c$) for crack propagation can be estimated by assuming that it is equal to half the critical value of the crack-opening displacement. On this basis, $\rho_c$ is given by

$$\rho_c = \frac{G_{Ic}}{2\sigma_y} \tag{2}$$

where $\sigma_y$ is the yield stress (*10*). The calculated values of $\rho_c$, shown in Figure 7, clearly reveal the operation of crack-tip blunting in the RTPMMA materials. The values of $\rho_c$ for the 2L materials are much higher than those for the 3L and 4L materials with similar $V_p$ because of their much lower $\sigma_y$ values.

Figure 8 shows the variation of impact $G_{Ic}$ with $V_p$ for each type of RTPMMA material. The performance of the 2L materials is relatively poor; their impact $G_{Ic}$ values are substantially lower than those obtained from low-strain-rate fracture testing (cf. Figures 6b and 8). The plateau value of $G_{Ic}$ observed for these materials at $V_p \sim 0.20$ is a factor of about 2.7 higher than the matrix PMMA impact $G_{Ic}$ (0.64 kJ/m$^2$), which is less than half of

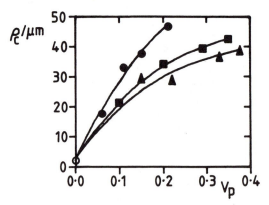

*Figure 7. Effects of toughening-particle structure and rubbery-phase volume fraction ($V_p$) on critical crack-tip radius ($\rho_c$). Key:* ●*, 2L;* ■*, 3L; and* ▲*, 4L.*

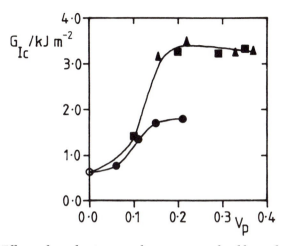

*Figure 8. Effects of toughening-particle structure and rubbery-phase volume fraction ($V_p$) on impact values of fracture energy ($G_{Ic}$). Key: ●, 2L; ■, 3L; and ▲, 4L.*

the enhancement achieved at the low strain rate. In contrast, the 3L and 4L materials remain tough under impact conditions and have essentially identical $G_{Ic}$ values at a given $V_p$. For these materials, impact $G_{Ic}$ values are slightly higher than the corresponding low-strain-rate values, but represent a slightly smaller enhancement (factors of 5.0–5.4 for $0.15 < V_p < 0.37$) over $G_{Ic}$ for the matrix PMMA.

There is considerable current interest in evaluating the importance of matrix ligament thickness ($\tau$), in particular the proposal of Wu (*14, 15*) that tough behavior will be observed only if $\tau$ is below a critical value ($\tau_c$) that is a characteristic property of the matrix material (for a given mode, rate, and temperature of deformation). The results presented here do not include sufficient data at low $V_p$ to enable brittle–ductile transitions to be evaluated precisely. However, they do show that under the same loading conditions the region of $V_p$ in which $G_{Ic}$ increases most rapidly is approximately the same for each type of RTPMMA material. These regions ($0.05 < V_p < 0.08$ at the low strain rate and $0.09 < V_p < 0.12$ under the impact conditions) were used to calculate corresponding ranges for $\tau_c$ by assuming a cubic lattice of toughening particles (*15*):

| RTPMMA Material | Low Strain Rate ($\mu m$) | Impact ($\mu m$) |
|---|---|---|
| 2L | $0.23 > \tau_c > 0.17$ | $0.15 > \tau_c > 0.12$ |
| 3L and 4L | $0.33 > \tau_c > 0.24$ | $0.22 > \tau_c > 0.18$ |

These estimates of $\tau_c$ are sufficiently accurate to reveal that, in contrast to the simple prediction of Wu, $\tau_c$ for the 2L materials is smaller than $\tau_c$ for the 3L and 4L materials. Therefore, $\tau_c$ is not completely insensitive to the structure and properties of the rubbery phase; the 2L particles are less efficient than 3L and 4L particles at toughening the matrix PMMA. The lower values of $\tau_c$ under the impact conditions are expected because of the increases in $E$ and $\sigma_y$ that result from the increase in strain rate (approximately 4 orders of magnitude) compared to the low-strain-rate testing.

**Fractography**. Major differences between the fracture surfaces of the unmodified matrix PMMA and the RTPMMA materials were revealed by SEM. Figure 9 shows micrographs that are typical of the fracture surfaces of tensile and fracture-test specimens of the matrix PMMA and the RTPMMA materials. The characteristic banded appearance of the matrix PMMA fracture surfaces arises from intermittent craze-branching at the crack tip (16). In distinct contrast, the fracture surfaces of the RTPMMA materials are much rougher. Most significantly, RTPMMA materials show holes and dome-like features with a range of diameters. Some of these features are of similar diameter to the toughening particles, but others are much smaller. Also, the number of these features increases as $w_p$ increases. These observations lead to the general conclusion that the holes and domelike features result from debonding and internal cavitation of the toughening particles. The domelike features are possibly caused by debonding at internal interfaces. This interpretation would also provide an explanation of the stress-whitening observed in the deformed RTPMMA materials.

## Conclusions

The results reported in this chapter show that the principal mechanism of deformation operating in the RTPMMA materials is shear yielding with debonding and cavitation of the toughening particles, and that this deformation results in crack-tip blunting. The tensile tests indicate that debonding and cavitation of the toughening particles just precede the formation of shear bands, which presumably are initiated by the voids created. These conclusions are in accord with previous observations on similar RTPMMA materials (12, 13, 17–19).

The properties of the 2L materials are generally inferior to those of the 3L and 4L materials. The 2L particles are less effective at toughening the matrix PMMA and cause greater reductions in $E$, $\sigma_y$, and $\sigma_f$ at a given value of $V_p$, as compared with 3L and 4L particles. The similar properties of the 3L and 4L materials indicate that the internal rubbery core in the 4L particles has no major effect on the deformation behavior of the blends.

On the basis of detailed investigations of deformation processes occurring just ahead of the crack tip in rubber-toughened polymers, Yee (20) proposed

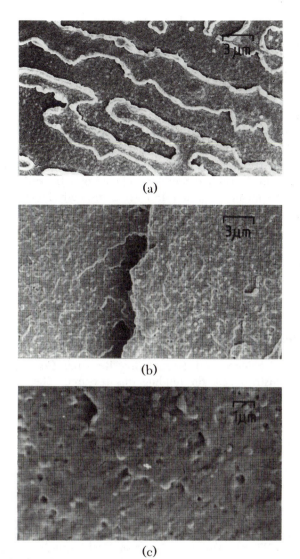

(a)

(b)

(c)

*Figure 9. SEM micrographs of fracture surfaces from impact testing: a, matrix PMMA showing features of breakdown of a single craze at the crack tip; b and c, 4L17, typical of the RTPMMA materials, shows holes and domelike features (small microcracks (about 1 μm long) caused by radiation damage can be seen at higher magnification).*

that certain conditions must be satisfied for maximum toughening. First, the toughening particles should be much smaller than the size of the crack tip and plastic zone. Secondly, they should debond or cavitate at stresses just below that for failure of the matrix material, thereby relieving triaxial tension and initiating the formation of shear bands. The RTPMMA materials broadly satisfy these conditions, although the design of the toughening particles could be modified to optimize the stress at which debonding–cavitation occurs.

Finite element (FE) analysis (21) shows that for the 2L materials the hydrostatic tensile stress in the rubber phase is approximately 0.02–0.03 times the applied stress, corresponding to about 1 MPa at the yield point. Preliminary FE calculations for the 4L materials (F. J. Guild, Queen Mary College, London, United Kingdom, unpublished work) revealed that the stress concentration factors are greatly increased by constraints resulting from the internal glassy layer. The stress concentration factor for the rubbery annulus is 0.21–0.25 and for the rubbery core is 0.10–0.13. The hydrostatic tensile stresses in the rubbery annulus and core of a 4L particle are therefore predicted to be about 8–10 and 4–5 MPa, respectively, at the yield point. Hence, the 4L and 3L particles are more likely to cavitate than the 2L particles. This observation may explain the generally inferior performance of the 2L materials. Further optimization of the toughening particle morphology is possible, therefore. Together with controlled changes in the degree of cross-linking and graft-linking within the particles, this optimization may lead to additional increases in toughness.

## Acknowledgments

We thank the Science and Engineering Research Council and Imperial Chemical Industries Ltd. for funding this research. The assistance of Pat Hunt, Bill Jung, Roy Moore, and Felicity Guild is gratefully acknowledged.

## References

1.  Imperial Chemical Industries Ltd., Br. 965,786, 1964.
2.  Imperial Chemical Industries Ltd., Br. 1,093,909, 1967.
3.  Rohm and Haas, Br. 1,340,025, 1973.
4.  Rohm and Haas, Br. 1,414,187, 1975.
5.  E. I. du Pont de Nemours and Company GB 2,039,496A, 1980.
6.  Williams, J. G., *EGF Newsletter*, Winter 1988/89, 8, 14 (Copies available from Prof. J. G. Williams, Department of Mechanical Engineering, Imperial College, London)
7.  Turner, S. *Mechanical Testing of Plastics*, Plastics and Rubber Institute: London, 1983.
8.  Marshall, G. P.; Williams, J. G.; Turner, C. E. *J. Mater. Sci.* **1973**, 8, 949.
9.  Williams, J. G. *Fracture Mechanics of Polymers*; Ellis Horwood: Chichester, England, 1984.
10. Kinloch, A. J.; Young, R. J. *Fracture Behaviour of Polymers*; Applied Science: London, 1983.
11. *Polymer Handbook*, 2nd Ed.; Brandrup, J.; Immergut, E. H., Eds.; Wiley-Interscience: New York, 1975.
12. Hooley, C. J.; Moore, D. R.; Whale, M.; Williams, M. J. *Plast. Rubb. Process. Appl.* **1981**, 1, 345.

13. Bucknall, C. B.; Partridge, I. K.; Ward, M. V. *J. Mater. Sci.* **1984**, *19*, 2064.
14. Wu, S. *Polymer*, 1985, 26, 1855.
15. Wu, S. *J. Appl. Polym. Sci.* **1988**, *35*, 549.
16. Doyle, M. J. *J. Mater. Sci.* **1983**, *18*, 687.
17. Milios, J.; Papanicolaou, G. C.; Young, R. J. *J. Mater. Sci.*, **1986**, *21*, 4281.
18. Frank, O.; Lehmann, J. *Colloid Polym. Sci.* **1986**, *264*, 473.
19. Shah, N. *J. Mater. Sci.* **1988**, *23*, 3623.
20. Yee, A. F., Presented at the 3rd European Symposium on Polymer Blends, Cambridge, England, July 1990.
21. Guild, F. J.; Young, R. J. *J. Mater. Sci.* **1989**, *24*, 2454.

RECEIVED for review March 6, 1991. ACCEPTED revised manuscript March 9, 1992.

# Toughened Semicrystalline Engineering Polymers

## Morphology, Impact Resistance, and Fracture Mechanisms

**E. A. Flexman**

**Du Pont Polymers, Experimental Station, E. I. du Pont de Nemours and Company, P.O. Box 80323, Wilmington, DE 19880-0323**

*Major factors that contribute to maximized impact resistance of semicrystalline engineering polymers include increased matrix molecular weight, minimal rubber glass-transition temperature, and optimal rubber-particle size, which depends on matrix type. Additional constraints are posed by the limitations of available materials and commercial requirements. Expanding the understanding of impact-modified crystalline engineering polymers will further extend their property ranges and utility. This chapter describes the phenomenology of certain factors, illustrates a new technique to measure rubber particle size, and contrasts the fracture mechanisms of poly(butylene terephthalate), nylon 66, and polyacetal.*

IMPACT MODIFICATION OF POLYMERS has been practiced for more than 60 years (*1*), but the mechanisms of rubber toughening are only partially understood. Knowledge of these rubber-toughening mechanisms has enabled improvement of semicrystalline engineering polymers to the highly toughened grades broadly used today.

Polymers are labeled engineering polymers because their physical properties allow them to be used as structural components in demanding applications. The most significant properties of engineering polymers are strength, stiffness, heat resistance and toughness. To ensure commercial viability of any polymer system, cost and the many aspects of processibility and aesthetics are

0065-2393/93/0233-0079$07.50/0

also important. The sum of these individual properties and considerations in any one polymer is what is finally purchased in the marketplace. Trade-offs such as stiffness–toughness, toughness–flow, and heat resistance–cost often limit the spectrum of properties available in a single polymer.

Toughness is one of the most difficult properties to reduce to a single parameter to aid designers in polymer selection. The components of strength, stiffness, rate of loading, crack-propagation resistance, and elongation vary dramatically with orientation, test temperature, part design and geometry, and molding conditions. In this chapter, toughness will be defined as a resistance to crack propagation. The notched Izod or Charpy tests are frequently used to determine crack propagation resistance because of their simplicity, convenience, and acceptance by the polymer-producing and -using community. Although these notched impact-resistance tests are inherently flawed because a basic material property is not measured, they are considered the most severe of the toughness tests.

The quest to improve polymer crack resistance in general, and in engineering polymers specifically has motivated the development of much new technology. Although the engineering polymers are all tough in an unnotched situation, the presence of molding flaws, environmental abuse, or poor design often act as stress risers to cause premature failure of molded objects. The commercial success of rubber-toughened versions of commodity polymers has been another factor in spurring the development of impact-modified versions of engineering polymers. As the value of toughened polymers became apparent to the ultimate customers, many of the lessons learned with styrenics, acrylates, and poly(vinyl chloride) have been applied successfully to engineering polymers.

The toughened semicrystalline engineering-polymer systems discussed in this chapter include polyesters, polyamides, and polyacetals, all of which are underrepresented in the literature of toughened plastics. A variety of reasons contribute to this lack of available information:

1. The relatively recent commercial appearance of these toughened engineering-polymer systems makes them less familiar to researchers.

2. Because these polymers are crystalline, they are opaque as a matrix. This opacity minimizes the usefulness of common optical techniques. The solvent techniques used with readily dissolved amorphous polymers are also more difficult to apply.

3. Additional levels of complexity are present in controlling the amount and type of crystallinity and, thus, the overall matrix properties (2–4).

4. Because of molecular orientation and microdifferences in thermal history, crystalline morphological variations commonly ex-

ist within a molded specimen. Interactions of a growing crack with these various crystalline structures must be considered (5, 6).

5. Because of higher shrinkage, specimens must be injection-molded to obtain optimum properties. This requirement is a significant barrier because few academic laboratories have such facilities.

6. Corporate secrecy as well as difficulty separating scientific content from commercial desires and legal requirements in patent publications have been problematic.

## *Poly(butylene terephthalate)*

Core-shell impact modifiers, as well as functionalized polyolefins and thermoplastic elastomers, are used commercially (7–9) to impact-modify poly(butylene terephthalate) (PBT). Blends of PBT with polycarbonate are also commercially important. Such blends are usually toughened with core-shell rubber tougheners (10). PBT, like most polymers, demonstrate an inverse relationship between matrix molecular weight and amount of impact modifier necessary to obtain ductile breaks. Higher molecular weight polymer requires less impact modifier, a combination that results in a stiffer product. However, higher molecular weight grades become increasingly difficult to injection-mold, and thus, commercial constraints intrude on ideal technical solutions.

Optimization of particle size is also an important factor in toughening PBT. Dispersed-phase particle size in polymer blends traditionally has been established by transmission electron microscopy (TEM) or scanning electron microscopy (SEM). Although the electron microscope has been a powerful tool in the study of toughened-polymer morphology for many years, it can have significant limitations. A relatively low number of particles can be observed simultaneously; rarely more than ~ 100. Because of the thickness of the slice and depth-of-field considerations, it is difficult to determine whether a particular particle is sectioned through its equator or through an edge, and thus measurements of its size and size distribution are questionable. Indeed, because most particles will not be sliced through the equator, a systematic error occurs toward a smaller average particle-size value than is truly present in the specimen. Selective stains often must be developed to observe particular phases. Although orientation of the particles caused by injection molding is an expected effect, unstable flow that causes portions of the section to be oriented perpendicular to the plane of the slice will display the particles smaller than they really are. Finally, perception factors cause larger and more distinct particles to be given more attention than small, less distinct particles in a visual evaluation of dispersion.

Many of the issues concerned with TEM measurement can be circumvented by the use of an alternative method of establishing domain dimensions. A particularly useful system for determining rubber particle size is a small angle X-ray scattering (SAXS) technique that was developed by Wilson (11) to measure a variety of size-related aspects of submicrometer particles in polymeric systems in which a sufficient electron-density difference exists. An ultrahigh resolution diffractometer (Bonse–Hart) allows resolution of particles and clusters of particles of 1 μm or smaller. The SAXS technique scans ~ $10^{12}$ particles and requires only 1 h for both acquisition and analysis with specifically developed software.

To obtain particle-size and size-distribution information from a SAXS measurement, the results are modeled by a log-normal distribution of independent spherical particles. Analysis of the invariant (the intensity times the square of the angle) vs. the angle 2θ leads to a size distribution. This size distribution may then be characterized by its median, mode, mean, and a distribution breadth factor. Figure 1 illustrates the general shape of and the type of information contained in the plots obtained by this method. Particles larger than several micrometers cannot be discerned because of interference from the beam. Smaller sizes occur at larger angles of 2θ. Because the SAXS

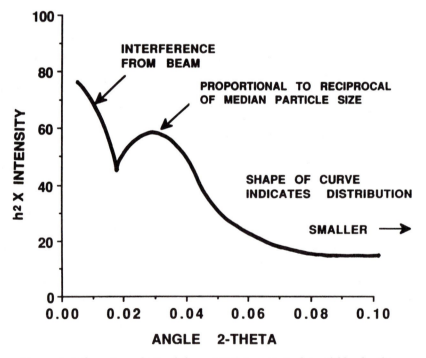

*Figure 1. Information obtained from SAXS invariant plots of blends of two polymers with different electron densities.*

method emphasizes the maximum particle width, it gives values slightly larger than values obtained by TEM.

The use of SAXS to study impact-modified blends of PBT is further illustrated in Figures 2 and 3. In Figure 2, the median particle size of a variety of blends toughened by a reactive polyolefin is plotted vs. their Izod values at $-30$ °C. Because a ductile–brittle transition is present, the Izod values between 4 and $\sim 12$ ft-lb/in. represent *average* values with a bimodal distribution of 4 or lower and 12 or higher. For a ductile response to the Izod test at $-30$ °C, PBT polymers of this molecular weight toughened by this polyolefin system (*12*) require a median particle size of $\sim 0.16$ μm or less. The actual particle size distribution of the impact-modifier particles in the impact-modified blends are of greater interest than their median size alone. Two distributions chosen to illustrate the actual particle sizes in a toughened PBT at the ductile–brittle transition of 0.16 μm and in a blend with a very small particle-size median of 0.08 μm (Figure 3).

## *Aromatic Polyester Fiber*

Poly(ethylene terephthalate) (PET) is chemically and physically very similar to PBT. A higher melting point, which provides a higher use-temperature advantage, is the primary difference. The negative side of the higher melting point attribute is a higher processing temperature, which is especially signifi-

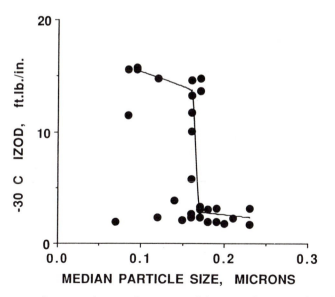

*Figure 2. Median particle size of impact-modifier particles in toughened PBT blends.*

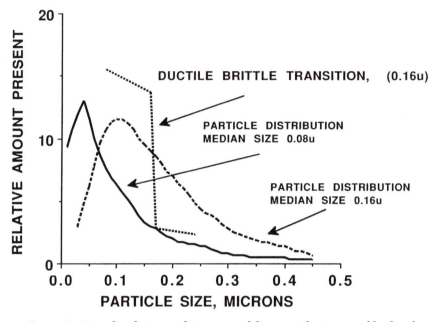

*Figure 3. Size distributions of impact-modifier particles in two blends of toughened PBT. One blend is at the median size for ductile–brittle transition at −30 °C; the other is smaller. Median size and distribution were measured by SAXS.*

cant in terms of toughening less thermally stable elastomers. Toughening systems commonly employed in PBT undergo thermal degradation at PET processing temperatures unless increased levels of antioxidant and additional thermal stabilization techniques are used. A commercially important advantage of PET is a lower price, which results from raw material economies of scale and the availability of scrap and recycled material.

Serious issues in the study of PET toughness arise from crystallization that is slower and less complete than that of PBT under molding conditions that have commercial utility. Crystallization packages that compensate for this PET characteristic are available, but their interaction with thermally stable rubber-toughening systems is complicated and will not be reported here.

## Nylon 66

In the Izod test, ductile breaks have been achieved in nylon 66 with a wide variety of impact modifiers (13). Nylon 66 was the first polyamide offered in a grade toughened beyond its ductile–brittle transition (14). The ductile–brittle transition in 1/8-in. dry-as-molded nylon 66 bars tested at room temperature occurs between 6 and 10 ft-lb/in. Although averages may be reported in this range, individual breaks are never observed with values between 6 and 10 ft-lb/in., and breaks between the broader range of 5 to 12 ft-lb/in. are rare.

In the initial scouting phase of the development work for highly toughened nylon 66, Izod measurements were made by cutting a 5-in. flex bar in half and testing both pieces. Observations indicated that an intermediate level of toughening occurred with certain impact modifiers when all the Izod values from one half of the bar (usually the end near the gate) were brittle and all those from the other half (the far end) were ductile. Figure 4 illustrates the Izod values along a bar that was notched every 1/8 in. from near the gate to near the far end. The term "supertough" was applied when both ends gave ductile breaks. Because the total amount of rubber present is one of the most important factors in toughening a matrix, simply adding more impact modifier often gives supertough blends. Figure 5 illustrates how increasing the level of impact modifier from 20% (used in Figure 4) to 30% gives a "supertough" material. Because additional impact modifier reduces

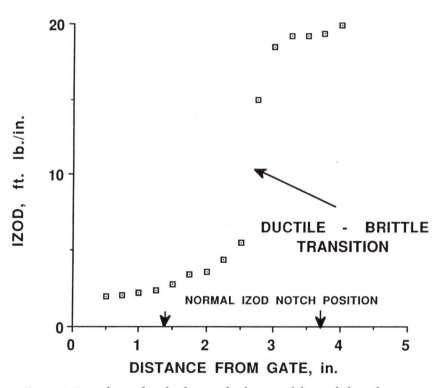

*Figure 4. Dependence of Izod values on the distance of the notch from the gate of an injection-molded bar of impact-modified nylon 66 blended with rubber of intermediate ability to toughen.*

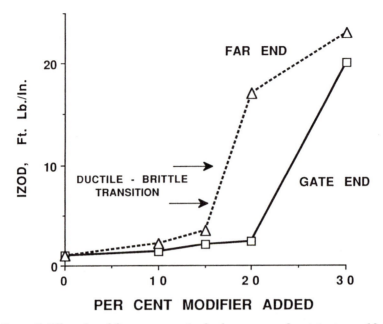

*Figure 5. Effect of modifier content on Izod values measured on injection-molded bars cut in half. The near-end values are notched 1.25-in. from the gate and the far-end values are notched 3.75 in. from the gate.*

other desirable properties of the nylon blend, such as stiffness, it is usually not practiced.

Rubber particle size proved to be the underlying cause of the brittle near-end–ductile far-end Izod values. When the nylon matrix was removed with formic acid, SEM observation of the impact-modifier particles (Figure 6) indicated that some of the particles near the gate were larger and highly elongated and thus much less efficient in their ability to toughen (15). Because 20% of this particular impact modifier produced a blend with bars at the ductile–brittle transition, the effect readily was detected and observed. The concentration of larger particles near the gate is believed to be an artifact of the injection-molding process, probably because of particle coalescence in the surface layers as the surface freezes and fresh melt is pushed beneath the surface. Coalescence has been investigated in other nongrafting rubber-toughened semicrystalline polymer systems (16).

Near-far ductile–brittle transition in a single bar can be corrected by several techniques, which include using a mixture of impact modifiers. In the experimental blend used in Figure 7, modifier A is similar to the modifier in Figure 4 in that both brittle near-end and ductile far-end Izod values are produced. At the 20% level, modifier B is less effective and gives brittle breaks on both ends of the bar. Figure 7 illustrates that, at certain compositional ratios of these impact modifiers, the partial replacement of the more

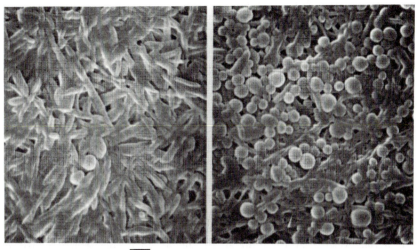

*Figure 6. Impact-modifier particles from the near (left) and far (right) ends of an injection-molded bar. The nylon matrix was removed by formic acid. (Micrographs supplied by P. N. Richardson.)*

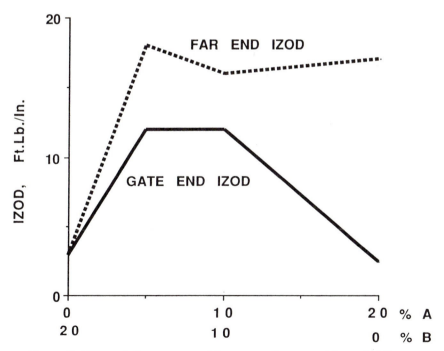

*Figure 7. Effect of the impact-modifier ratio of two modifiers of different toughening ability on Izod values from notches cut near and far from the gate of injection-molded specimens.*

effective impact modifier by a less efficient one results in ductile breaks all along the bar.

The SAXS technique was developed during the precommercial phase of Du Pont's supertough nylon product development. Application of this technique addressed many questions about the final blend morphology and rubber particle size. Figure 8 illustrates an early analysis of a blend with a brittle Izod value near the gate and a ductile Izod value at the far end of the bar. Although quantification of the median particle size was not available at that time, the results clearly showed that the particles in the near end were larger than those in the far end. In the rare instances where the reverse impact results (ductile breaks near the gate and brittle breaks at the far end) occurred, the SAXS analysis continued to show that tougher portions of the bar resulted from smaller particles (Figure 9).

Both compositional and processing effects can be monitored by SAXS. Figure 10 illustrates the results of changing only the molecular weight of a pair of toughened blends. Although the higher molecular weight matrix is expected to yield a higher Izod value, the SAXS scan demonstrates that higher shear yields smaller rubber particles that contribute an additional increment to the blend toughness. Figure 11 illustrates how different extru-

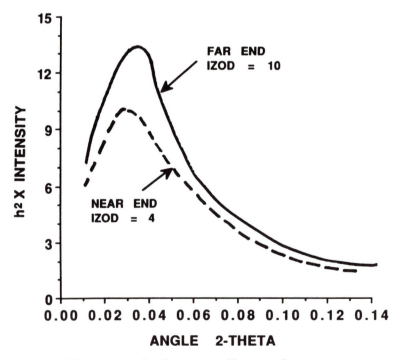

*Figure 8. SAXS invariant plot illustrating rubber-particle size at two positions along an injection-molded bar spanning a ductile–brittle transition.*

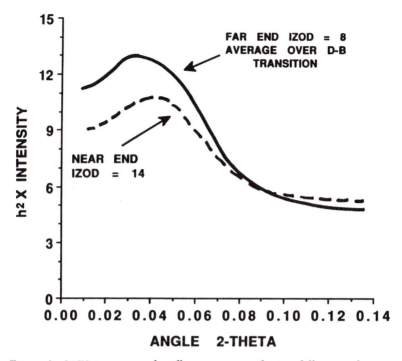

*Figure 9. SAXS invariant plot illustrating particle-size difference along an injection-molded bar with the less common feature of a near-end Izod value toughener than the far-end Izod value.*

sion rates affect minor differences in particle size and distribution. Many other significant variables can also be studied by this technique.

A further demonstration of the utility of the SAXS technique to measure the particle-size distribution is the characterization of the domain sizes of multicomponent blends. Figure 12 is a TEM of a toughened polyarylate dispersed in a toughened nylon (*17*). The large light domains are the toughened polyarylate, and lighter toughener particles can be seen within them. The small light particles are those of the nylon toughener.

Figure 13 illustrates the size distributions of all three discontinuous domains in the TEM of Figure 12. The size distributions of these three distinct types of domains are superimposed in this figure. The SAXS measurements were made on the toughened nylon and polyarylate polymers prior to blending. It was assumed that little change in toughener morphology occurred during combination of the two toughened polymers. Solvolytic removal of polyarylate from a microtomed slice, followed by image analysis on the resulting SEM photomicrographs of the cavities enabled measurement of the dispersed polyarylate domains. Consequently, estimates of the sizes

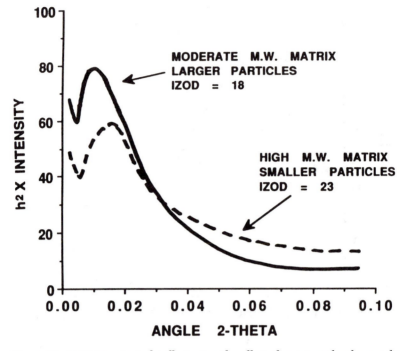

*Figure 10. SAXS invariant plot illustrating the effect of matrix molecular weight on blend particle size.*

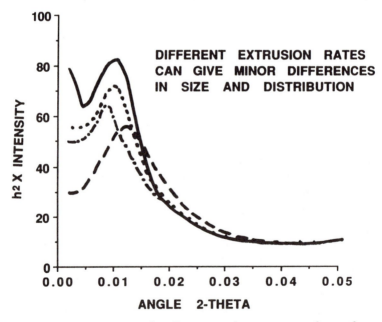

*Figure 11. SAXS invariant plot illustrating the sensitivity of particle size distribution to processing variables.*

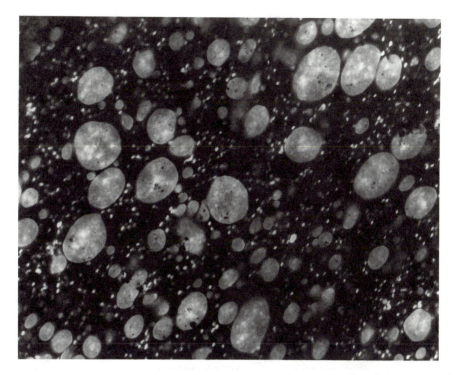

*Figure 12. Transmission photomicrograph of toughened polyarylate dispersed in toughened nylon.*

and size distributions of the three different types of polymeric particles in this complex blend can be quantitatively established and compared.

## Nylon 6

In relation to nylon 66, the lower melting point of nylon 6 affords the advantage of lower processing temperature for toughening with less thermally stable elastomers. This advantage is balanced by a disadvantage in terms of upper use temperature. Nylon 6 absorbs more water than nylon 66, which aids toughening by plasticizing the matrix and increasing matrix ductility. The benefits of increased water absorption are counterbalanced by a greater reduction in modulus and dimension changes. A difference in the response of oriented fibers of nylon 6 and nylon 66 to sinusoidal strain also has been reported (*18*).

As with PBT, both core-shell-type impact modifiers and functionalized elastomers are used to toughen nylon 6 (*6*). Soft polyolefins are also used (*19, 20*). Borggreve and Gaymans (*21*) recently published an excellent series of papers on toughening nylon 6 with functionalized elastomers, and a comprehensive treatment appeared in Borggreve's thesis (*21*).

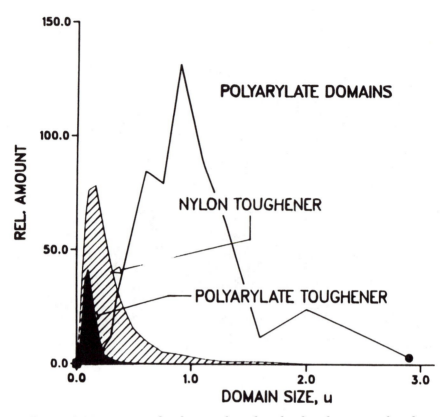

*Figure 13. Domain size distribution of toughened polyarylate in toughened nylon. Impact-modifier size distributions were measured by SAXS. Polyarylate was measured by SEM after solvolytic removal.*

## Polyacetal

Because of high-level crystallinity, sensitivity to chemically induced degradation, and lack of reactive functional groups, polyacetal homopolymer was the last engineering polymer offered in a high-impact grade (22). Polyacetal is similar to PBT and nylon 66 in that the rubber particle size and rubber glass-transition temperature $(T_g)$ should be minimized (23) to maximize the Izod test results. Figure 14 is an idealization of those interacting parameters. The Izod axis was truncated at ~ 30 ft-lb/in. because at greater values the bars merely bend out of the pendulum's path and little or no tearing results.

Unlike nylon 66 and PBT, in polyacetal a ductile–brittle transition does not occur with this category of toughening because the morphology of the supertough grade is that of a co-continuous network (24–26). The gradation in cross-sectional rubber-domain size from very fine at the surface to coarser

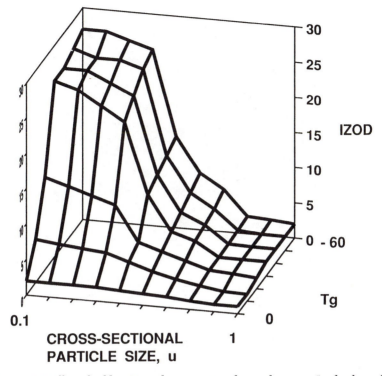

*Figure 14. Effect of rubber* T$_g$ *and cross-sectional particle size on Izod values of impact-modified polyacetal.*

in the center may be the primary cause for lack of an observed ductile–brittle transition. A similar gradation in morphology and orientation of the matrix molecules from perhaps an extended chain morphology on the surface to spherulitic crystalline morphology in the center also may be a factor. A ductile–brittle transition was observed, however, in other modified polyacetal systems at low loading rates (27). Figure 15 illustrates the rubber morphology in highly toughened polyacetal. Similar complicated rubber morphology interacting with a changing crystalline morphology in impact-modified polypropylene has been reported (16).

Particle size measurements were made by analysis of TEM photomicrographs of various toughened polyacetal blends. Because of the complicated nature of the co-continuous morphology, bars were cut perpendicular to the flow direction and electron photomicrographs of the region in the center of the bar were taken. This center location means that the largest particles are observed. Image analysis of these TEMs provides a cross-sectional particle size that can be plotted vs. Izod values as illustrated in Figure 16. The correlation was improved by consideration of the interparticle distance using

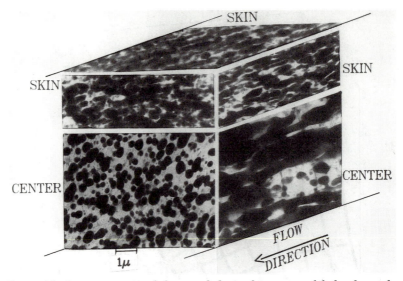

*Figure 15. Composite view of the morphology of impact-modified polyacetal.*
*(Micrographs supplied by S. A. Barenberg.)*

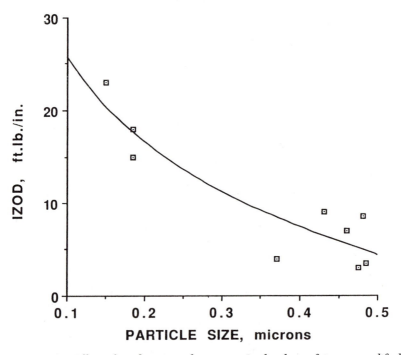

*Figure 16. Effect of median particle size on Izod values of impact-modified*
*polyacetal. Particles are viewed by TEM perpendicular to the flow direction in*
*the center of injection-molded bars.*

a nearest-neighbor approach with image analysis of the same photographs (Figure 17).

## Comparison of Fracture Mechanisms

Tensile elongation is one of the least severe toughness test because no notches are present, the injection-molded specimens are tested in their optimum (oriented) direction, and a moderate rate of loading usually is used. The mode of plastic deformation in tension differs between the untoughened PBT, nylon 66, and polyacetal matrix polymers. All of the higher molecular weight grades of these matrix polymers normally elongate greater than 50%. Polyacetal gradually thins over the entire gauge length of the dumbbell-shaped specimen bar, whereas PBT and nylon 66 form a stable neck. This stable neck can grow and, depending on the composition of the bar and presence of other materials, can result in considerable strain hardening.

A second toughness difference between these matrix polymers is that ductile Izod breaks eventually occur with increasing temperature as nylon approaches its $T_g$, but for polyacetal, although above its $T_g$, such breaks do not occur, even up to 110 °C. This difference cannot be ascribed to a molecular-weight effect because the polyacetal (M.W. ≃ 65,000) has several

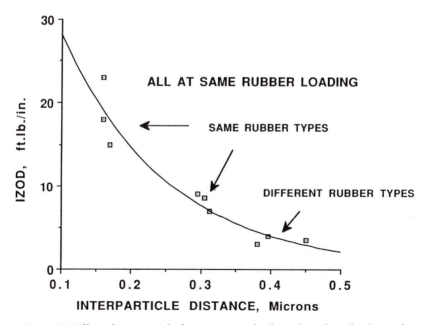

*Figure 17. Effect of interparticle distance on Izod values of toughened polyacetal. Particles are viewed by TEM perpendicular to the flow direction in the center of injection-molded bars.*

times the molecular weight of the nylon (M.W. ≃ 17,000). Polyacetal also has a higher Izod value at room temperature compared to dry as-molded nylon. The higher level of crystallinity of the polyacetal is a possible explanation for this behavior. Details of molecular morphology, such as fewer interlamellar tie molecules, may also have an effect. The presence of strong hydrogen bonding in the nylon is an additional possible cause for this difference. Conversely, polyacetal has a greater resistance to crack growth under conditions of repeat impact (fatigue). This difference is maintained in the toughened versions of these two polymers (28).

Volumetric-strain measurements are one way to obtain detailed information about deformation modes in these polymeric systems. The volumetric-strain test, developed by Bucknall (29), measures longitudinal and transverse strains during a tensile test. Figure 18 shows volumetric-strain results for the three matrix materials. A flat curve represents shear-yielding processes, and a line with a slope of unity represents dilatational processes. This test shows that nylon 66 and PBT deform primarily by shear yielding, whereas polyacetal primarily forms voids in tension. Because polyacetal is a high-strength material even at 50–80% elongation, crazing with fibril formation may occur, although no visual evidence has been observed.

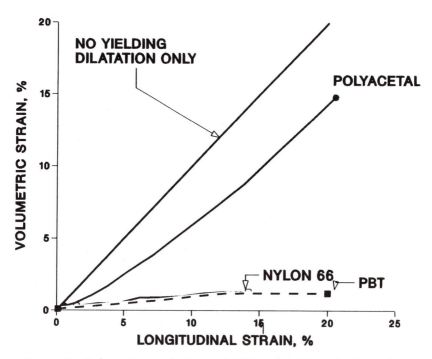

Figure 18. Deformation mechanisms of PBT, nylon 66, and polyacetal as determined by volumetric-strain measurements in tension.

The deformation mechanisms of nylon 66 and PBT change with the addition of toughener (*30–35*), as do the deformation mechanisms in nylon 6 (*36*). In nylon 66, the additional toughener increases the level of dilatation (*14*) as shown in Figure 19. Dilatation is also added to the allowed modes of deformation in the highly toughened grade PBT (Figure 20). In toughened polyacetal, the level of shear yielding increases (Figure 21). The apparent role of the toughener in these systems is to allow the "missing" deformation mechanism to occur. The interaction of dilatation and shear yielding becomes synergistic in terms of the total energy needed to break a specimen. In toughened nylon 66 vs. its matrix resin, energy dissipation during impact in the form of heat was proposed as a fracture mechanism and was measured (*37*). The J-integral technique of fracture mechanics was applied to the toughened nylon (*38*) and standard fracture mechanics was applied to nylons and polyesters (*39*).

With the addition of toughener, the response of the toughened polyacetal in tensile tests changed. The material now forms a stable neck that grows with increased loading, which is consistent with the volumetric-strain results.

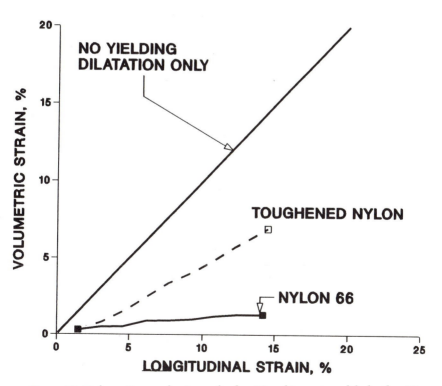

*Figure 19. Deformation mechanisms of nylon 66 and impact-modified nylon 66 as determined by volumetric-strain measurements in tension.*

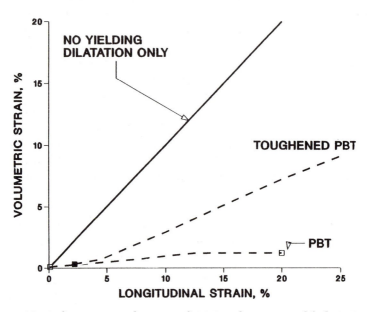

Figure 20. *Deformation mechanisms of PBT and impact-modified PBT as determined by volumetric-strain measurements in tension.*

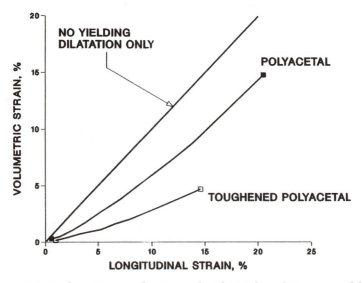

Figure 21. *Deformation mechanisms of polyacetal and impact-modified polyacetal as determined by volumetric-strain measurements in tension.*

Previous SEM studies in which the rubber was removed solvolytically after stretching the bar (22) also illustrate that the deformation mode of the polyacetal matrix has changed from brittle to ductile. This brittle–ductile change also was noted at other laboratories (40–42).

The SAXS technique can be employed to investigate the fracture mechanism of the engineering polymers. Void dimensions from the stress-whitened impact region of a highly toughened polyamide specimen are smaller in comparison with the original rubber particle size distribution as shown in Figure 22. Because the two distributions are quite parallel, the comparison results suggest that the cavitation arises from within the particles. The voids have a much lower density than the impact modifiers and, because the technique depends on an electron-density difference, the resulting intensity is much greater and was normalized for this comparison. Cavitation within the polyamide particles was demonstrated elsewhere by TEM (43), as well as in toughened nylon 6 (44).

The dilatational processes previously reported in polyacetal (26) were observed by SAXS elsewhere (45, 46) and in our laboratory. Figure 23 illustrates the invariant treatment results from observation of a bar of high-molecular-weight polyacetal strained to 40% at a rate of 2 in./min and held in position by clamps. Clamping was necessary because the cavities disappear very rapidly when the specimen is allowed to relax, unlike the cavities in the toughened nylon. Both the transverse and machine directions (melt-flow and tensile-stress directions) were scanned.

The curves in Figure 23 indicate a bimodal distribution of features with a significant difference in electron densities. Figure 24 shows the size distribution of these features. Although it is possible that two populations of cavities are observed, it is more likely that the smallest features are craze-spanning fibrils because the load-bearing ability of the bar at 40% elongation is undiminished at 10,000 psi.

Further observation of Figures 23 and 24 indicates that the large population viewed in the transverse direction is about three times the size of the population viewed in the machine direction, which seems reasonable at a 40% strain level. The small-size population viewed in the machine direction has twice the intensity of the large-size population in the transverse direction, and this finding indicates that more electron density differences are being measured in this dimension. The size distribution in the machine direction is much narrower, as might be true of extended fibrils viewed from the side rather than from the long axis that includes base areas that are not fully drawn.

## Summary

Various factors that contribute to the toughening of impact-modified semicrystalline engineering polymers were discussed. Optimum impact-modifier particle size was a persistent theme. The utility of a new SAXS method

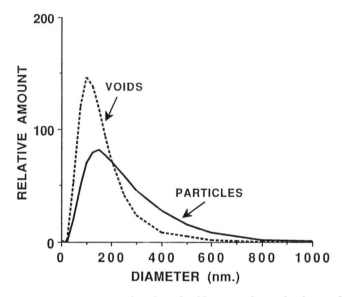

*Figure 22. SAXS comparison of voids and rubber particles in the distorted zone of a broken Izod test bar of highly impact-modified nylon 66.*

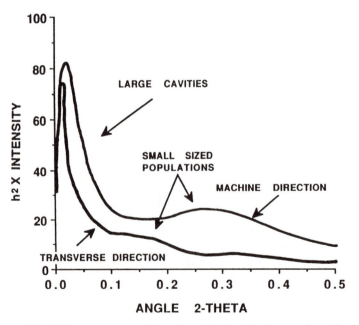

*Figure 23. SAXS comparison of voids in high-molecular-weight polyacetal strained to 40% and clamped.*

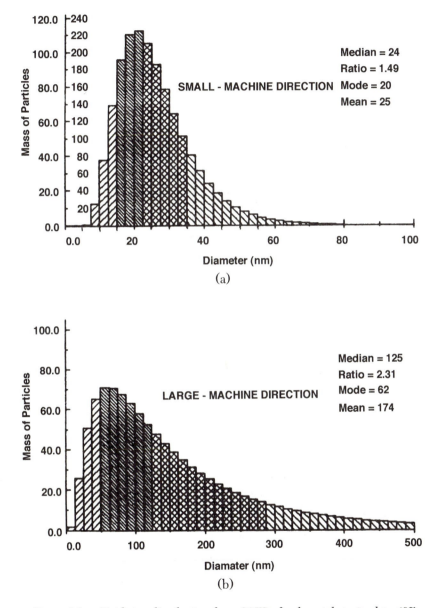

*Figure 24.    Void size distribution from SAXS of polyacetal strained to 40% and clamped. Parts a and b are scans looking in the direction of melt flow and tensile strain. Part a shows the distribution of small features (possibly fibrils); part b shows the distribution of large voids. Parts c and d are scans that are perpendicular to the bar and direction of strain. Part c shows the distribution of small features (possibly fibrils); part d shows the distribution of large voids.*
*Continued on next page.*

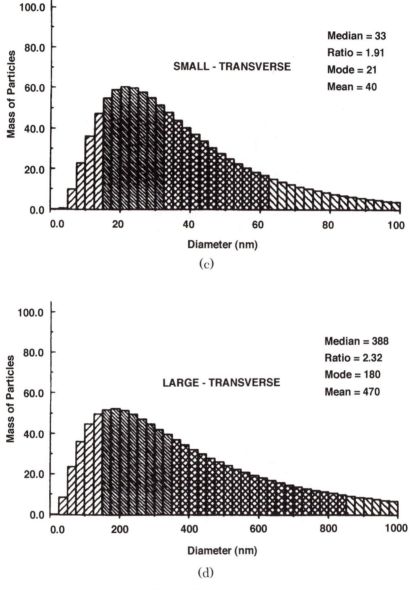

Figure 24. Continued

to measure the size and size distribution of particles in rubber-toughened polymer blends was illustrated. Fracture mechanisms of polyacetal, PBT, and nylon 66, and their highly toughened grades were discussed. The SAXS technique and volumetric-strain measurements were used to characterize the deformation mechanisms.

The preceeding discussion concentrated on the toughened versions of single-matrix polymer systems. Reports on commercially important blends of crystalline and amorphous polymers, such as PBT and polycarbonate (*10*) and nylon 66 and poly(phenylene oxide) (*47–49*), have been referenced.

## *References*

1. Ostromislensky, I. U.S. Patent 1,613,673, 1927.
2. Katti, S. S.; Schultz, J. M. *Polym. Eng. Sci.* **1982**, *22*, 1001.
3. Boyd, R. H. *Polymer* **1985**, *26*, 323.
4. Mandelkern, L. *Polym. J.* **1985**, *17*, 337.
5. Schultz, J. M. In *Fourth Cleveland Symposium on Macromolecules: Irreversible Deformation*; Case Western Reserve University: Cleveland, OH, 1983.
6. Kinloch, A. J.; Young, R. J. *Fracture Behavior of Polymers*; Applied Science Publishers: London, 1983.
7. Literature on Paraloid Impact Modifiers, Rohm & Haas Company, Philadelphia, PA.
8. Epstein, B. N. U.S. Patent 4,172,859, 1979.
9. Deyrup, E. J. U.S. Patent 4,753,980, 1988.
10. Hobbs, S. Y.; Dekkers, M. E. J.; Watkins V. H. *J. Mater. Sci.* **1988**, *23*, 1219; *Polym. Bull.* **1987**, *17*, 341.
11. Wilson, F. C. Presented at Material Resources Society Symposium on Polymer-Based Molecular Composites, Boston, MA, Nov. 1989. See also U.S. Patent 4,536,541, 1985.
12. Deyrup, E. J.; Flexman, E. A.; Howe, K. L. U.S. Patent 4,912,167, 1990.
13. Epstein, B. N. U.S. Patent 4,174,358, 1979.
14. Flexman, E. A. *Polym. Eng. Sci.* **1979**, *19*, 564. Also Presented at the Plastics and Rubber Institute Conference on Toughening of Plastics, London, July 5, 1978.
15. Richardson, P. N., private communication.
16. Karger-Kocsis, J.; Csikai, J. *Polym. Eng. Sci.* **1987**, *27*, 241.
17. Flexman, E. A.; Sosnowski, J. J. European Patent Publication 291, 997, November 1988.
18. Prevorsek, D. C.; Kwon, Y. D.; Sharma, R. K. *J. Appl. Polym. Sci.* **1980**, *25*, 2063.
19. Chuang, H.-K.; Han, C. D. *J. Appl. Polym. Sci.* **1985**, *30*, 2457.
20. Willis, J. M.; Favis, B. D. *Polym. Eng. Sci.* **1988**, *28*, 1416.
21. Borggreve, R. J. M., Thesis, University of Twente, 1988. Published in serial as: Borggreve, R. J. M.; Gaymans, R. J.; Schuijer, J.; Ingen Housz, J. F. *Polymer* **1987**, *28*, 1489; Borggreve, R. J. M.; Gaymans, R. J.; Luttmer, A. R. *Makromol. Chem. Symp.* **1988**, *16*, 195; Borggreve, R. J. M.; Gaymans, R. J. *Polymer* **1989**, *30*, 63; Borrgreve, R. J. M.; Gaymans, R. J.; Schuijer, J. *Polymer* **1989**, *30*, 71; Borggreve, R. J. M.; Gayman, R. J.; Eichenwald, H. M. *Polymer* **1989**, *30*, 78.
22. Flexman, E. A. *Mod. Plastics*, February **1985**, 72. Also *Society of Plastic Engineers Annual Technical Conference*; Society of Plastic Engineers: Brooksfield Center, CT, 1984; p 558.
23. Flexman, E. A. U.S. Patent 4,804,716, 1989.
24. Pecorini, T. J.; Manson, J. A.; Hertzberg, R. W. *ACS Polm. Prepr.* September **1988**, 136.
25. Wadha, L. H.; Dolce, T. J.; LaNieve, H. L. *Society of Plastic Engineers Annual Technical Conference*; Society of Plastic Engineers, Inc.: Washington, DC, 1984; p 558.

26. Flexman, E. A. *ACS Polym. Prepr.* September **1988**, 189.
27. Chang, F.; Yang, M. *Polym. Eng. Sci.* **1990**, *30*, 543.
28. Adams, G. C. ASTM Standard Technical Publication 936; American Society for Testing and Materials: Philadelphia, PA, 1987; p 281.
29. Bucknall, C. *Toughened Plastics*; Applied Science Publishers: London, 1977; p 195.
30. Wu, S. *J. Appl. Polym. Sci.* **1988**, *36*, 549.
31. Hobbs, S. Y.; Bopp, R. C.; Watkins, V. H. *Polym. Eng. Sci* **1983**, *23*, 380.
32. Ramsteiner, F.; Heckmann, W. *Polym. Chem.* **1985**, *25*, 199.
33. Speroni, F.; Castoldi, E.; Fabbri, P.; Casiragli, T. *J. Mater. Sci.* **1989**, *24*, 2165.
34. Dekkers, M. E. J.; Hobbs, S. Y.; Watkins, V. H. *J. Mater. Sci.* **1988**, *23*, 1225.
35. Bucknall, C. B.; Heather, P. S.; Lazzeri, A. *J. Mater. Sci.* **1989**, *24*, 2255.
36. Sunderland, P.; Kausch, H.; Schmid, E.; Arber, W. *Makromol. Chem. Makromol. Symp.* **1988**, *16*, 365.
37. Wu, S. *J. Polym. Sci. Phys.* **1983**, 699.
38. Huang, D. D. ASTM Standard Technical Publication 1114; American Society for Testing and Materials: Philadelphia, PA, 1991; p 290.
39. Schultz, J. M.; Fakirov, S. *Solid State Behavior of Linear Polyesters and Polyamides*; Prentice-Hall: Englewood Cliffs, NJ, 1990.
40. Kloos, F.; Wolters, E. *Kunstoffe* **1985**, *75*, 735.
41. Chiang, W.; Lo, M. *J. Appl. Polym. Sci.* **1988**, *36*, 1685.
42. Chiang, W.; Huang, C. *J. Appl. Polym. Sci.* **1989**, *37*, 951.
43. Wood, B. A. In *Proceedings of the 47th Annual Meeting of the Electron Microscopy Society of America*; Electron Microscopy Society of America: Woods Hole, MA, 1989; p 352.
44. Speroni, F.; Castoldi, E.; Fabbri, P.; Casiragli, T. *J. Mater. Sci.* **1989**, *24*, 2165.
45. Wendorff, J. H. *Prog. Colloid Polym. Sci.* **1979**, 135.
46. Wendorff, J. H. *Polymer* **1980**, *21*, 553.
47. Sue, H.-J.; Yee, A. F. *J. Mater. Sci.* **1989**, *24*, 1447.
48. Hobbs, S. Y.; Dekkers, M. E. J. *J. Mater. Sci.* **1989**, *24*, 1316.
49. Gallucci, R. R. *Society of Plastic Engineers Annual Technical Conference*; Society of Plastic Engineers, Inc.: Washington, DC, 1986; p 48.

RECEIVED for review March 6, 1991. ACCEPTED revised manuscript March 31, 1992.

# Deformation and Fracture Toughness in High-Performance Polymers

## Comparative Study of Crystallinity and Cross-Linking Effects

Ruth H. Pater[1,] Mark D. Soucek[1], and Bor Z. Jang[2]

[1] NASA Langley Research Center, Mail Stop 226, Hampton, VA 23665-5225
[2] Materials Science and Engineering, 201 Ross Hall, Auburn University, Auburn, AL 36849

*A systematic study was made of 10 principal high-performance thermoplastics and two semiinterpenetrating polymer networks (semi-IPNs). The fundamental tendency to undergo localized crazing or shear banding, as opposed to a more diffuse homogeneous shear-yielding deformation, was evaluated. Amorphous thermoplastics exhibited crazing as the primary mode of deformation. In contrast, semicrystalline materials displayed both crazing and shear banding. Increasing the crystallinity increased diffuse shear yielding at the expense of craze growth. Another effect was an enlargement of the deformation zone. Some ordered polymers showed only diffuse shear yielding, whereas others displayed a combination of weak crazes and diffuse shear yielding. For a semi-IPN, increasing the degree of cross-linking decreased crazing, deformation zone size, and fracture toughness of an amorphous thermoplastic. Thus, crystallinity acts like cross-linking in reducing crazing, but, exerts the opposite effect on changing the size of the deformation zone. These results suggest that the reduction in fracture toughness by crystallinity is mainly due to decreased crazing, whereas reduction by cross-linking arises from both decreased crazing and diminished deformation zone.*

**H**IGH-PERFORMANCE THERMOPLASTICS ARE EMERGING as an important class of engineering materials. Thermoplastics are being considered for use as composite matrices, adhesives, molded articles, films, and coatings for a wide variety of aerospace structural, electronic–electrical, and automotive applications. Most of these applications utilize the materials' excellent resistance to impact and microcrack. The majority of these polymers are semicrystalline in nature. It is of theoretical and technological importance to understand the relation between fracture toughness and crystallinity in these polymers.

Extensive studies of polyether ether ketone (PEEK) have shown that fracture toughness strongly depends on crystallinity. Lee et al. (1) showed that the mode-I fracture toughness ($K_{Ic}$) of PEEK 150P decreased by almost a factor of 3 with a crystallinity increase from 27 to 43 wt %. Similarly, Friedrich, et. al. (2) showed a 40% reduction in $K_{Ic}$ for a crystallinity change from 15 to 35% in PEEK 450G. Chu's study (3) related decreased molecular weight, larger spherulites, and more perfect crystal lamellae to a reduction in fracture toughness. Currently, PEEK is the only high-performance thermoplastic that has received considerable attention. Whether the same relation holds for other high performance semicrystalline polymers remains to be seen.

Why does an inverse relation exist between fracture toughness and crystallinity? Despite a theme variation, earlier studies offer no explanation for the relation. Evaluation of crystallinity and deformation may answer the question and provide additional knowledge that would have important implications for designers, users, and manufacturers of these polymers.

Like crystallinity, cross-linking significantly affects fracture toughness of a toughened polymer, such as a semiinterpenetrating polymer network (semi-IPN). The origin of the inverse relationship between fracture toughness and cross-linking is unclear. Thus, there also exists a need to understand the role of cross-linking in deformation and fracture mechanisms in a toughened polymer. Also of great interest is the comparison of crystallinity and cross-linking effects on the fracture mechanisms, because such a comparative study under well-controlled conditions has not been undertaken for high-performance polymers.

Crazing and fracture toughness are closely related. Crazing and shear banding are the most important deformation mechanisms in glassy thermoplastics (4, 5). Another deformation mode is diffuse shear yielding (4, 5). Whether crazing or shear banding can be a primary mode of plastic deformation in a highly cross-linked thermoset resin, such as epoxies, remains controversial. However, both crazing and shear banding have been suggested as the dominant energy-dissipating mechanisms in rubber-toughened thermoset resins (4, 6–8), although doubts about these claims have been raised (5, 9–12). Other proposed mechanisms include particle deformation and crack bridging (9, 10, 13), cavitation-induced shear deformation or stress relief (12, 14–21), crack pinning (22–25), and crack-tip blunting (26). Several review articles have been published on this subject (4, 5, 9, 12, 16, 26, 27).

The materials selected for this study include 10 principal high-performance thermoplastics that are under experimental and developmental evaluation and are commercially available. The chemical structures of the thermoplastics are shown in Chart I, except for the structure of New TPI, which remains undisclosed for proprietary reasons. Table I lists the neat resin thermal and fracture-toughness properties of the thermoplastics as reported in the cited literature. The materials chosen for study cover a broad area of thermoplastic chemistry varying from poly(arylene ether ketone) to polysulfone to polyimide, with variations in their connecting groups. Some of the materials are related to each other as structural isomers. A wide spectrum of crystalline morphology is represented, ranging from a very high degree of crystallinity to a totally amorphous structure. Thus, these materials present a good opportunity for a systematic study of crystallinity and deformation. Included in this study are two newly developed semi-IPNs called LaRC-RP41 and LaRC-RP40. Schemes I and II show their syntheses. By combining easy-to-process but brittle PMR-15 with tough but difficult-to-process LaRC-TPI or NR-150B$_2$, the resultant semi-IPNs exhibit an attractive combination of properties including significantly improved fracture toughness over the unmodified thermoset resin (fracture energy $G_{Ic}$ 476 vs. 87 J/m$^2$ in LaRC-RP41, for example) (28–30). Changing the composition allows the cross-linking and fracture toughness to be varied and, thus provides an opportunity to study the role of cross-linking in plastic deformation and fracture toughness in a semi-IPN environment.

## Experimental Details

**Materials**.  PEEK (100 μm), New TPI (20 μm), polyaryl sulfone (PAS, 80 μm), polyether sulfone (PES, 140 μm), and Upilex (80 μm) films were commercially obtained and used as received. The film thickness is indicated in the parentheses. LaRC-CPI (30 μm), LaRC-I-TPI (40 μm), and polyimide sulfone (PISO$_2$, 80 μm) films were previously prepared and are reported in references 31, 32, and 33, respectively. The films of LaRC-TPI, NR-150B$_2$, and their semi-IPNs were prepared in this study as follows: The LaRC-TPI polyamic acid solution, which had a 30-wt % solids concentration in N,N-dimethylacetamide (DMAc), was purchased from Mitsui Toatsu Chemicals, Chiyoda-Ku, Tokyo, Japan. The NR-150B$_2$ monomer precursor solution with 54-wt % solids in N-methylpyrrolidone (NMP) was supplied by Du Pont. The PMR-15 molding powder used to cast films was obtained from the monomethyl ester of 5-norbornene-2,3-dicarboxylic acid (NE), 4,4'-methylenedianiline (MDA), and dimethyl ester of 3,3',4,4'-benzophenonetetracarboxylic acid (BTDE). A 50-wt % methanol solution of the monomer mixture at a molar ratio of NE:MDA:BTDE = 2.00:3.09:2.09 was stirred at room temperature for 0.5 h to give a clear dark-brown solution. This solution

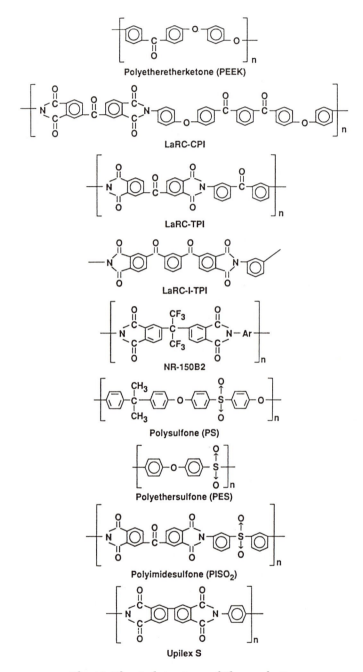

*Chart I. Chemical structures of Thermoplastics.*

**Table I. Thermal and Fracture-Toughness Properties
of High-Performance Thermoplastics**

| Polymer | Supplier | $T_g$ (°C) | $T_m$ (°C) | $G_{Ic}$ ($J/m^2$) | Ref. |
|---------|----------|------------|------------|---------------------|------|
| PEEK | ICI | 143 | 343 | 4025 | 37 |
| LaRC-CPI | NASA–Langley | 222 | 350 | 6620 | 31 |
| LaRC-TPI | Mitsui Toatsu | 252 | 348 | 1768 | 41 |
| LaRC-I-TPI | NASA–Langley | 259 | none | — | 32 |
| New TPI | Mitsui Toatsu | 250 | 388 | — | 42 |
| NR-150B$_2$ | Du Pont | 360 | none | 2398 | 42 |
| PES (Radel A400) | Amoco | 220 | none | 3500 | 37 |
| PES (Victrex 4100G) | ICI | 230 | none | 1925 | 37 |
| PISO$_2$ | HiTech Services | 273 | none | 1400 | 37 |
| Upilex S | ICI/Ube Chem. | 328 | — | — | 43 |

NOTE: The dashes indicate that data were unavailable.

was concentrated at 80 °C in a nitrogen atmosphere for 2 h and further dried at 150 °C in air for 1.5 h.

A resin solution with 20-wt % solids concentration in NMP was centrifuged and the decantate was cast onto a glass plate by using an 80 μm doctor blade and dried in a well-ventilated and dust-free chamber for 48 h to remove excess solvent. The film was thermally converted into the polyimide by heating at 80, 120, 150, 210, 250, 300, and 350 °C for 1 h at each temperature in air and then cooled to room temperature at a very slow rate (2–3 °C/min) to reduce residual thermal stress. The film was removed from the plate by soaking in water overnight. The thickness of the obtained film varied from 20 to 50 μm.

**Film Characterization.** Rectangular 13-mm-wide strips were cut from each material with a pair of sharp scissors. Each strip specimen was supported on light-weight cardboard and cut with a razor blade to create the edge notch. This procedure gave a crack that appeared sharp under the microscope (34).

A plane-stress fracture toughness test was conducted on the rectangular specimens with a single-edge-notch (SEN) geometry. A microscope stage-top miniature testing system equipped with a 25-lb (100-N) load cell and a variable speed motor was used to deform the specimen in a tensile mode. The load-displacement curve of each test was recorded on a chart recorder. The crack-tip region was observed at 110× or 220×, and the onset of the crack propagation was noted and marked on the chart record. This visual determination technique gave an acceptable degree of consistency (34).

The deformation zone near a crack tip could be observed in situ continuously or intermittently at various stages of loading. Photomicrographs were obtained by stopping the cross-head at various extents of plastic

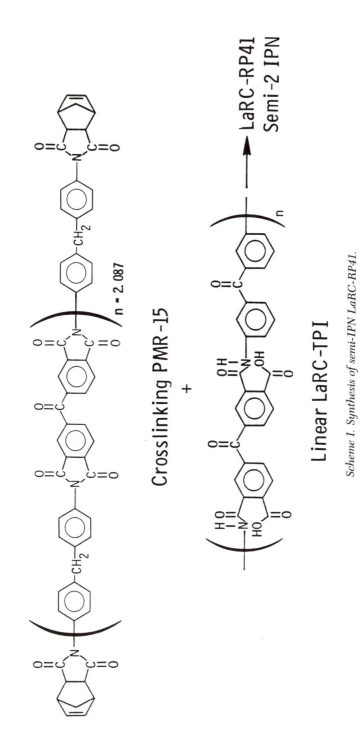

Crosslinking PMR-15

+

Linear LaRC-TPI

→ LaRC-RP41
Semi-2 IPN

n = 2.087

*Scheme I. Synthesis of semi-IPN LaRC-RP41.*

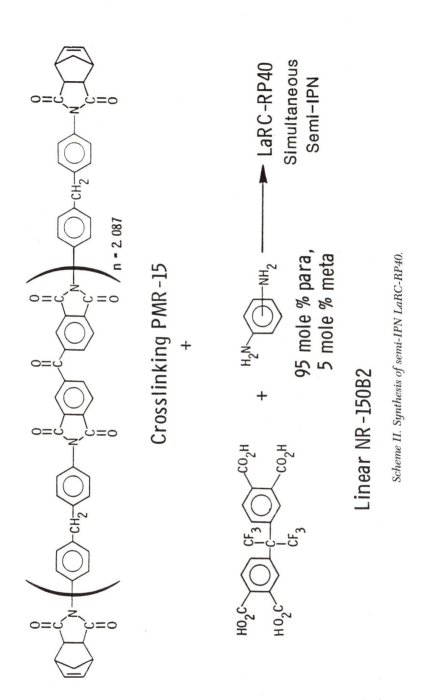

Crosslinking PMR-15

+

Linear NR-150B2

95 mole % para,
5 mole % meta

LaRC-RP40
Simultaneous
Semi-IPN

n = 2.087

*Scheme II. Synthesis of semi-IPN LaRC-RP40.*

deformation. Deformation zones of select samples were also further examined on a scanning electron microscope (SEM).

The plane-stress fracture toughness (or stress-intensity factor, $K_{Ic}$) was obtained from eq 1 in reference 35:

$$K_{Ic} = YS\sqrt{a}$$

$$Y = [3.94(2w/\pi a)\tan(\pi a/2w)]^{1/2}$$

$$S = F/[(w - a)b]$$

where $w$ is the specimen width, $b$ is the specimen thickness, $a$ is the notch depth, $F$ is the load at which crack propagation begins, and $Y$ is a geometric factor. The differential scanning calorimetry (DSC) analysis was carried out with 2–10-mg film samples on a Du Pont Model 940 calorimeter at a heating rate of 5 °C/min. From the DSC thermogram, the apparent glass-transition temperature ($T_g$), crystallization temperature ($T_c$), and melting temperature of a crystalline phase ($T_m$) were determined. The wide-angle X-ray diffraction was measured on films with the X-ray generator operated at 45 kV and 40 mH. The intensity of counts taken every 0.01° ($2\theta$) was recorded on hard disk for the angular range 10–40° ($2\theta$). Typical intensities ranged between 600 to 2500 counts/s.

## Results and Discussion

**Crystallinity Effects.** *PEEK*. Figure 1 illustrates the DSC curves for the three samples of PEEK. The as-received material exhibited a $T_g$ at 149 °C, a $T_c$ at 178 °C associated with crystallization, and a $T_m$ at 336 °C corresponding to the melting of the crystals. The heat of fusion ($\Delta H_f$) was 32.9 J/g. Annealing this sample at a temperature (200 °C) above its $T_g$ for 1.5 h enhanced $\Delta H_f$ to a value of 37.2 J/g. However, annealing at a temperature (325 °C) slightly below its $T_m$ for 2 h had a more pronounced effect on $\Delta H_f$ as well as on the crystalline structure: the $\Delta H_f$ value increased to 45.0 J/g and the crystals became more perfect, as indicated by the increased sharpness of the $T_m$ peak. Blundell and Osborn (36) reported a value of 130 J/g for $\Delta H_f$ for the 100% crystalline PEEK. By using this value, the degree of crystallinity for these three samples was estimated to be 25, 29, and 35%, respectively. The X-ray diffraction data shown in Figure 2 confirm this trend.

Figure 3 shows the deformation behavior of the three samples viewed through a polarizing light microscope. When loaded in a tensile mode, the as-received specimen showed crazes at the early stage of loading (Figure 3a). Continued loading turned the deformation mode into one of a more localized nature. Both crazes and shear bands dominated the crack-tip deformation zone. Crazes appeared to grow at a slightly faster rate than shear bands and

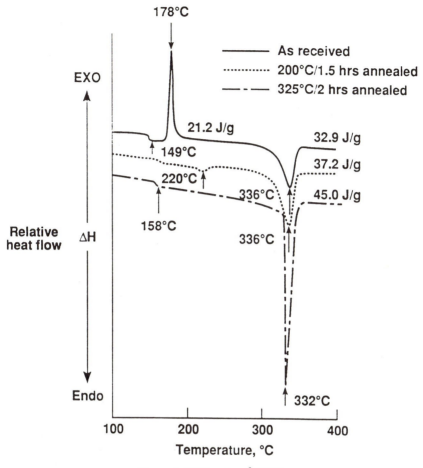

*Figure 1. DSC scans of PEEK.*

could be observed at the leading edge of the growing deformation zone. At a higher magnification (Figure 4), two packets of shear bands intersecting each other at an angle of 30–60° were observed. Comparison of Figure 3a with Figures 3b and 3c readily demonstrates that increasing the crystallinity dramatically decreases crazing and shear banding but increases diffuse shear yielding. Thus, crystallinity caused a transition in the deformation mechanism from a localized to a more homogeneous mode. Another effect is an enlargement of the deformation zone as shown in Figure 5. The deformation zone ahead of the crack tip as well as the fracture surfaces near the crack-tip regions were affected by the crystallinity as illustrated in Figure 6.

**LaRC-CPI.**   Compared to PEEK, LaRC-CPI has a higher $T_g$ (226 °C vs. 149 °C), a higher $T_m$ (350 °C vs. 336 °C), and, more importantly, a higher

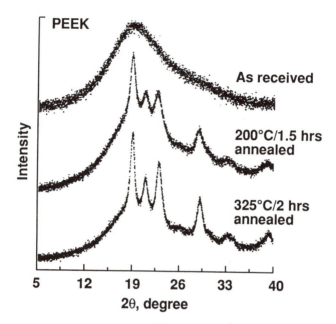

Figure 2. X-ray diffractograms of PEEK.

 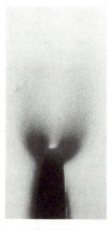 

(a)  As received PEEK        (b)  200°C/1.5 hrs           (c)  325°C/2 hrs
                                  annealed                      annealed

Figure 3. Polarized optical micrographs of the crack-tip deformation zones in PEEK.

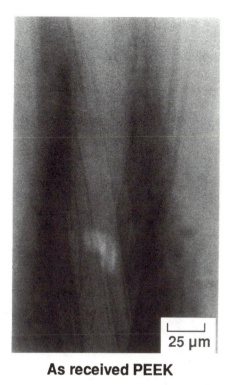

**As received PEEK**

*Figure 4. Higher magnification view of Figure 3a.*

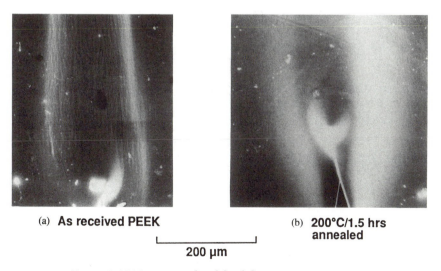

(a) **As received PEEK**          (b) **200°C/1.5 hrs annealed**

⊢————————⊣
**200 µm**

*Figure 5. SEM micrographs of the deformation zones in PEEK.*

(a)  **As received PEEK**

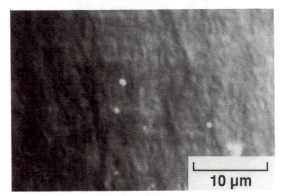

(b)  **200°C/1.5 hrs annealed**

(c)  **325°C/2 hrs annealed**

Figure 6. SEM micrographs of the fracture surfaces near the crack-tip regions
in PEEK.

$G_{Ic}$ [6620 J/m$^2$ (*31*) compared to 4025 J/m$^2$ (*37*)]. The X-ray diffraction patterns shown in Figure 7 suggest that the as-received film as well as the annealed samples had a high degree of crystallinity. The DSC curves for the three LaRC-CPI film samples are illustrated in Figure 8. Annealing the as-received sample at 200 °C for 1.5 h increased the crystallinity somewhat ($\Delta H_f$ increased from 17.8 to 18.5 J/g) and induced more perfect crystals, as evident from the increased sharpness of the $T_m$ peak. A pronounced increase in the crystallinity was obtained when the as-received material was annealed at a temperature of 327 °C, close to its $T_m$. The as-received film showed crazes extending a considerable distance ahead of the crack tip (shown in Figure 9). Comparison of Figure 10a with Figure 10c shows that the crystallinity in LaRC-CPI exerted the same effects seen in the previously discussed PEEK material. The apparent crazes were dramatically reduced, and the deformation zone enlarged as the crystallinity increased. Diffuse shear yielding became the primary mode of deformation for the sample with the highest crystallinity.

***LaRC-TPI.***   The crystalline behavior of the chemically imidized LaRC-TPI 1500 series was extensively investigated by Hou and Bai (*38*), who showed, as indicated in Figure 11, that the area under the $T_m$ peak increased as the area under the $T_g$ peak decreased with increasing annealing time. Hou and Bai also indicated that, unlike PEEK, a LaRC-TPI polymer is not readily recrystallized after melting the initial crystalline phase. Annealing a LaRC-TPI

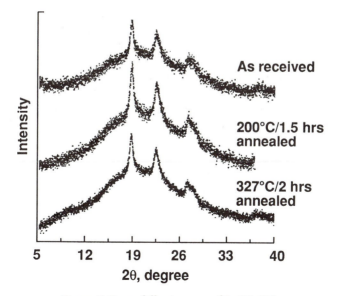

*Figure 7. X-ray diffractograms of LaRC-CPI.*

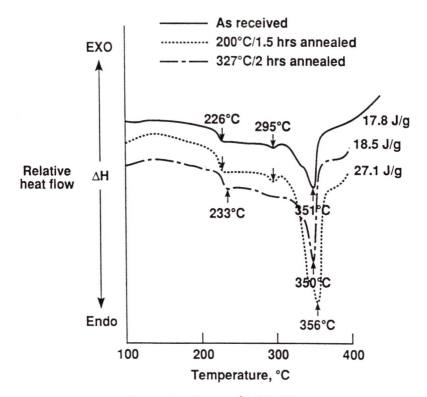

*Figure 8. DSC scans of LaRC-CPI.*

specimen at temperatures above 320 °C resulted in a fully amorphous structure as determined by DSC. Similarly, thermally imidizing LaRC-TPI also results in a fully or nearly fully amorphous material, as evidenced from the following results.

Three film samples of the thermally imidized LaRC-TPI were prepared in this study. One film was obtained by treating the 316 °C cured material at 200 °C for 1.5 h. The second film was slowly cured at 80, 120, 150, 210, 250, 300, and 350 °C for 1 h at each temperature. The details of the preparation were described in the Experimental Details section. The third sample, designated 316 °C cured material, was prepared by heating the film at 100, 200, and 316 °C for 1 h at each temperature. Figure 12 shows the DSC traces for these three samples, all of which exhibited a $T_g$ at temperatures of 255–262 °C, but no $T_m$ (327 and 341 °C from reference 38) could be found. The absence of a semicrystalline structure is also confirmed by the X-ray data shown in Figure 13. Because there is a variation in the area under the $T_g$ peak (*see* $\Delta H_f$ values in Figure 12) and considering the results of the study by Hou and Bai (*38*) discussed earlier, it is possible that the three samples

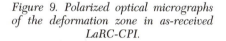
25 μm

*Figure 9. Polarized optical micrographs of the deformation zone in as-received LaRC-CPI.*

contained crystals that were too small or too few in number to be detected by both the DSC and X-ray diffraction measurements, but that do exert effects on the deformation behavior.

Figure 14 illustrates that the 350 °C cured sample has crazes and that shear bands dominate the crack-tip deformation zone. A closer examination of the right and left arm regions as indicated by arrows (Figures 14a and 14c) reveals the presence of two packets of shear bands intersecting at an angle of 30–60° to each other. The shear-banding phenomenon was quite pronounced in this sample. The SEM micrographs of Figure 15 compare the craze-growth behavior of the three films. The 200 °C annealed specimen has many crazes that grew to a great length without diminishing their width (Figure 15a). The 350 °C cured sample (Figure 15b) also shows well-developed crazes that display smaller width and length than in the previous sample. In comparison to the other two samples, the 316 °C cured material has the least amount of crazes but considerably more shear bands.

**LaRC-I-TPI.**   LaRC-I-TPI is a structural isomer of LaRC-TPI. A comparison of their chemical structures can be made from Chart I. Both materials behave alike in many aspects of their physical and mechanical properties that have been compared in a study by Pratt and St. Clair (32). The thermally imidized film of LaRC-I-TPI shows a $T_g$ at 251 °C (Figure 16), which is lower than a reported value (32) of 259 °C. The absence of a semicrystalline structure is evident from the DSC (Figure 16) and the X-ray data (Figure 17). This amorphous polymer displays deformation behavior

(a) **As received LaRC-CPI**      (b) **200°C/1.5 hrs annealed**

├─────────────────┤
       **200 μm**

(c) **327°C/2 hrs annealed**

*Figure 10. SEM micrographs of the crack-tip deformation zones in LaRC-CPI.*

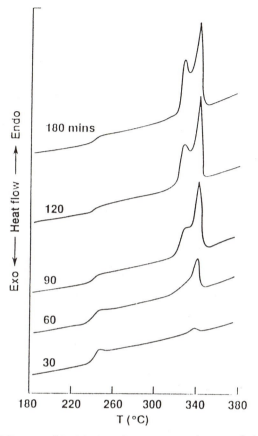

*Figure 11. DSC scans of LaRC-TPI 1500 series samples annealed at 310 °C for various times. (Reproduced with permission from reference 38. Copyright 1990.)*

(Figure 18a) analogous to the 350 °C cured LaRC-TPI sample (Figure 14). Both crazes and shear bands dominate the crack-tip deformation zone as shown at a higher magnification in Figure 18b.

**New TPI**[3]. Hou and Reddy (39) reported that the commercially obtained New TPI powder had no detectable $T_g$ but two $T_m$s at 354 and 384 °C with a $\Delta H_f$ value of 46–49 J/g. They also showed five reflections in the X-ray diffraction pattern. The present New TPI film (also obtained from a commercial source) shows a $T_g$ at 260 °C, a $T_c$ at 308 °C, and a $T_m$ at 383 °C (Figure 19). A $T_g$ of 250 °C and a $T_m$ of 388 °C have been reported (Table I). The X-ray pattern shown in Figure 20 suggests an amorphous structure for

---

[3] New is part of the name of a specific compound. For proprietary reasons the specific compound cannot be divulged.

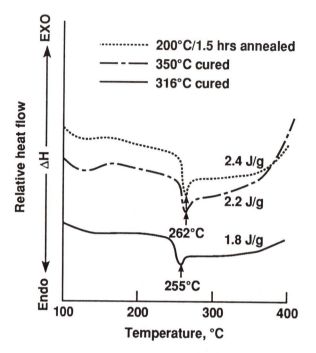

*Figure 12. DSC scans of LaRC-TPI.*

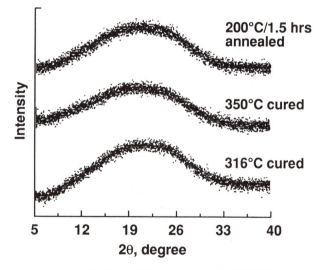

*Figure 13. X-ray diffractograms of LaRC-TPI.*

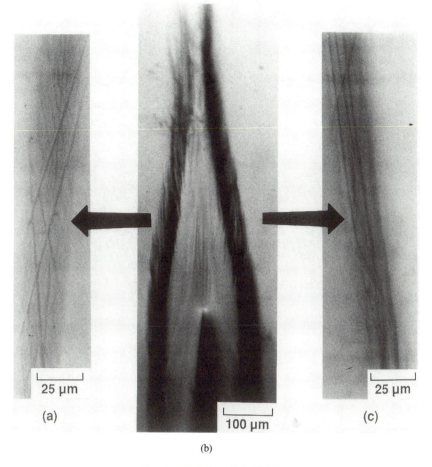

(a)                                    (b)                                    (c)

25 μm          100 μm          25 μm

**As received LaRC-TPI**

*Figure 14. Polarized optical micrographs of 350 °C cured LaRC-TPI. Parts a and c are higher magnification views of the regions indicated by arrows.*

the present New TPI. The deformation and fracture behavior of this material (Figure 21) resembles that of LaRC-TPI and LaRC-I-TPI. A higher magnification view of one of the arms (Figure 22) illustrates the presence of well-developed localized crazes and shear bands that grew to a considerable length without diminishing their widths.

*Polyimide Sulfone.* The polyimide sulfone ($PISO_2$) film was fully amorphous as evidenced by the DSC data shown in Figure 23, and its $T_g$ occurred at 271 °C, which is close to a value of 273 °C reported by St. Clair and Yamaki (33). Among the 10 thermoplastics studied, the present $PISO_2$

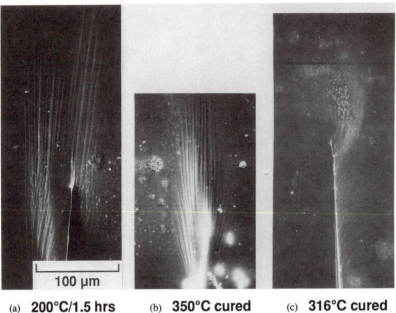

(a) **200°C/1.5 hrs annealed**     (b) **350°C cured**     (c) **316°C cured LaRC-TPI**

*Figure 15. SEM micrographs of the crack-tip deformation zones in LaRC-TPI.*

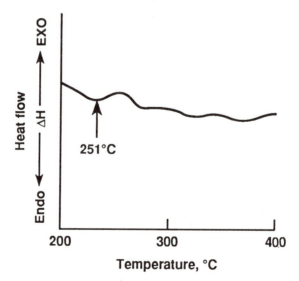

*Figure 16. DSC scan of LaRC-I-TPI.*

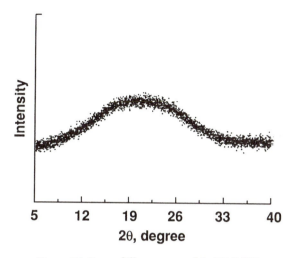

*Figure 17. X-ray diffractogram of LaRC-I-TPI.*

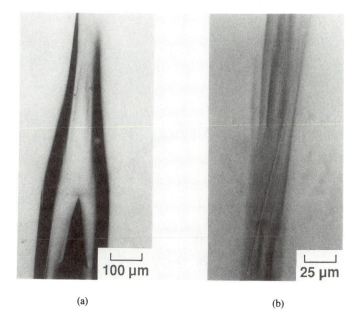

(a)                                          (b)

*Figure 18. Polarized optical micrographs of the crack-tip deformation zone in a, LaRC-I-TPI and b, crazes near the leading edge of a grown deformation zone.*

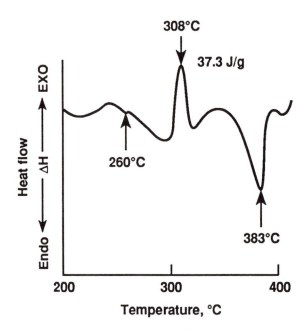

Figure 19. DSC scan of New TPI.

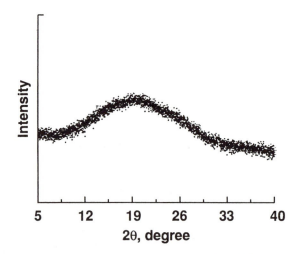

Figure 20. X-ray diffractogram of New TPI.

exhibits the most pronounced crazes, which are very wide and extend to long distances, as can readily be seen in Figures 24 and 25.

**NR-150B$_2$.** An NR-150B$_2$ polymer is generally thought of as an amorphous polymer, because of the presence of a bulky hexafluoroisopropylidene

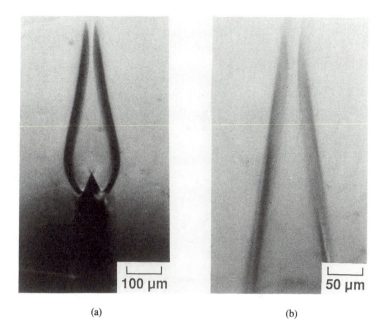

(a)                                          (b)

*Figure 21. Polarized optical micrographs of the crack-tip deformation zone in New TPI.*

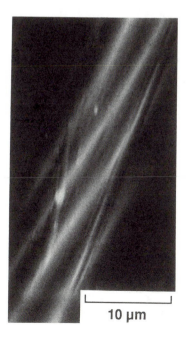

*Figure 22. SEM micrographs of the deformation zone showing shear bands in New TPI.*

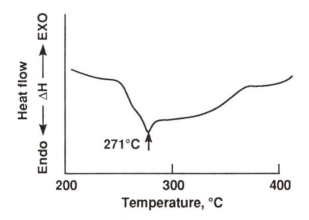

Figure 23. DSC scan of polyimide sulfone.

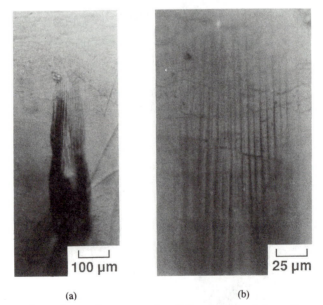

(a)                                          (b)

Figure 24. Polarized optical micrographs of the crack-tip deformation zone in
polyimide sulfone at different magnifications.

group that tends to prevent the formation of a crystalline structure. The DSC
scan of the prepared NR-150B$_2$ film shows a sharp thermal transition peak at
353 °C (Figure 26). However, the X-ray diffraction pattern shown in Figure
27 clearly indicates that some ordered structure, if not a semicrystalline
structure, is present. Possibly a thermal transition due to crystal melting
occurs at a temperature above its decomposition temperature and, thus,

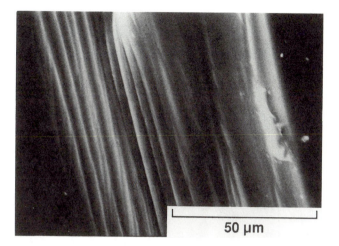

*Figure 25. SEM micrograph of the deformation zone in polyimide sulfone showing extended crazes.*

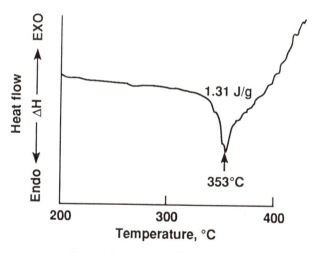

*Figure 26. DSC scan of NR-150B$_2$.*

cannot be observed in the DSC scan. The peak at 353 °C is most likely due to a $T_g$, because a $T_g$ of 360 °C has been reported (Table I). Figure 28 illustrates diffuse shear yielding as the principal mode of deformation. The deformation zone initiated with two primary arms, each at an angle of approximately 45° with respect to the crack-tip propagation direction. These two arms gradually reoriented almost parallel to the tip growth direction. At the root of the crack-tip, only a heavily deformed shear-yielding zone is observed. No crazes

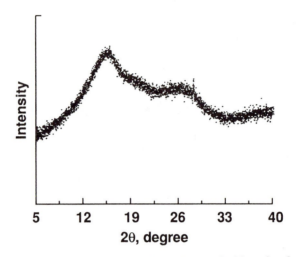

Figure 27. X-ray diffractogram of NR-150B$_2$ showing highly ordered structure.

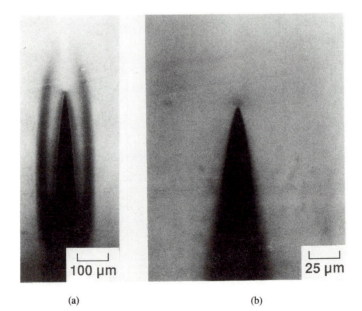

(a)                                              (b)

Figure 28. Polarized optical micrographs of the crack-tip deformation zone in
NR-150B$_2$.

or shear bands are discernible even at a higher magnification (Figure 29). Most likely the NR-150B$_2$ material had a high entanglement density, which is responsible for the diffuse shear yielding.

*Polyaryl Sulfone (PAS).* The commercial polyaryl sulfone film exhibits multiple thermal transitions in the DSC curve shown in Figure 30. In conjunction with the X-ray data of Figure 31, the peaks at 194 and 230 °C may be associated with $T_g$s, whereas the peaks occurring at 331, 403, and 470 °C are associated with $T_m$s. A $T_g$ of 220 °C was reported (Table I). The transition peak at 331 °C has a very small $\Delta H_f$ value (0.4 J/g) and its assignment as a $T_m$, rather than a $T_g$, is consistent with the chemical structure shown in Chart 1. A combination of shear banding and diffuse shear yielding characterizes the deformation behavior as readily seen in Figure 32, particularly at higher magnification (Figure 32b).

*Polyether Sulfone (PES).* Analogous to the behavior of PAS, the commercially obtained polyether sulfone (PES) film showed multiple thermal transitions in the DSC trace (Figure 33) and an ordered structure in X-ray

**50 µm**

*Figure 29. SEM micrograph of the deformation zone in NR-150B$_2$.*

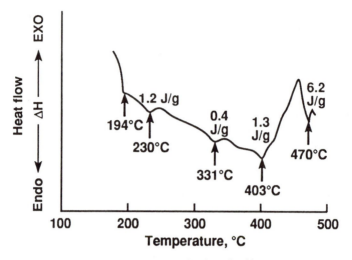

Figure 30. DSC scan of polyaryl sulfone.

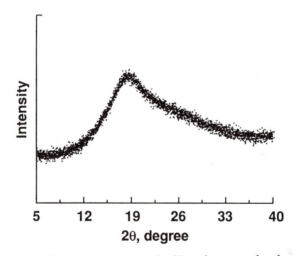

Figure 31. X-ray diffractogram of polyaryl sulfone showing ordered morphology.

diffraction (Figure 34). The peak at 235 °C can be assigned to a $T_g$, whereas the peaks at temperatures above 400 °C are most likely caused by different forms of the crystalline structure. As shown in Figure 35, the crack-tip deformation zone shows poorly developed crazes that are short and thin. A considerable amount of homogenous diffuse shear yielding is also observed. The inability to grow extended crazes is seen in Figure 36. One possible explanation for this behavior is that a large number of small crystallites act like "cross-links" to prevent molecules from becoming highly oriented, a condition necessary for crazing.

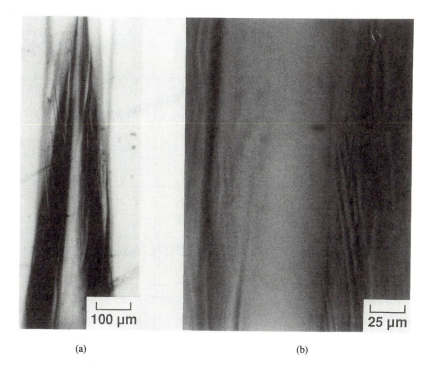

100 µm                                                    25 µm

(a)                                                        (b)

*Figure 32. Polarized optical micrographs of the deformation zone in polyaryl sulfone at different magnifications.*

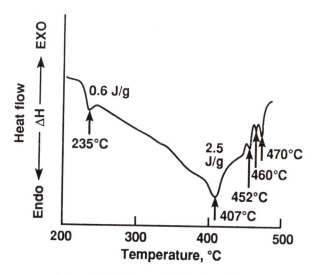

*Figure 33. DSC scan of polyether sulfone.*

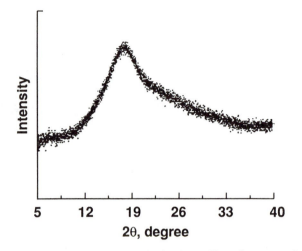

*Figure 34. X-ray diffractogram of polyether sulfone showing ordered morphology.*

**Upilex.**   The Upilex sample had two $T_g$s at 230 and 333 °C and a $T_m$ at 403 °C (Figure 37). A $T_g$ of 328 °C has been reported (Table I). The crystalline structure of Upilex is confirmed by the X-ray data that show at least six reflections (Figure 38). Upilex shows the same deformation behavior as the NR-150B$_2$ specimen presented earlier. As shown in Figure 39, its deformation is dominated by diffuse shear yielding without any traces of crazing or shear banding. The similarity in the fracture behavior between the NR-150B$_2$ and Upilex materials is unexpected because of the difference in the degree of rigidity in their molecular structures. The NR-150B$_2$ contains a hexafluoroisopropylidene connecting group, which acts as a flexibilizing unit, whereas the Upilex has a stiff rigid biphenyl bridge between rigid imide rings. The Upilex has a greater probability of forming a crystalline phase. Further study is needed to clarify this point.

**Cross-Linking Effects**.   Figure 40 illustrates the deformation zone patterns for the series of semi-IPNs prepared from PMR-15 and LaRC-TPI (LaRC-RP41). Another series of the semi-IPNs comprising PMR-15 and NR-150B$_2$ (LaRC-RP40) was also studied. Because many variables, such as processing conditions, specimen thermal history, testing conditions, and temperature, affect the results, a comparative study was made under carefully controlled conditions. All the specimens were prepared and tested under identical conditions as described in the Experimental Details section. Increasing the thermosetting component concentration probably increases the cross-link density in the semi-IPN system. Comparison of Figure 40a with Figures 40b–40f shows that an increase in cross-linking resulted in a size reduction of

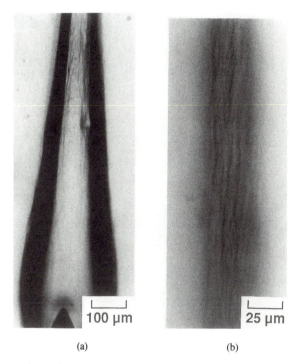

(a)                                        (b)

*Figure 35. Polarized optical micrographs of the deformation zone in polyether sulfone at different magnifications.*

*Figure 36. SEM micrograph of the deformation zone in polyether sulfone.*

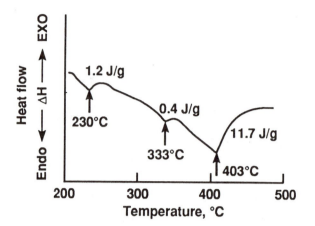

Figure 37. DSC scan of Upilex.

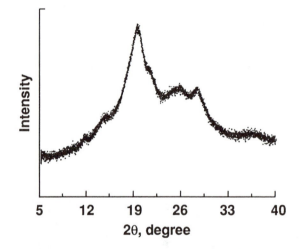

Figure 38. X-ray diffractogram of Upilex.

the deformation zone as well as a shift from primarily crazing deformation to only diffuse shear-yielding deformation as the weight percentage of LaRC-TPI to PMR-15 varied from 100 to 0%. A dramatic decrease in the deformation zone size occurred with the composition containing 35 and 65 wt % of LaRC-TPI and PMR-15, respectively. With PMR-15 as the matrix and LaRC-TPI as the dispersed phase, diffuse shear yielding was the primary mode of plastic deformation as shown in Figures 40d–40f. Conversely, with LaRC-TPI as the matrix, both crazing and shear yielding were still the major mechanisms for energy dissipation in the semi-IPN (Figures 40a–40c). However, when LaRC-TPI was replaced by NR-150B$_2$, the resultant semi-

**Upilex**

*Figure 39. Polarized optical micrograph of the deformation zone in Upilex.*

IPN films exhibited only diffuse shear yielding. This diffuse shear-yielding behavior is understandable because both the pure NR-150B$_2$ and PMR-15 components showed only diffuse shear yielding.

The $K_{Ic}$ values for pure PMR-15, LaRC-TPI, and NR-150B$_2$ and their semi-IPNs are plotted in Figures 41 and 42. The low $K_{Ic}$ value for PMR-15 (0.2 MPa · m$^{1/2}$) shows that PMR-15 is a very brittle material because of its high cross-link density. By incorporating 25 wt % of a thermoplastic component (either LaRC-TPI or NR-150B$_2$), the $K_{Ic}$ value of PMR-15 can be more than doubled. The pure thermoplastic films are considerably tougher than PMR-15 and their semi-IPN counterparts, as expected. Although LaRC-TPI is tougher than NR-150B$_2$ in thin film form, the NR-150B$_2$ is more effective in toughening PMR-15. This situation may be due to the difference in phase morphology that facilitated operation of different toughening mechanisms or to the same mechanisms but to a different extent.

## Summary

High-performance thermoplastics exhibit three modes of plastic deformation: crazing, shear banding, and diffuse shear yielding. Semicrystalline polymers, such as PEEK and LaRC-CPI, show a combination of crazing and shear banding. When crystallinity is increased through annealing, diffuse shear yielding increases at the expense of craze growth. This process is accompa-

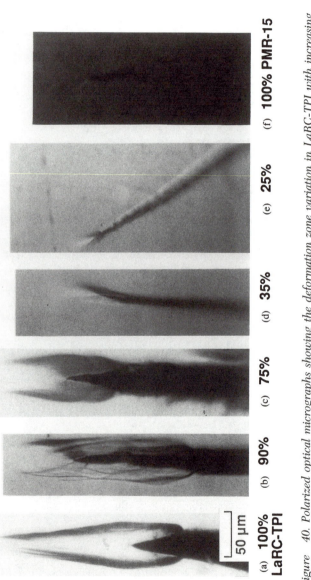

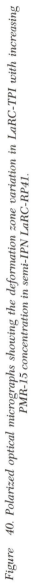

Figure 40. Polarized optical micrographs showing the deformation zone variation in LaRC-TPI with increasing PMR-15 concentration in semi-IPN LaRC-RP41.

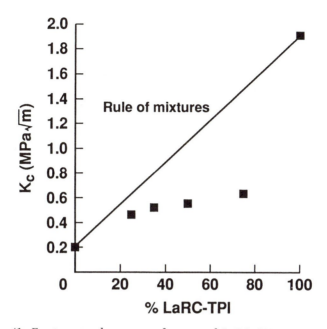

*Figure 41. Fracture toughness as a function of LaRC-TPI concentration in semi-IPN LaRC-RP41.*

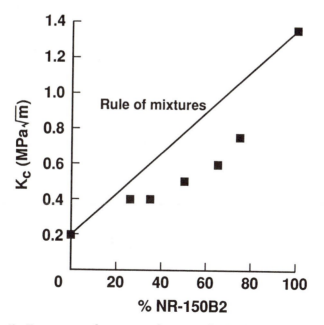

*Figure 42. Fracture toughness as a function of NR-150B$_2$ concentration in Semi-IPN LaRC-RP40.*

nied by enlargement of the deformation zone. For fully amorphous materials, like LaRC-TPI, LaRC-I-TPI, New TPI, and polyimide sulfone, crazing is the primary energy dissipation mechanism, but some shear bands are also found. Some of the ordered polymers, namely, NR-150B$_2$ and Upilex, show only diffuse shear yielding, whereas others, such as polyaryl sulfone and polyether sulfone, display weak crazing (short and thin crazes) combined with diffuse shear yielding.

Adding a small amount (25 wt %) of a thermoplastic component, either LaRC-TPI or NR-150B$_2$, significantly improves the fracture toughness of highly cross-linked and brittle PMR-15 polyimide (40). Pure PMR-15 film shows only diffuse shear yielding and has a very small deformation zone. This behavior is consistent with the limited extensibility of a highly cross-linked network structure. Increasing the cross-linking by increasing the thermoset component concentration causes a reduction in the deformation zone, diminished craze growth, and a shift from crazing to diffuse shear yielding in semi-IPNs of LaRC-TPI. Variations in fracture toughness are correlated with changes in both deformation mode and zone in the semi-IPN systems. The cross-linking plays a role similar to that of crystallinity in retarding craze growth and in promoting diffuse shear yielding. However, cross-linking and crystallinity exert opposite effects on the size of the deformation zone. The inverse relation between the crystallinity and fracture toughness is most likely due to decreased crazing, whereas the reduction in fracture toughness by cross-linking may arise from both decreased crazing and a diminished deformation zone.

## References

1. Lee, W. I.; Talbott, M. F.; Springer, G. S.; Berglund, L. A. *J. Reinf. Plast. Compos.* **1987**, *6*, 2–12.
2. Friedrich, K.; Carlsson, L. A.; Gillespie, J. W., Jr.; Karger-Kocsis, J. In *Thermoplastic Composite Materials*; Carlsson, L. A., Ed.; Elsevier Science Publishers, B. V.: New York, 1991; p 242.
3. Chu, J.-N. M.Sc. Thesis, University of Delaware, Newark, DE, 1988.
4. Bucknall, C. B. *Toughened Plastics*; Applied Science Publishers: London, 1977.
5. Kinloch, A. J.; Young, R. J. *Fracture Behavior of Polymers*; Applied Science Publishers: New York, 1983.
6. Sultan, J. N.; McGarry, F. J. *Appl. Polym. Symp.* **1971**, *16*, 127.
7. Sultan, J. N.; McGarry, F. J. *J. Polym. Sci.* **1973**, *13*, 29.
8. Bucknall, C. B.; Yoshii, T. *Br. Polym. J.* **1978**, *10*, 53.
9. Kunz-Douglass, S.; Beaumont, P. W. R.; Ashby, M. F. *J. Mater. Sci.* **1980**, *15*, 1109.
10. Kunz, S.; Beaumont, P. W. R. *J. Mater. Sci.* **1981**, *16*, 3141.
11. Kunz, S. Ph.D. Thesis, University of Cambridge, 1978.
12. Yee, A. F.; Pearson, R. A. In *Fractography and Failure Mechanisms*; Roulin-Moloney, A. C., Ed.; Elsevier Applied Science Publishers: London, 1989; p 291.
13. Evans, A. G. et al. *Acta Metall.* **1986**, *34*, 79–87.
14. Bascom, W. D. et al. *J. Mater. Sci.* **1981**, *16*, 2657.

15. Bitner, J. R. et. al. *J. Adhes.* **1982**, *13*, 3.
16. Donald, A. M.; Kramer, E. J. *J. Mater. Sci.* **1982**, *17*, 1765.
17. Hagerman, E. M. *J. Appl. Polym. Sci.* **1973**, *17*, 2203.
18. Yee, A. F. *J. Mater. Sci.* **1977**, *12*(8), 757.
19. Yee, A. F. et al. In *Toughened and Brittleness of Plastics*; Deanin, R. D.; Grugnola, A. M., Eds.; American Chemical Society: Washington, DC, 1976; p 97.
20. Kinloch, A. J.; Williams, G. J. *J. Mater. Sci.* **1980**, *15*, 987.
21. Bascom, W. D.; Cottington, R. L. *J. Adhes.* **1976**, *7*, 333.
22. Lange, F. F. *Phil. Mag.* **1970**, *22*, 983.
23. Evans, A. G. *Phil. Mag.* **1972**, *26*, 1327.
24. Evans, A. G. *J. Mater. Sci.* **1974**, *9*, 1145.
25. Green, D. J.; Nicholson, P. S. *J. Mater. Sci.* **1979**, *14*, 1413.
26. Cartwell, W. J.; Roulin-Moloney, A. C. In *Fractography and Failure Mechanisms*; Roulin-Moloney, A. C., Ed.; Elsevier Applied Science Publishers: London, 1989; pp 233–290.
27. Garg, A. G.; Mai, Y. W. *Comp. Sci. Technol.* **1988**, *31*(3), 179–223.
28. Pater, R. H. *Polym. Eng. Sci.* **1991**, *31*(1), 20–27.
29. Pater, R. H.; Morgan, C. D. *SAMPE J.* **1988**, *24*(5), 25–32.
30. Pater, R. H. In *Encyclopedia of Composites*; Lee, S., Ed.; VCH Publishers: New York, 1990; Vol. 2, pp 377–401.
31. Hergenrother, P. M.; Haven, S. J. *SAMPE J.* **1988**, *24*(4), 13–18.
32. Pratt, J. R.; St. Clair, T. L. *SAMPE J.* **1990**, *26*(6), 29–38.
33. St. Clair, T. L.; Yamaki, D. A. In *Polyimides*; Mittal, K. L., Ed.; Plenum Press: New York, 1984; pp 99–116.
34. Hinkley, J. A.; Mings, S. L. *Polymer* **1990**, *31*, 75–77.
35. Williams, J. G. *Fracture Mechanics of Polymers*; Applied Science Publishers: London, 1987; p 64.
36. Blundell, D. J.; Osborn, B. N. *Polymer* **1983**, *24*(8), 953–958.
37. Johnston, N. J.; Hergenrother, P. M. *Proc. 32nd SAMPE Symp.* **1987**, 1400–1412.
38. Hou, T. H.; Bai, J. M. *Proc. 35th SAMPE Symp.*, **1990**, *35*, 1594–1608.
39. Hou, T. H.; Reddy, R. M. *Report No. CR 187445*, National Aeronautics and Space Administration: Washington, DC, 1990.
40. Jang, B. Z.; Pater, R. H; Soucek, M. D.; Hinkley, J. A. *J. Polym. Sci. Polym. Phys. Ed.*, **1992**, *30*.
41. Johnston, N. J. Presented at the 4th Industry, Government Review of Thermoplastic Matrix Composites, San Diego, CA, February 9–12, 1987.
42. Johnston, N. J. Presented at the Interdisciplinary Symposium on Recent Advances in Polyimides and Other High Performance Polymers, Sponsored by American Chemical Society, San Diego, CA, January 22–25, 1990.
43. Kochi, M.; Horigome, T.; Mita, I.; Yokota, R. *Proc. 2nd Internat. Conf. Polyimides* **1985**, 454–468.

RECEIVED for review March 19, 1991 ACCEPTED revised manuscript March 13, 1992

# 6

# The Damage Zone in Some Ductile Polymers under a Triaxial Tensile Stress State

**A. Tse, E. Shin, A. Hiltner\*, and E. Baer**

**Department of Macromolecular Science and Center for Applied Polymer Research, Case Western Reserve University, Cleveland, OH 44106**

*Analysis of the damage zone at a semicircular notch clarifies the irreversible deformation behavior of some ductile polymers in a triaxial stress state. Plasticity concepts describe the thickness dependence of shear-yielding modes in polycarbonate, whereas impact-modified poly(vinyl chloride) deforms by a cavitation mechanism that is described by a critical volume strain. Contributions of both shear and cavitation mechanisms are observed in poly(vinyl chloride).*

**F**AILURE PROCESSES IN DUCTILE POLYMERIC MATERIALS depend on a number of complex interacting phenomena, some of which are not completely understood. One approach involves the study of the plastic deformation zone ahead of a stress concentrator. The sharp notch geometry is frequently used for triaxial tension observations because the stress state approximates triaxial tension at the tip of the propagating crack. On the other hand, a blunt notch, which results in large amounts of plastic yielding, is particularly attractive when the objective is to examine the prefailure damage zone while minimizing the tendency for crack growth. Specifically, the semicircular notch geometry is used to clarify the relationship of the stress state to the deformation modes as determined by specimen geometry, composition, applied stress, and test temperature.

This chapter discusses several types of damage zones that form ahead of

\* Corresponding author.

0065–2393/93/0233–0143$06.00/0

a semicircular notch under slow tensile loading. Polycarbonate (PC) represents a material that undergoes localized shear deformation (*1*). The importance of dilation in rubber-toughened materials is the motivation for studying polymer blends. Although most blends are opaque and thick sheets are not readily amenable to characterization by optical techniques, the damage zone can be studied in transparent or translucent blends that have been developed by matching refractive indices. An example is the translucent blends of poly(vinyl chloride) (PVC) with experimental chlorinated polyethylene (CPE) impact modifiers (*2, 3*). The contribution of dilation in glassy polymers is illustrated with PVC, which exhibits both shear yielding and cavitation (*4*).

## Material That Shear Yields—Polycarbonate

**Pressure-Dependent Core Yielding.** Polycarbonate is a ductile polymer at room temperature and exhibits shear yielding in tension and compression. The shear-yielding modes of polycarbonate were described by Ma et al. (*1*), who used a semicircular notch and tension deformation. The stress–displacement curve is independent of specimen thickness for thicknesses in the range of 1.2 to 6.5 mm (Figure 1). Yielding is observed first near position 1 on the stress–displacement curve and occurs at the notch root in the center of the specimen where the triaxial stress is highest. This mode of yielding, referred to as core yielding, is clearest in thicker sheets as two families of curving flow lines that grow outwardly from the notch (Figure 2).

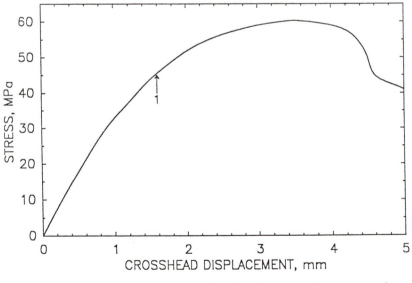

Figure 1. Stress–displacement curve of polycarbonate with a semicircular notch (*1*).

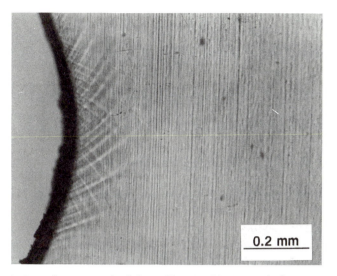

*Figure 2. Optical micrograph of the yield zone of 3.18-mm-thick PC at position 1 on the stress–displacement curve.*

Plane-strain core yielding at a semicircular notch of a perfectly plastic material is described by logarithmic spirals. The slip line field in the $xy$ plane is given by Hill (5) as

$$\theta = \pm \ln(r/a) + \text{constant} \tag{1}$$

in polar coordinates $(r, \theta)$, where $a$ is notch length. The signs indicate the two families of orthogonal $\alpha$ and $\beta$ slip lines that follow the directions of maximum shear stress. In this case, which is for a pressure-independent yield material, the slip lines intersect at 90°.

The slip lines of core yielding are easily observed with the optical microscope in thick sections of transparent glassy polymers that shear-yield, although the angle at which the two families of curves intersect is often less than 90°. This smaller intersection angle occurs when the yield stress is higher in compression than in tension, and it signifies the need for a pressure-dependent yield criterion to describe the increase in yield stress with increasing hydrostatic pressure. A widely used approach is the modified von Mises yield criterion with the form

$$k = k_0 - \mu \sigma_m \tag{2}$$

where $k$ is the shear-yield stress, $\sigma_m$ is the hydrostatic mean stress, and $k_0$ and $\mu$ are material parameters. A modification of equation 1 for a pressure-dependent yield material has been proposed (6):

$$\theta \cot \psi = \pm \ln(r/a) + \text{constant} \qquad (3)$$

where $2\psi$ is the angle between the slip lines and is related to the pressure dependency through $\mu = \cos 2\psi$. By using the published value of $\mu = 0.08$ for PC (7), the angle between $\alpha$ and $\beta$ slip lines is calculated to be 85°, which is in good agreement with the observed value of 86°. The slip lines of core yielding in PVC also intersect at an angle that is less than 90°. For this polymer, $\mu = 0.11$ (8), which gives an angle of 84°.

**Stress Instability and Plane-Strain Hinge Shear.** Plane-strain core yielding at the notch surface is kinematically constrained by the surrounding elastic material and cannot grow beyond a certain point. However, the external surface provides a larger degree of freedom for external yielding. The mode of macroscopic shear yielding is thickness dependent as is illustrated in Figure 3 by optical micrographs of four thicknesses of polycarbonate sheet. The plane-strain hinge shear mode initiates when the yield condition is achieved at the four points of shear stress concentration where the notch surface intersects the external surface above and below the $x$ axis. Hinge shear length increases with increasing stress and penetrates the entire thickness to form a pair of hinge shear bands that grow outwardly from the notch surface, one above and one below the core-yielding zone (Figure 4a).

In glassy polymers, hinge shear is observed as a pair of dark bands that grow outwardly from the notch at an angle that depends on the notch geometry. The angle of the hinge shear bands increases with increasing notch sharpness from about 40° with a semicircular notch (Figure 3a), which is close to the Tresca condition for the unnotched case, to 70°, which is the angle of maximum shear strain predicted analytically for the sharp notch geometry (9). Observation of the birefringent isostrain fringes that extend through the thickness at an angle of 90° to the plane of the sheet (1) confirm that the hinge shear mode is through thickness yielding.

Hinge shear is observed in ductile metals where it is identified with the sliding of linear dislocation arrays (10–12). Hinge shear together with core yielding is observed on single-edge notched specimens of silicon steel tested under plane-strain conditions (13) and also on Charpy specimens of high-nitrogen steel in a three-point flexural mode (14). In unnotched polycarbonate, hinge shear probably produces the characteristic epsilon damage zone that is observed during fatigue fracture (15–17). The dependence of hinge shear length on the applied tensile stress can be described by the Bilby, Cottrell, and Swinden (BCS) model (18)

$$L/a = 1/2\left[\sec(\pi\sigma/2\sigma_{ys}) - 1\right] \qquad (4)$$

where $L$ is the hinge length, $a$ is the notch length, and $\sigma$ and $\sigma_{ys}$ are the tensile stress and yield stress of the polymer, respectively.

**Stress Instability and Plane-Stress Intersecting Shear.**    If the specimen thickness is small, the through-thickness stress condition $\sigma_3 = 0$ will be approximately satisfied throughout the entire thickness; when the yield stress is achieved at the notch, yielding throughout the entire thickness occurs by slip along planes parallel to the $x$ axis that make an angle with the plane of the sheet (Figure 4b). Intersecting shear is clearly visible with transparent glassy polymers such as polycarbonate; the broad intersecting flow lines that extend through the thickness at an angle to the loading direction are seen when a cross section is viewed in the polarizing optical microscope (Figure 5). Although the intersecting shear angle is 45° for a material that obeys the Tresca criterion, the observed direction of yield in glassy polymers is inclined at more than 45°; in PC the intersecting shear angle is 53°. This deviation in the direction of the shear plane may result from volume expansion that occurs during plastic yielding of glassy polymers (*19*). Light scattered from the surface curvature created by the localized shear displacement causes intersecting shear to appear as a pair of parallel dark bands that grow outwardly from the notch (Figure 3d).

In cases that are neither plane stress nor plane strain, the dark bands of hinge shear are observed first, followed at higher stresses by the appearance of the second pair of dark bands created by intersecting shear, as illustrated with the two intermediate thicknesses, 4.70 and 3.14 mm, in Figures 3b and 3c. The spacing of the parallel bands depends on thickness and is equal to $t \cot 53°$ where $t$ is the sheet thickness.

Conditions for the transition from plane-strain hinge shear to plane-stress intersecting shear, obtained from a study of edge-slotted silicon steel, are described by Hahn and Rosenfield (*13*). Their study appears to be the only other instance in which this transition has been described. With the assumption that intersecting shear is constrained until the yield stress is achieved at a distance equal to one-half the sheet thickness above and below the crack plane, the condition for hinge shear dominance is

$$t/a > \sec\left(\pi\sigma/2\sigma_{ys}\right) - 1 \qquad (5)$$

It is also assumed that when the intersecting shear zone approaches a certain Dugdale-type shape, intersecting shear will predominate; specifically, when

$$t/a < (1/4)\left[\sec\left(\pi\sigma/2\sigma_{ys}\right) - 1\right] \qquad (6)$$

Intermediate stresses constitute a transition region. The analysis of Hahn and Rosenfield is applicable to polymers that shear-yield. In the case of polycarbonate, only hinge shear is observed below the lower limit of the transition region given by equation 5, whereas the onset of intersecting shear occurs at a stress within the transition region defined by equations 5 and 6.

## Materials That Cavitate—Blends of PVC with CPE

**Stress-Whitened Zone.**  Most toughened plastics are not readily amenable to studies of cavitation in thick sheets because the optimum rubber

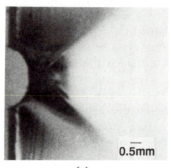

(a)

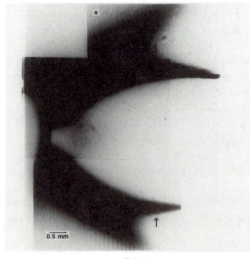

(b)

*Figure 3. Micrographs showing the yield zone of PC with a semicircular notch at four different thicknesses: a, 6.54 mm, 58.0 MPa; b, 4.70 mm, 58.8 MPa; c, 3.14 mm, 58.8 MPa; and d, 1.20 mm, 56.2 MPa (1).*

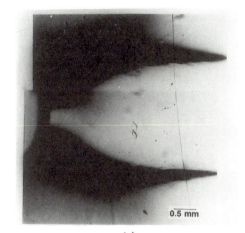

(c)

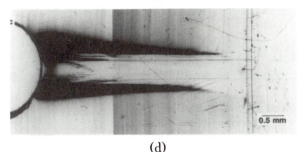

(d)

*Figure 3. Continued.*

particle size for toughening also produces the maximum light scattering and optical opacity so that stress-whitening is not readily discernible by optical techniques. Tse et al. (3) studied the damage zone of impact-modified PVC by using chlorinated polyethylene CPE impact modifiers with refractive indices that closely matched those of PVC. The stress–displacement curve of a translucent blend of PVC with 15 parts of the experimental CPE resin 5896 is shown in Figure 6. The first observed irreversible deformation is not core yielding; the first visible damage occurs near the linear limit of the stress–displacement curve as a small curved zone of stress-whitening ahead of the notch. As loading proceeds, the zone size increases, and the same crescent shape is maintained. At higher stresses the shape changes, and the damage zone becomes elongated ahead of the notch.

The stress-whitened zone occurs only in the center region of the sheet where plane-strain conditions prevail. For a sheet of finite thickness, the through-thickness stress is neither completely plane stress nor plane strain.

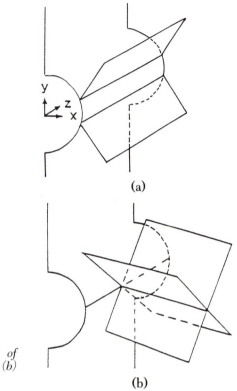

Figure 4. Schematic representation of hinge shear (a) and intersecting shear (b) ahead of a semicircular notch.

Most photoelastic experiments and finite element analyses indicate that the plane-strain condition is approximately realized through the thickness except for a region of rapidly decreasing triaxiality at the surface that constitutes approximately 10% of the thickness (20, 21). The observation that stress whitening does not extend to the surface but leaves a region of about 10% of the thickness that is visibly unaffected is confirmation that stress whitening occurs by a plane strain deformation mechanism.

Instead of a shear-yield condition producing the damage zone at the notch tip, as in PC, the damage zone described for PVC blends is determined by a cavitational condition. When the zone is small, the elastic stress distribution of Maunsell (22) is used to obtain the plane-strain stress condition at the zone boundary (3). The shape of the initial stress-whitened zone resembles a major principal stress contour, which is also curved near the notch root. However, the actual shape is not the same, and the major principal stress varies by approximately 10% around the boundary of the zone. Cavitation is frequently the cause of stress whitening in impact-

*Figure 5. Polarized optical micrograph of the yield zone of 1.20-mm-thick PC loaded to 58.5 MPa and viewed in the yz plane (1).*

modified polymers, and the boundary of the crescent-shaped zone in Figure 7 is described by a constant mean stress $\sigma_m = (\sigma_1 + \sigma_2 + \sigma_3)/3$, where $\sigma_1$, $\sigma_2$, and $\sigma_3$ are the major principal, minor principal, and through-thickness stresses, respectively. Figure 8 illustrates how the mean stress at the zone boundary remains almost constant as the load is increased; the variation in the major principal stress is also shown.

Cavity nucleation in ductile metals is most often described by a critical strain condition rather than a critical stress condition (23). A principal strain criterion has been suggested for polymer crazing, which is also a cavitational process (24, 25). The volume strain calculated from the bulk modulus for the blend described in Figure 7 is 0.8%, which is significantly lower than the 1.2% reported for polycarbonate (26). A lower critical volume strain for cavitation can explain why PVC is more amenable than PC to conventional methods of impact modification that rely on cavitation as the mechanism of toughening.

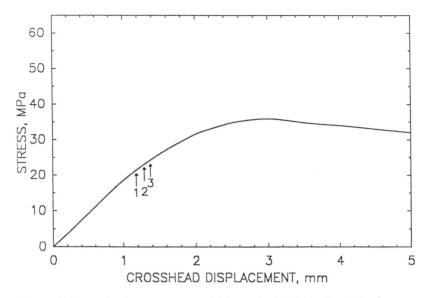

*Figure 6. Stress–displacement curve of 3.6-mm-thick PVC with 15 phr (parts per hundred parts resin) experimental resin 5896 (3).*

**Composition and Temperature Effects**. Numerous variables have been examined for their effect on the damage zone of PVC blends (3); these include the ratio of the notch radius to sheet thickness, temperature in the subambient range, amount of impact modifier from 5 to 20%, and composition of the CPE impact modifier, which altered the glass transition of the rubber as well as other properties. In all cases, the damage zone is a crescent-shaped stress-whitened zone that is described by a constant mean stress. Although the value of the mean stress obtained from an analysis of the damage zone may vary, the volume strain obtained from the mean stress and the calculated bulk modulus is always about 0.8% for PVC. An example of this condition in the study cited (3) is that either increasing the amount of impact modifier (Table I) or increasing the temperature caused the mean stress at the zone boundary to decrease (Table II). This effect leads to the conclusion that cavitation is controlled by a critical volume strain that is a material parameter of the resin (in this case PVC).

Assuming an elastic stress distribution to obtain the condition at the boundary of the damage zone is justifiable only when the zone is small. As the zone increases in size, stress redistribution because of the presence of the zone cannot be neglected. When stress redistribution occurs, the crescent-shaped stress-whitened zone elongates in the direction of the minor principal stress with a shape that is qualitatively described by the elastic–plastic boundary (27, 28).

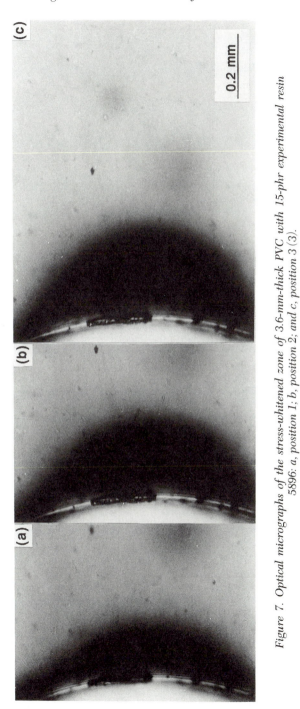

*Figure 7. Optical micrographs of the stress-whitened zone of 3.6-mm-thick PVC with 15-phr experimental resin 5896: a, position 1; b, position 2; and c, position 3 (3).*

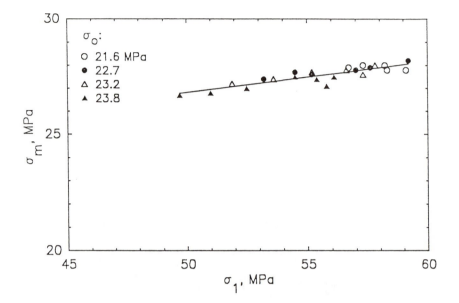

*Figure 8. Mean stress along the boundary of the stress-whitened zone of 3.6-mm-thick PVC with 15 phr experimental resin 5896 at several remote stresses (3).*

**Table I. Mean Stress and Volume Strain of PVC with Experimental Resin 5896(3)**

| Experimental Resin 5896 (phr) | Elastic Modulus[a] (GPA) | Yield Stress[a] (MPa) | $\sigma_m$ (MPa) 1.6 mm | 3.6 mm | Volume Strain[a](%) |
|---|---|---|---|---|---|
| 5 | 3.2 | 57 | 34 | 33 | 0.7 |
| 10 | 2.6 | 51 | 29 | 29 | 0.8 |
| 15 | 2.5 | 49 | 26 | 28 | 0.8 |
| 20 | 2.3 | 44 | 20 | 23 | 0.7 |

[a]At 10 mm/min, crosshead speed.

**Table II. Mean Stress and Volume Strain of 1.6-mm-Thick PVC with 10-phr Experimental Resin 5896 at Different Temperatures (3)**

| Temperature (°C) | Modulus (GPa) | Mean Stress (MPa) | Volume Strain (%) |
|---|---|---|---|
| 23 | 2.6 | 29 | 0.7 |
| 0 | 3.5 | 37 | 0.8 |
| −20 | 3.7 | 48 | 0.9 |
| −30 | 4.7 | 54 | 0.8 |
| −40 | 4.9 | 59 | 0.9 |

## Material That Both Shear-Yields and Cavitates— Poly(vinyl chloride)

**Initial Damage Zone.**    Poly(vinyl chloride) undergoes localized shear deformation, but readily cavitates when modified with a second phase (*3, 4*). Both shear and cavitational mechanisms contribute to the damage zone that forms in PVC ahead of a semicircular notch as described by Tse et al. (*4*). The first visible deformation occurs near position 1 on the stress–displacement curve in Figure 9, with the appearance of two families of intersecting slip lines that characterize core yielding. Core yielding of PVC is completely analogous to that of polycarbonate except for the slight difference in the angle at which the slip lines intersect. This discrepancy is caused by the larger pressure dependency of yielding in PVC. However, in PVC a region of stress whitening, which is not presence in polycarbonate, appears near the tip of the slip line zone (Figure 10). As the load increases, the size of the stress-whitened zone increases in length and width, but the position of the boundary closest to the notch remains fixed at approximately 0.11 mm from the notch tip.

The location of the stress-whitened zone relative to the notch tip, and the size of the cavities that cause light to be scattered, are apparent on a fracture surface created by brittle fracture through the preexisting damage zone (Figure 11). The core-yielding zone is closest to the notch and appears as two families of ridges that intersect at an approximately 90° angle. A region that is profusely cavitated is located approximately 0.1 mm from the notch and

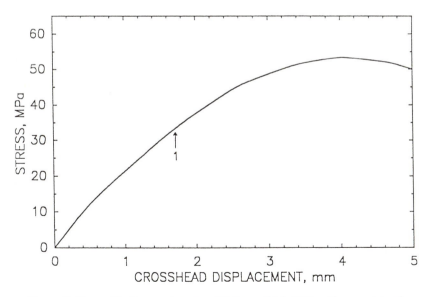

*Figure 9. Stress–displacement curve of 3.6-mm-thick PVC with a semicircular notch (4).*

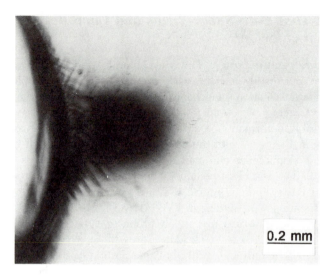

*Figure 10. Optical micrograph of the damage zone of 3.6-mm-thick PVC at position 1 on the stress–displacement curve (4).*

adjacent to the core-yielding zone. The region of the surface occupied by the 1-μm voids coincides with the location of the stress-whitened zone, and it can be distinguished from the areas further from the notch and closer to the edges that are featureless.

The critical volume strain condition for stress-whitening of PVC blends is thought to apply to PVC as well; additives in the PVC formulation act as the second phase that promotes cavitation (29). However, in this case, the remote stress required for shear yielding is achieved at the notch tip before the remote stress for cavitation is achieved, and the initial irreversible deformation occurs by a shear mode, specifically core yielding, rather than by cavitation. Once a plastic yield zone forms, the resulting stress redistribution creates a gradually increasing stress at the tip of the zone as it grows away from the notch root. In PVC, the additional stress concentration is sufficient for the critical cavitational condition to be achieved at the tip of the core-yielded zone before the stress instability is reached. The stress-whitened zone is viewed as being superimposed on an elastic–plastic interface so that the part nearest to the notch lies within the plastic-yielded zone and the part furthest from the notch is in the surrounding elastic region.

Although no exact solution is possible for the plastic–elastic stress distribution at a semicircular notch, some approximate calculations of the stress state at the boundary of the stress-whitened zone suggest that, like the PVC blends, stress whitening of PVC follows a constant mean stress condition and, furthermore, the corresponding critical strain is the same as PVC blends.

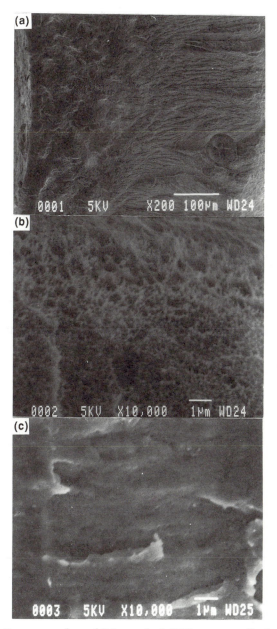

*Figure 11. Fracture surface of 3.6-mm-thick PVC cryogenically fractured in a brittle manner through the damage zone: a, the core-yielding zone at the notch root; b, high magnification view of the stress-whitened zone; and c, high magnification view of the region beyond the stress-whitened zone (4).*

The expression for the plastic mean stress, obtained from the stress distribution for the slip line field modified for a pressure-dependent material, is (6, 30)

$$\sigma_m = (k_0/\mu)\left\{1 - [1/(1 + \mu)](r/a)^{-2\mu/(1+\mu)}\right\} \tag{7}$$

and the corresponding volume strain is about 0.8%. The plastic–elastic stress distribution at a notch is amenable to finite element analysis, and estimates of the major principal stress suggest that a constant stress condition also exists at the point in the elastic region where the boundary of the stress-whitened zone intersects the $x$ axis.

**Stress Instability.** Although sheet thickness has little or no effect on the appearance of the core-yielding and stress-whitened zones of PVC, macroscopic shear yielding occurs by thickness-dependent shear modes analogous to those described by polycarbonate. The broad flow lines of plane-stress intersecting shear are observed superimposed on the stress-whitened zone in thinner sheets, whereas plane strain through thickness yielding on inclined planes above and below the notch, referred to as hinge shear, predominates in thicker sheets. As with polycarbonate, transitional behavior is observed with intermediate thicknesses.

The characteristics of hinge and intersecting shear modes in PVC closely conform with those established for polycarbonate, and shear yielding of PVC appears superimposed on the cavitation mechanism when the instability condition is achieved. The confined nature of the stress-whitened zone in PVC, and the fact that the zone does not extend to the notch surface where it could have a significant notch blunting effect, is probably the reason the presence of the stress-whitened zone has little or no effect on the shear-yielding modes.

## Conclusions

A comparison of the damage zone that forms in tension ahead of a semi-circular notch has been made between materials that typically shear-yield and those that exhibit stress whitening. The response to the triaxial tensile stress state created by a stress concentrator is controlled by material parameters that determine the critical condition for each mechanism of irreversible deformation. Analysis of the damage zone in polycarbonate, PVC, and rubber-modified PVC leads to the following conclusions:

1. Plasticity concepts describe the thickness dependence of shear-yielding modes in polycarbonate. The transition from plane-strain hinge shear on inclined planes above and below

the notch to plane-stress intersecting shear through the thickness follows conditions suggested for ductile metals.

2. The cavitation mechanism can be studied in plane strain with translucent blends of PVC with experimental CPE resins. The initial small crescent-shaped zone is controlled by a critical volume strain that is independent of composition and temperature.

3. Both shear and cavitation processes are observed in PVC. The initial irreversible deformation occurs by core-yielding, but the resulting stress intensification is sufficient for cavitation to subsequently initiate at the tip of the plastic zone.

## Acknowledgments

The work was generously supported by the Dow Chemical Company and the National Science Foundation, Polymers Program (DMR 87–13041).

## References

1. Ma, M.; Vijayan, K.; Im, J.; Hiltner, A.; Baer, E. *J. Mater. Sci.* **1989**, *24*, 2687.
2. Tse, A.; Laakso, R.; Hiltner, A.; Baer, E. *J. Appl. Polym. Sci.* **1991**, *42*, 1205.
3. Tse, A.; Shin, E.; Laakso, R.; Hiltner, A.; Baer, E. *J. Mater. Sci.* **1991**, *26*, 2823.
4. Tse, A.; Shin, E.; Hiltner, A.; Baer, E. *J. Mater. Sci.* **1991**, *26*, 5374.
5. Hill, R. *The Mathematical Theory of Plasticity*; Clarendon: Oxford, England, 1950.
6. Narisawa, I.; Ishikawa, M.; Ogawa, H. *J. Mater. Sci.* **1980**, *15*, 2059.
7. Christiansen, A. W.; Baer, E.; Radcliffe, V. *Phil. Mag.* **1971**, *24*, 451.
8. Yuan, J.; Hiltner, A.; Baer, E. *J. Mater. Sci.* **1983**, *18*, 3063.
9. Rice, J. R. *J. Mech. Phys. Solids* **1974**, *22*, 17.
10. Orowan, E. In *Proceedings of the International Conference on Physics*; Cambridge University Press: London, 1934; Vol. 2, p 81.
11. Cottrell, A. H. In *The Mechanical Properties of Matter*; Wiley: New York, 1964; p 348.
12. Clarke, F. J. P.; Sambell, R. A.; Tattersall, H. G. *Phil. Mag.* **1962**, *7*, 393.
13. Hahn, G. T.; Rosenfield, A. R. *Acta Metall.* **1965**, *13*, 293 .
14. Wilshaw, T. R.; Pratt, P. L. *J. Mech. Phys. Solids* **1966**, *14*, 7.
15. Mills, N. J.; Walker, N. *J. Mater. Sci.* **1980**, *15*, 1832.
16. Takemori, M. T.; Kambour, R. P. *J. Mater. Sci.* **1981**, *16*, 1108.
17. Takemori, M. T.; Matsumoto, D. S. *J. Polym. Sci. Polym. Phys. Ed.* **1982**, *20*, 2027.
18. Bilby, B. A.; Cottrell, A. H.; Swinden, K. H. *Proc. Roy. Soc. London* **1963**, A272, 304.
19. Whitney, W.; Andrews, R. D. *J. Polym. Sci.* **1967**, *C16*, 2981.
20. Dixon, J. R. In *Physical Basis of Yield and Fracture: Conference Proceedings*; Institute of Physics and the Physical Society: London, 1967; Vol. 1, p 6.
21. Villarreal, G.; Sih, G. C. *Mechanics of Fracture 7*; Sih, G. C., Ed.; Marinus Nijhoff: The Hague, Netherlands, 1981; p 253.

22. Maunsell, F. G. *Phil. Mag.* **1936**, *21*, 765.
23. Goods, S. H.; Brown, L. M. *Acta Metall.* **1979**, *27*, 1.
24. Wang, T. T.; Matsuo, M.; Kwei, T. K. *J. Appl. Phys.* **1971**, *42*, 4188.
25. Miltz, J.; DiBenedetto, A. T.; Petrie, S. *J. Mater. Sci.* **1978**, *13*, 1427.
26. Kambour, R. P.; Vallance, M. A.; Farraye, E. A.; Grimaldi, L. A. *J. Mater. Sci.* **1986**, *21*, 2435.
27. Theocaris, P. S. *J. Appl. Mech.* **1962**, *29*, 735.
28. Allen, D. N. de G.; Southwell, R. *Phil. Trans. Roy. Soc. London* **1950**, *242*, 379.
29. Vincent, P. I.; Willmouth, F. M.; Cobbold, A. J. In *Preprints of the 2nd International Conference on Yield, Deformation and Fracture of Polymers*; The Plastics Institute: Cambridge, England, 1973; p 5/1.
30. Kitagawa, M. *J. Mater. Sci.* **1982**, *17*, 2514.

RECEIVED for review March 6, 1991. ACCEPTED revised manuscript March 25, 1992.

# TOUGHENED THERMOSETS

# Additive Effects on the Toughening of Unsaturated Polyester Resins

Laurent Suspene[1,3], Yeong Show Yang, and Jean-Pierre Pascault[2]

[1] Cray Valley–Groupe TOTAL, Centre de Recherches et de Technologies de Verneuil, Ponc Alata, 60550 Verneuil-en-Halatte, France
[2] Laboratoire des Matériaux Macromoléculaires, URA–Centre National de la Recherche Scientifique 507, Bât. 403, Institut National des Science Appliqúees de Lyon, 69621 Villeurbanne Cédex, France

*An elastomer additive, carboxy-terminated acrylonitrile–butadiene copolymer, was used for toughening in the free radical cross-linking copolymerization of unsaturated polyester (UP) resins. For molded parts, Charpy impact behavior was generally enhanced and the number of catastrophic failures was reduced. The miscibility and interfacial properties of additive and resin blends play important roles in the toughening process. Phase-diagram studies showed that the elastomer additive is immiscible with the UP resin and is phase-separated from the resin matrix during curing. This phase-separation phenomenon is similar to that in the low-profile mechanism of UP resins. Additive-resin system miscibility greatly influences curing morphology. Microvoids occurred in the additive phase of cured resin because of shrinkage stress. The intrinsic inhomogeneity of the polyester network and the existence of microvoids in the final product limit the toughening effect of additives on unsaturated polyester resins.*

UNSATURATED POLYESTER RESINS ARE VITAL for composite applications. They are particularly useful in sheet-molding compounds (SMC) and bulk-molding compounds (BMC) for manufacturing automotive parts. Many researchers (1–9) have studied the curing kinetics of the bulk copolymerization

[3] Current address: Cook Composites and Polymer, 217 Freeman Drive, P.O. Box 996, Post Washington, WI 53074.

0065–2393/93/0233–0163$07.50/0

of styrene–unsaturated polyester (ST–UP) resins. Like most of the thermosetting matrix, unsaturated polyester resins are blended with several additives to improve their properties (10). For example, the high polymerization shrinkage of the UP resin may cause molding problems such as poor surface quality, warpage, sink marks, internal cracks, blisters, and dimension control (11, 12). Most of these problems are eliminated by the use of low-profile additives (LPA) such as poly(methyl methacrylate) (PMMA) and poly(vinyl acetate) (PVAc) to compensate for resin shrinkage (13–18).

Blending UP resins with additives also improves their mechanical properties. As with many thermoset polymers, the UP resins are limited by their brittleness, especially when good impact behavior is required. The impact strength and fracture properties of UP resins can be reinforced by means of an elastomeric additive. Such a component leads to a randomly dispersed rubbery phase in the material and creates a high dissipation energy for impact failures. This technique is commonly and successfully applied to epoxies by using random copolymers (19). The mechanisms of phase separation during the processing of epoxy resins have been widely studied (20). In the UP resins, the toughening effect of an elastomeric additive may depend on phase separation to provide the final morphology and to control the rubber-particle size and volume fraction.

This chapter proposes a phase-separation mechanism leading to the particular morphology obtained with UP resins. The toughening effect of UP resin with elastomers may be well explained with the mechanism. "Low-profile" systems were used for a better explanation of phase separation. Finally, the toughened systems were applied in toughening studies.

## Experimental Procedure

**Materials.** Two types of SMC-grade polyester resins (containing M1, M2, and M3 unsaturated polyester (UP) prepolymers, respectively) were used in this study. Both M1 and M2 are 1:1 copolymers of maleic anhydride and propylene glycol. They have similar number-average molecular weights, with an average of about 10 vinylene groups per molecule. However, the M1 prepolymer has a broader molecular weight distribution, higher acid index, and lower hydroxyl index than the M2 prepolymer. M3 is a copolymer of isophthalic acid, maleic anhydride, propylene glycol, and diethylene glycol (0.4:0.6:0.74:0.45 in mole ratio). The molecular characteristics of these UP resins are given in Table I.

The low-profile additive under investigation was PVAc (LP40A, Union Carbide, $M_n$ = 40,000, 40 wt % styrene solution). A reactive liquid rubber, a carboxyl-terminated poly(butadiene-co-acrylonitrile) (CTBN 1300 × 8, from BFGoodrich), was used as an elastomer.

### Table I. Molecular Characteristics of M1, M2, and M3 Unsaturated Polyester Prepolymers

| Prepolymer | $M_n^a$ | $M_w/M_n^a$ | $I_A^b$ | $I_{OH}^b$ |
|-----------|---------|-------------|---------|-----------|
| M1 | 1689 | 7.48 | 30 | 50 |
| M2 | 1480 | 3.14 | 20 | 60 |
| M3 | 2100 | 12 | 10 | 71 |

[a] Number-average and weight-average molecular weights were measured by a differential refractometer detector.
[b] Acid index ($I_A$) and hydroxyl index ($I_{OH}$), milligrams of KOH per gram of resin.
SOURCE: Reproduced with permission from reference 13. Copyright 1991 Butterworth–Heinemann Ltd.

An adduct epoxy-terminated poly(butadiene-*co*-acrylonitrile) (ETBN), was prepared by reacting an excess of epoxy prepolymer diglycidyl ether of bisphenol A (DGEBA) with the reactive rubber CTBN 1300 × 8. It was synthesized by the condensation of carboxyl chain ends on epoxy groups at 150 °C under vacuum without catalyst. The excess of DGEBA was eliminated by successive precipitations in the selective solvent ethanol. This synthesis was designed to refunctionalize the rubber and to obtain a DGEBA–CTBN–DGEBA triblock copolymer. The number-average molecular weight and the adduct density were calculated by using a method described elsewhere (*21*). This DGEBA–CTBN adduct has a number-average molecular weight, $M_n$, of 3840; its density, ρ, is 0.979.

**Instrumentation and Measurement. Cloud Point and Phase Diagram.** A liquid mixture could be transparent or opaque (cloudy), depending on the system composition and the system temperature. Theoretically, if the refractive indices of the mixing components are sufficiently different, a transparent mixture indicates a miscible one-phase system; a cloudy mixture always indicates an immiscible two-phase system. A *cloud point* is a point at which the system changes from transparent to cloudy or from cloudy to transparent as the composition or the temperature is changed.

The cloud points of styrene-UP resins, with or without LPA, were obtained by observing the transparency changes of the system when a styrene monomer is added at constant temperature (i.e., isothermal cloud point) or when the temperature of a constant-composition mixture is raised (i.e., nonisothermal cloud point). The measurements were taken under moderate agitation.

Totally immiscible blends became a two-phase system separated by a distinct interface. The equilibrium phase diagrams were determined by composition analysis of each phase. The time required to reach equilibrium varied from days to weeks, depending on the total polymer concentration and the proximity to the critical point (*22*). When the mixture was equilibrated, the two phases could be sampled. The amount of CTBN in each phase was determined by quantitative analysis with gel permeation chromatography (GPC) (22) or by elemental analysis of nitrogen content. The amount of PVAc was determined by acidic titration and, finally, the amount of styrene was determined by an evaporation method (23, 24).

*Chromatography.* A gel permeation chromatograph (Perkin Elmer, series 10) was used to measure the polymer molecular weight at room temperature with 50-, 100-, 500-, and 1000-Å columns combined with a mix column of 500-Å average size. A dual detector system consisting of an ultraviolet (UV) detector and a differential refractometer was used to measure the molecular weights of all the species in the systems studied. The UV detector used a monochromatic light at 254 nm (for vinylene groups) to measure the molecular weight of unsaturated polyester resin without the interference of PVAc or CTBN. All the GPC curves were analyzed by using the calibration curve obtained with standard samples of monodispersed polystyrene to estimate the molecular weight.

*Morphology and Impact Testing.* The samples containing PVAc were molded at 150 °C with 70 bar of pressure for 4 min with 1.5% *t*-butyl perbenzoate (TBPB) as high-temperature initiator. The samples containing CTBN were molded in a closed mold for 1 h at 150 °C. Methyl ethyl ketone peroxide (MEKP) was used as the initiator.

The molded samples were broken into several pieces by impact. One piece was etched in acetone for 30 min to dissolve the soluble materials on the fracture surface. The piece was then gold-coated, and the fracture surface was viewed by a scanning electron microscope (SEM). In addition, another piece was microtomed and stained by $RuCl_4$ for transmission electron microscope (TEM; Philips, EH 31) observation.

Unnotched impact test specimens were milled from 50-× 10-× 4-mm plates. A Charpy three-point bend pendulum (O. Wolpert, DDR) with a maximum energy of 4 J (pendulum striking velocity was 2.4 m/s) was used for impact measurement. The loading span was 40 mm, and the tests were carried out at room temperature (23 °C). Twenty-five specimens of each composition were tested.

## Results

**Phase Diagrams.** Figure 1 shows a typical triangle diagram of a ternary system of styrene–unsaturated polyester–low-profile additive (ST–UP–LPA). On the diagram UP represents M1 prepolymer and PVAc is used as LPA. The boxes show isothermal cloud points of the ternary system measured at 23 °C. An isothermal ternary cloud point curve (CPC) could be obtained by connecting all the isothermal cloud points; it forms an envelope.

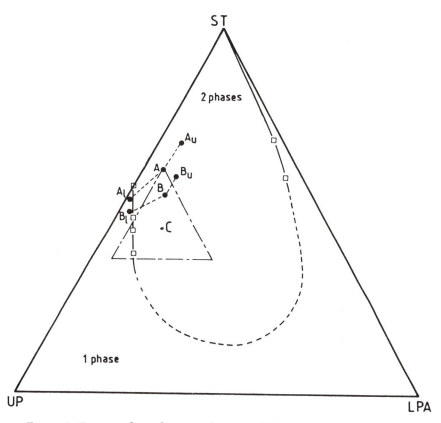

*Figure 1. Ternary phase diagram of styrene–M1 prepolymer–PVAc ternary systems. Key: □, experimental cloud point of styrene–M1–PVAc at 23 °C; solid lines and dashed lines, isothermal CPC at 23 °C; broken triangle, range of industrial formulations for low-profile UP resin; points A and B, compositions studied; subscripts u and l, upper and lower phases, respectively; and point C, composition chosen for molding studies. (Reproduced with permission from reference 13. Copyright 1991 Butterworth–Heinemann Ltd.)*

Inside the envelope is a two-phase region in which the system is cloudy after mixing; phase sedimentation will occur after storage. Outside the envelope is a one-phase region in which all the mixtures are miscible and transparent at the specific temperature.

A mixture that contains a large amount of UP or LPA is highly viscous and the corresponding cloud point is difficult to observe. For that reason the isothermal CPC was not experimentally completed. The bottom part of the curve is shown as a dashed line instead of a solid line. However, the bottom point of the envelope probably does not reach the UP–LPA line because the UP prepolymer is always miscible with PVAc. For the styrene–M1 prepolymer system a cloud point exists at 23 °C. Thus, the M1 prepolymer is only partially miscible with styrene (25, 26).

The broken-line triangle represents the formulation range of typical industrial UP resins. For M1 prepolymer, most of the formulations are not miscible and will separate into two phases after a period of storage. For example, point A represents a formulation of 5 wt % PVAc, 33 wt % M1, and 62 wt % styrene.

The mixture, which is cloudy after mixing, forms two phases in a test tube after a 2-day storage at room temperature. Its upper phase contains around 60% of the total volume. The compositions of the upper and the lower phases are determined by gas chromatography (GC) for styrene content and by acid titration for UP content. The LPA content is calculated with the measured styrene and UP contents. The upper phase is shown as point $A_u$ on Figure 1; the lower phase is shown as the point $A_l$. Similarly, formulation B (9 wt % PVAc, 36 wt % M1, and 55 wt % styrene) also forms two phases that are shown as points $B_u$ (upper phase) and $B_l$ (lower phase).

Points A, $A_u$, and $A_l$ (or B, $B_u$, and $B_l$) are not on a straight line because of data-manipulating deviation in the molecular-weight calculation (27). For both formulations, the upper phases are rich in styrene monomer and PVAc; in contrast, the lower phases are rich in M1 prepolymer.

Figure 2 shows GPC molecular weight distribution (MWD) curves of the prepolymers and the LPA contained in the $A_u$ and $A_l$ phases. The GPC curves of the M1 prepolymer are also shown for comparison. Their GPC–RI (refractive index) curves are shown in Figure 2a. The styrene peaks are not labeled in the figure. Clearly, the upper phase ($A_u$) contains PVAc, and the lower phase ($A_l$) shows an insignificant amount of PVAc. The curves confirm the previous result of the composition analysis. Figure 2b shows the GPC–UV curves of the $A_u$ and $A_l$ phases. Because they are measured by UV detector at 254 nm, the PVAc interference has been eliminated. The styrene peaks are not shown in the figure, either. Therefore, all the curves represent the MWD of the UP prepolymer only. The three MWD curves are clearly not identical.

The measured molecular weights of UP prepolymers in phases A, $A_u$, and $A_l$ are shown in Table II. The polydispersities of the UP prepolymers remaining in the upper and lower phases are not the same and are different from those of the original M1 prepolymer. The $A_l$ (lower) phase contains more high-molecular-weight prepolymer (short elution time, $M_w = 34,277$ g/mol) than the $A_u$ (upper) phase ($M_w = 5693$ g/mol). In other words, the high-molecular-weight prepolymer is fractionated by the PVAc and likely to remain in the lower UP-rich phase.

Theoretically, a phase diagram drawn through phase points such as $A_u$, $B_u$, $A_l$, and $B_l$ should be identical with the isothermal CPC. Koningsveld (28) revealed that phase separation in a polydispersed polymer solution would be accompanied by a phenomenon that causes a change of polydispersity (shown in Figure 2 and Table II) and then causes the phase diagram to

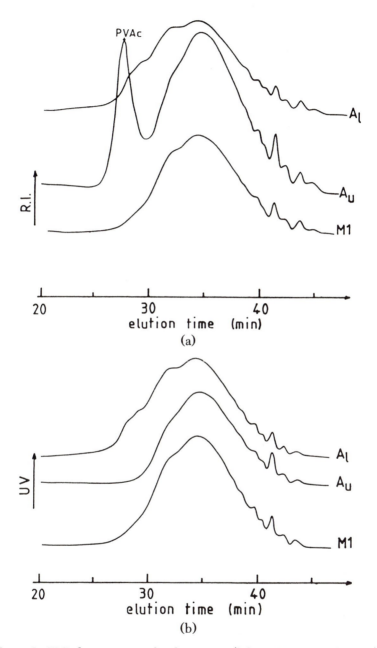

*Figure 2. GPC chromatograms for the point A (below, MI-contained system), the upper phase $A_l$ (above), and the lower phase $A_u$ (middle) in Figure 1. Key: a, recorded by RI detector; and b, recorded by UV detector at 254 nm. (Reproduced with permission from reference 13. Copyright 1991 Butterworth–Heinemann Ltd.)*

deviate from the CPC. That is why points $A_u$, $B_u$, $A_l$, and $B_l$ are not located on the isothermal CPC of Figure 1.

The miscibility of the UP prepolymer with styrene monomer could be changed by changing the structural characteristics of the UP prepolymer. Figure 3 shows a comparison of the miscibility of M1 and M2, the two prepolymers without LPA, in styrene. The nonisothermal CPC is drawn from the cloud points (or cloud temperatures) observed by changing the mixture temperature at each composition. A two-phase region is under the CPC and a one-phase region is above it. The curves are similar to an upper critical

### Table II. Molecular Weights of Unsaturated Polyester Prepolymers

| UP Prepolymers | $M_n$ | | $M_w$ | |
| --- | --- | --- | --- | --- |
| | $RI^a$ | $UV^b$ | $RI^a$ | $UV^b$ |
| Phase A$^c$ (M1) | 1,689 | 1,942 | 12,630 | 11,322 |
| Phase A$_u$ | — | 1,823 | — | 5,693 |
| Phase A$_l$ | 2,005 | 1,990 | 22,475 | 34,277 |

$^a$ Measured by the differential refractometer detector.
$^b$ Measured by the ultraviolet detector.
$^c$ Phases as shown in Figure 1.

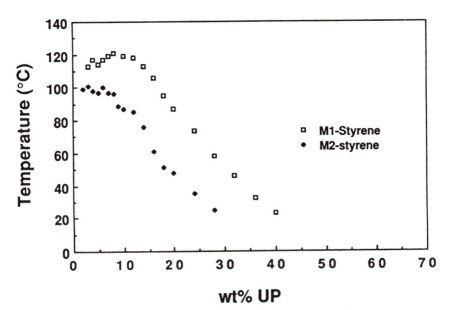

*Figure 3. Nonisothermal CPCs for styrene–UP prepolymer binary systems. (Reproduced with permission from reference 13. Copyright 1991 Butterworth–Heinemann Ltd.)*

solution temperature curve of most of the binary polymer solutions (29). To obtain a miscible resin, either the temperature or the UP content must be increased. However, the increased temperature is not practical for storage; the altered UP content would increase the resin viscosity and then limit the introduction of fillers and fibers.

As shown in Figure 3, the M2 prepolymer is more miscible with styrene than the M1 prepolymer because the nonisothermal CPC of M2 is lower than that of M1. At room temperature the cloud point of M2 is around 72 wt % styrene; that of M1 is around 60 wt % styrene. That comparison implies that the M2 prepolymer can allow higher styrene extent at room temperature and still remain in the one-phase region. The difference is believed to result from the molecular-weight polydispersity and acid and hydroxyl indices of the prepolymers, as shown in Table I.

The styrene miscibility difference between M1 and M2 also significantly influences the phase diagrams of the ternary (ST–UP–LPA) system shown in Figure 4, in which the LPA is PVAc. The isothermal CPCs (at 23 °C) of M1

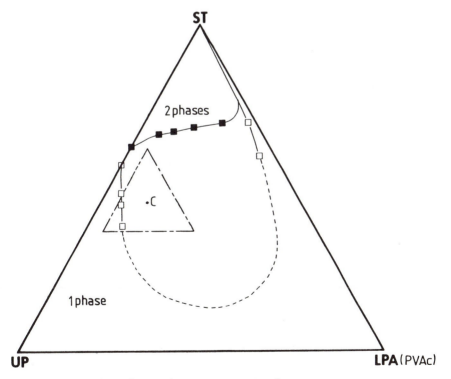

*Figure 4. Isothermal CPCs for styrene–UP prepolymer–PVAc ternary systems at 23 °C. Key: □, styrene–M1–PVAc; ■, styrene–M2–PVAc; and point C, composition chosen for the molding studies. (Reproduced with permission from reference 13. Copyright 1991 Butterworth–Heinemann Ltd.)*

and M2 are shown as the envelopes noted by open and filled squares, respectively. Again, the dashed part of the envelope reflects incomplete experimentation because of high viscosity. As in Figure 3, M2 shows a smaller two-phase region than M1. Apparently a UP prepolymer that has better miscibility with styrene in a binary system will also show better miscibility with PVAc in the ternary system.

Figure 5 shows the ternary phase diagram of a styrene–unsaturated polyester–elastomer system in which the UP is M3 prepolymer and CTBN 1300 × 8 is used as an elastomer. The diagram shows equilibrium tie lines for all the mixture compositions considered. The initial blend separated into two distinct phases. The upper or minor phase is mostly composed of CTBN and styrene. The lower or major phase is mostly composed of UP prepolymer and styrene.

Contrary to the previous systems, the M3 prepolymer in this case appears miscible with the styrene monomer. Moreover, the shape of the diagram

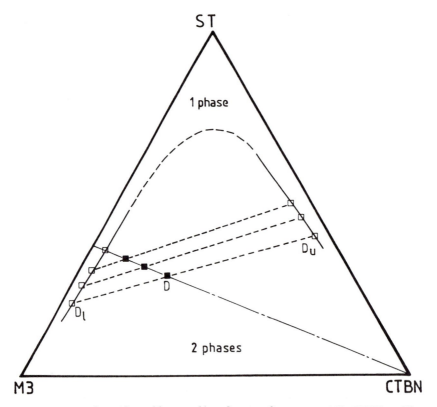

*Figure 5. Isothermal equilibrium phase diagram for styrene–M3–CTBN at 25 °C, determined by quantitative GPC measurements. Key: ■ , starting mixture composition; □ , separated phase composition. (Reproduced with permission from reference 27. Copyright 1990 John Wiley & Sons, Inc.)*

tends to indicate that the polyester prepolymer is immiscible with CTBN. This observation is correlated to the fact that the calculation of the Flory–Huggins interaction parameter $\chi$, by using van Krevelen's tables (*30*), gives 0.5 for the M3–styrene mixture and 0.18 for the CTBN 1300 × 8–styrene mixture. The coexistence curve is asymmetric, and the tie lines slope downward toward the UP–styrene axis representing the binary system with the larger interaction parameter. This observation agrees qualitatively with other experimental findings (*31*).

The phase diagram shown in Figure 6 was obtained by varying the styrene content in the mixture while keeping the ratio of M3 prepolymer:CTBN constant at 50:50. As previously, the mixtures separated into two distinct phases: a concentrated UP-rich bottom major phase and a diluted CTBN-rich top minor phase. This diagram has no tie lines. Because of the fractionation, the phase diagram obtained can be considered the sum

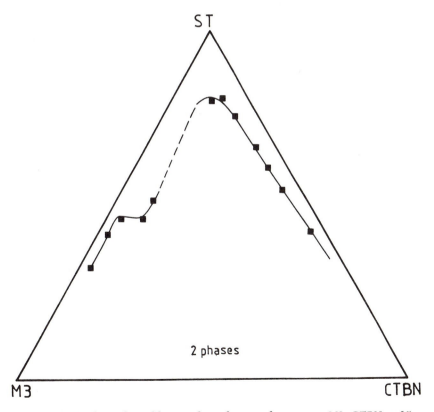

*Figure 6. Isothermal equilibrium phase diagram for styrene–M3–CTBN at 25 °C, determined by elemental analysis. Key: points, experimental phase composition determined by elemental analysis; curve, CPC curve. (Reproduced with permission from reference 27. Copyright 1990 John Wiley & Sons, Inc.)*

of $n \times m$ binodal curves, where $n$ and $m$ are the number of different molecular weight species of M3 prepolymer and CTBN, respectively (32).

The UV chromatograms of the top and bottom phases are presented in Figure 7. They clearly show that the higher molecular weight polyester molecules were in the UP-rich bottom (major) phase. On the contrary, the lower molecular weight polyester species were in the CTBN-rich top (minor) phase. This fractionation phenomenon was enhanced when polymer concentration decreased. Quantitative GPC analysis was not used to determine each phase composition because the detector response was not proportional to the polyester concentration. Nevertheless, the problem was solved by using elemental analysis.

**Morphology**.   Ternary systems of 45.4 wt % ST, 41.3 wt % UP, and 13.3 wt % PVAc were molded. The morphology of the molded samples was observed by SEM and TEM. The SEM micrographs of the fracture surfaces of the ST–M1–PVAc and ST–M2–PVAc systems after 30 min of etching in acetone are shown in Figures 8a and 8b.

The figures showed a fracture morphology typical of cured UP resins with PVAc (11, 12, 15–17). The nodules shown on the figures are cross-linked polyester microparticles that form a continuous macronetwork that is insoluble in most organic solvents. The PVAc, which is distributed around the microparticles, forms the other continuous phase. The PVAc phase on the fracture surface, however, is washed out by etching in acetone. In Figure 8b the size of the microparticles is singly distributed around 1–2 μm. Figure 8a shows a double size distribution of the nodules, one larger than 10 μm and one less than 2 μm. The distinction results from the miscibility difference between M1 and M2, which will be further discussed later in this chapter.

Figure 9 shows the TEM micrograph of the molded sample of UP resin (ST + M2) with PVAc, which is similar to the sample shown in Figure 8b. The cross-linked UP phase exists in the form of nodules, shown as the grey microparticles in Figure 9, and forms a polymeric network through linking the nodules. Figure 9 differs from Figure 8b by clearly showing the PVAc phase, which is stained as the dark area on the photo. The PVAc phase exists around the polyester nodules and forms a second continuous phase, as mentioned.

On the fracture surfaces of the specimens containing the M3 resin and 5 phr of CTBN, presented in Figures 10a and 10b, the rubber-phase particles are dispersed in the continuous UP matrix. As shown in Figure 10b, the rubbery nodules are burrowed after etching with acetone; nevertheless, the substructure of linked microgels remains the same as that shown in Figure 8b. This similarity indicates a complex morphology inside the rubber phase. This observation is in agreement with other experimental findings (33). The elastomer particle-size distribution obtained (1–40 μm) is presented in

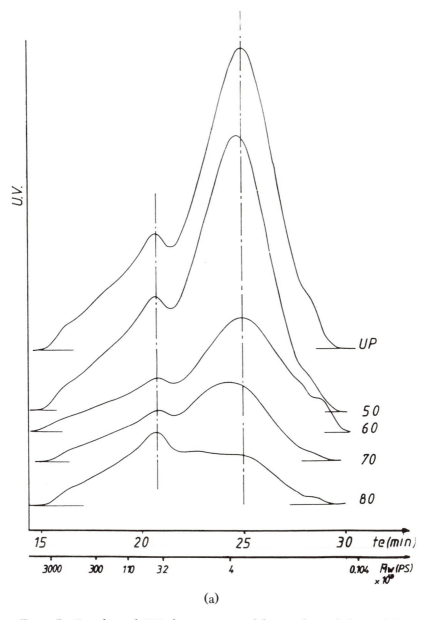

(a)

*Figure 7a. Top phase of GPC chromatograms of the two demixed phases of the styrene–M3–CTBN system with various amounts of styrene (50, 60, 70, and 80 wt %). (Reproduced with permission from reference 27. Copyright 1990 John Wiley & Sons, Inc.)*

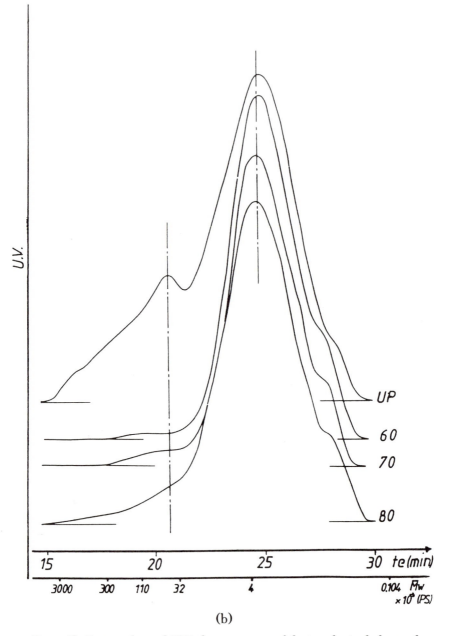

(b)

*Figure 7b. Bottom phase of GPC chromatograms of the two demixed phases of the styrene–M3–CTBN system with various amounts of styrene (50, 60, 70, and 80 wt %). (Reproduced with permission from reference 27. Copyright 1990 John Wiley & Sons, Inc.)*

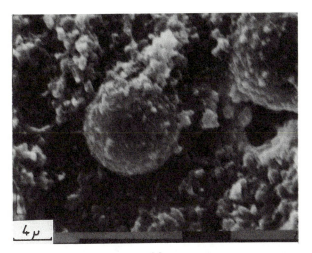

(a)

(b)

*Figure 8. SEM micrographs of molded styrene–UP–LPA ternary systems. Key: a, styrene–M1–PVAc; and b, styrene–M2–PVAc. Both photos show the same magnification. (Reproduced with permission from reference 13. Copyright 1991 Butterworth–Heinemann Ltd.)*

Figure 11a. The mean diameter was about 12 μm, with a large concentration at about 8-μm diameter.

A way to reduce particle size is to reduce the interfacial tension between phases. Size reduction was achieved with the introduction of the ETBN triblock copolymer DGEBA–CTBN–DGEBA into the CTBN elastomer

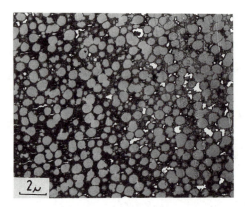

*Figure 9. TEM micrograph of molded styrene–M2–LPA ternary system. (Reproduced with permission from reference 13. Copyright 1991 Butterworth–Heinemann Ltd.)*

(24). The ETBN triblock copolymer, 10% by weight, was added to pure CTBN 1300 × 8 to have a 90:10 rubber-blend mixture. The morphology of the molded sample containing 5 phr of this blend is presented in Figure 12. As in the pure CTBN, the ETBN rubber-phase particles are dispersed in the UP matrix. Again, those particles present a continuous substructure of cross-linked polyester nodules embedded in an elastomeric continuous phase. But in this case, the elastomer particles are in the 1–15-μm size range (Figure 11b). The mean diameter was about 5 μm with a narrow distribution (±2.7 μm). The addition of ETBN clearly reduced the elastomer particle size by controlling the interfacial tension between the two initially separated phases.

**Impact Testing**. Figures 13a, 13b, and 13c show the impact behaviors of the pure M3 unsaturated polyester matrix, the UP matrix with 5 phr of CTBN 1300 × 8 elastomer, and that with 5 phr of the rubber blend (90 wt % CTBN and 10 wt % ETBN), respectively.

Impact energy is represented as a histogram of frequencies because, as will be seen later, the scatter in this test is an important parameter. *Impact energy* is the absorbed energy of the specimen divided by its cross-sectional area. For the system with elastomer modification, the fracture-impact energy per unit area, $R_s$ (Table III), was not really enlarged, and the scatter of the results was the same (standard deviation about 2 kJ/m²). In fact, the mean diameter of the dispersed phase, the width of the distribution, and the low fracture-impact energies are reduced.

This behavior can be explained by the shape of the particle size distribution curve (Figures 11a and 11b). Generally, improvements in the fracture impact of thermoset resin can be obtained by dispersing elastomer particles with diameters from 0.5 to 5 μm in the blends (34, 35). In our case, the

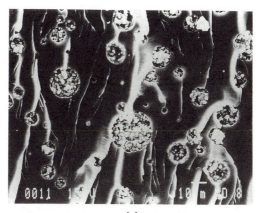

(a)

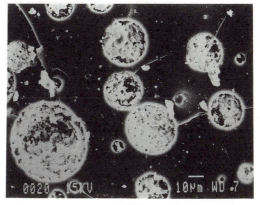

(b)

*Figure 10. SEM micrographs of molded styrene–M3–CTBN ternary systems. Key: a, unetched; and b, etched in acetone. Both photos have the same magnification. (Reproduced with permission from reference 24. Copyright 1990 Society of Plastics Engineers.)*

smaller particle size caused by the addition of the triblock copolymer to the rubber enhanced the fracture-impact energy. The smaller particle size may produce more microcracks in the UP matrix (cf. Figures 10 and 12).

The fracture surface is rough around the rubbery nodules, and impact behavior arises from crack deviation. Thus, the energy for breaking increases with increasing energy dissipation through the creation of surfaces (35). In the micrograph in Figure 10 (5 phr of pure CTBN 1300 × 8) the secondary cracks are less numerous, and the presence of large particles leads to zones where cracks can pass through easily. Stresses created under loading become concentrated in these regions and induce catastrophic failures.

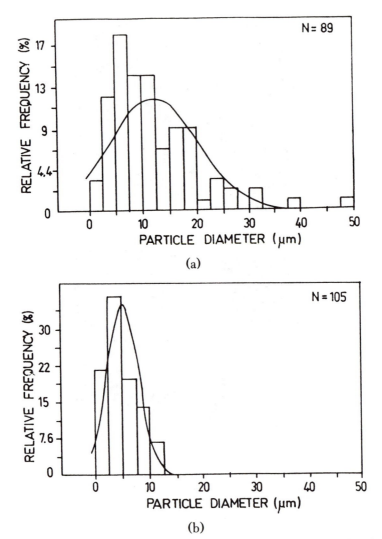

*Figure 11. Elastomer particle size distribution. Key: a, styrene–M3–CTBN ternary system; and b, styrene–M3–rubber (90 wt % CTBN, 10 wt % triblock copolymer ETBN) system. (Reproduced with permission from reference 24. Copyright 1990 Society of Plastics Engineers.)*

These results agree with the data of Crosbie and Phillips (36). They studied the impact behaviors of the UP resin by adding an experimental reactive liquid rubber (37) that is very similar to the ETBN used in our study. No significant change in impact energy was found as the rubber content was increased. They also found that, for such toughened polyesters,

(a)

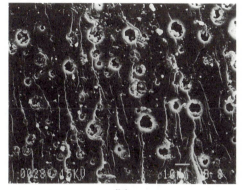

(b)

*Figure 12. SEM micrographs of molded styrene–M3–rubber (90 wt % CTBN, 10 wt % triblock copolymer ETBN) ternary systems. Key: a, unetched; and b, etched in acetone. Both photos have the same magnification. (Reproduced with permission from reference 24. Copyright 1990 Society of Plastics Engineers.)*

large improvements were observed only at slow testing speeds. Little or no improvement was seen with high-rate impact.

## Discussion

For simplicity, the relationships of phase diagram and morphology of the polyester–additive blends will be discussed by using the system of ST–UP–LPA, in which UP represents M1 or M2 prepolymer and PVAc is used as LPA. A ternary phase diagram (e.g., Figure 1) is used as a setup for a nonreactive system without initiators.

During the reaction, because of the increase of polymer content and the decrease of monomer content, the system's time- and conversion-dependent phase diagram is too complicated to obtain. However, a prereaction phase diagram still can be correlated to the morphology of molded blends.

In Figure 4, point C for the ST–M1–PVAc system is located in a two-phase region, although it is a one-phase mixture for the ST–M2–PVAc system. Figure 8 clearly shows that, after curing, the ST–M2–PVAc system

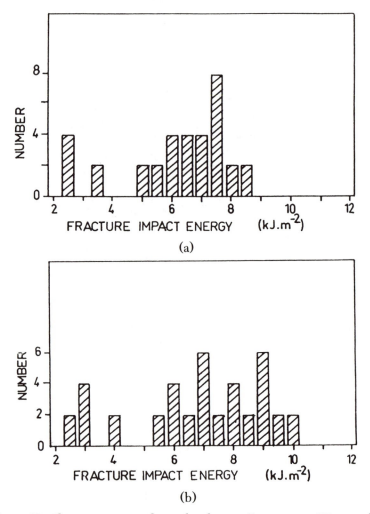

(a)

(b)

*Figure 13. Charpy impact toughness distribution. Key: a, pure M3 resin; b, styrene–M3–CTBN; and c, styrene–M3–rubber (90 wt % CTBN, 10 wt % triblock copolymer ETBN). (Reproduced with permission from reference 24. Copyright 1990 Society of Plastics Engineers.)*

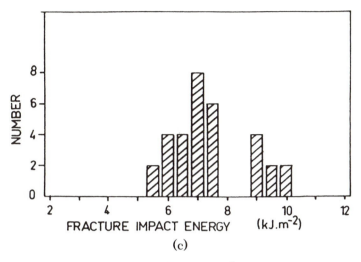

Figure 13. Continued

**Table III. Charpy Impact Toughness of the Various Materials**

| Material | Fracture-Impact Energy $R_s$ $(kJ/m^2)$ | Minimum $R_s$ $(kJ/m^2)$ | Maximum $R_s$ $(kJ/m^2)$ | Standard Deviation |
|---|---|---|---|---|
| Pure M3 | 6.2 | 2.5 | 8.5 | 1.8 |
| M3 + 5 phr CTBN | 6.8 | 2.5 | 10 | 2.2 |
| M3 + 5 phr elastomer blend | 7.4 | 5.5 | 10 | 1.3 |

SOURCE: Reproduced with permission from reference 24. Copyright 1990 Society of Plastics Engineers.

reveals a more homogeneous nodule distribution than the ST–M1–PVAc system. It indicates that the initial system miscibility, which is modified by prepolymer characteristics as shown on the ternary phase diagram, greatly influences the final product morphology.

According to a phase equilibrium theory, a mixture that is located in the two-phase region of a phase diagram eventually will sediment into two distinct phases. Point A, for example, in Figure 1 forms an upper phase, $A_u$, and a lower phase, $A_l$, before curing. The $A_u$ phase is rich in styrene and PVAc, and the $A_l$ phase is rich in UP prepolymer. Nevertheless, the UP prepolymer is not homogeneously distributed in both phases because it is not a monodispersed prepolymer but a prepolymer mixture with a broad MWD. Figure 2 reveals that the lower phase contains more high-molecular-weight UP molecules than the upper phase. These results indicate that, when a phase separation takes place in the initial ST–UP–LPA system, the high-molecular-weight UP molecule is likely to remain in the UP-rich minor phase, although the LPA remains in the styrene–LPA-rich major phase.

Adding an initiator and increasing the temperature starts the polymerization of the ST–UP–LPA system. The polymerization of UP resin is a free radical chain-growth reaction. At the beginning of the reaction, some high-molecular-weight polymer chains are formed by the free radical growth. As a result of the intramolecular cross-linking reaction, the high-molecular-weight polymer chains immediately become tiny, highly intra-cross-linked particles. Those highly intra-cross-linked, high-molecular-weight polymer particles are called *microgels* (38–41). The microgels may or may not be miscible in the system, depending on the conversion-dependent phase diagram during curing.

If the microgel is considered a swollen high-molecular-weight UP molecule, the addition of microgel is expected to increase the polydispersity of the UP component and then to change the CPC on the phase diagram. For example, point C in Figure 4 for ST–M2–PVAc is initially located in the one-phase region When the reaction takes place microgels are continuously formed, so M2 polydispersity is increased with polymerization. Consequently, the CPC envelope moves downward and becomes larger. Eventually the expanding CPC envelope may cover point C. At that point the system becomes immiscible and will be phased out. When phase separation occurs, according to the initial phase diagram, the LPA remains in the styrene–LPA-rich phase; the high-molecular-weight microgels precipitate and form the UP-rich phase. This fractionation effect is caused by LPAs. For low-profile systems the phase separation takes place at the beginning of the reaction, when the conversion is less than 1% (13).

When phase separation occurs, the microgels precipitate to form the UP-rich phase. Then these microgels concentrate and coagulate to form larger particles (called microparticles (14, 15) or nodules), as shown in Figure 8. The nodule, composed of a number of microgels depending on the system, is a basic cross-linking unit of a macronetwork.

As the reaction goes on, the styrene and the UP prepolymer in the styrene–LPA-rich phase continuously polymerize to form new microgels, which are converted to more nodules. The nodules are then cross-linked to form a macronetwork structure. When the reaction is complete, only LPA is left in the styrene–LPA-rich phase. Therefore, LPA forms a second continuous phase that surrounds the macronetwork of polyester resin. This morphology can easily be observed in Figures 8b and 9.

The mechanism of the ST–M1–PVAc system is similar to that of ST–M2–PVAc, but a little different. In Figure 4, point C of the ST–M1–PVAc system is located in the two-phase region. Before the reaction, the system exists as two phases even at 100 °C. When the reaction begins, the initial styrene–LPA-rich phase follows the mechanism described for the ST–M2–PVAc system. The UP nodules dispersed in this phase show a size distribution similar to those in the ST–M2–PVAc system (cf. Figures 8a and 8b). The initial UP-rich phase dispersed in the initial styrene–LPA-rich phase would follow a similar mechanism. However, because of the low LPA content in this UP-rich phase, the fractionation effect is not significant. Each dispersed UP-rich microdomain finally forms a large nodule in the network structure. This morphology appears in Figure 8a as the larger nodules.

The formation mechanism of the morphology of the system containing CTBN is different from that of the system containing LPA. A comparison of the phase diagrams of Figures 1 and 5 clearly shows the miscibility of the systems when LPA and CTBN are not initially the same. With CTBN, the blends are always located in a two-phase region, even at high molding temperature. The ST–M3–CTBN system, contrary to that of ST–M1–PVAc, shows a continuous UP-rich phase in which CTBN is completely excluded. The dispersed CTBN-rich phase also contains styrene and UP prepolymer. During the reaction, the continuous phase follows a pure UP resin reaction mechanism and then forms a "flakelake" morphology (*39, 41*). The dispersed phase, on the other hand, follows the ST–M2–PVAc mechanism. Consequently, it forms a nodule network dispersed in a continuous bulk UP phase (*14, 17*), similar to that shown in Figure 8b.

Exactly the same morphologies are observed for many additives like PMMA (*39*), polystyrene (PS) (*12*), polyethylene (PE), or some styrene–butadiene copolymers (*42*) that are not initially miscible with UP resins. Such additives are commonly used for shrinkage compensation or for toughening.

It is generally believed that plastics can be toughened by means of randomly dispersed elastomer particles (*35*). The morphology of the elastomeric phase (such as shape, size, and volume fraction) has been widely correlated to mechanical improvements for a great number of materials such as epoxies (*20*) or thermoplastics (*35*). However, the improvement of impact strength obtained by introducing an elastomer into a UP matrix ($< 10\%$) is not important compared to that obtained with an epoxy matrix ($\sim 100\%$) (*19, 20*). Similarly, creation of chemical bonding between elastomer and UP resin using methacrylate end-capped CTBN (*43*) or isocyanate end-capped polybutadiene (*44, 45*) was attempted as a way to increase fracture-impact energy. The resulting morphologies were similar to those achieved with CTBN, and the impact strength was in the same order of magnitude. This unexpected behavior of impact improvements with UP resin may be explained by its unusual morphology.

For a reactive material such as polyester resin, the system is initially in a liquid state and becomes solid during curing. The final morphology is generally fixed by the polymerization kinetics and the thermodynamics. When poly(epichlorohydrin) is used as an additive (*46*), the UP resin–additive mixture is miscible before reaction. Because strong hydrogen-bonding interactions formed in those systems (*47*) lead to a strong thermodynamic effect, the mixture remains one phase after cross-linking. In this case little improvement of impact strength is achieved, and this improvement is associated with the decrease of $T_g$.

With LPAs such as PVAc and polyurethane as additives, the mixture is initially miscible. Because of competition between thermodynamics and polymerization kinetics, it finally becomes either a co-continuous two-phase morphology for a UP–PVAc system or a relatively homogeneous but still two-phase microscale morphology for a UP–polyurethane system. Even with co-continuous morphologies, the use of a polyurethane elastomer does not lead to a significant improvement of impact properties (48). PVAc is different because it is vitreous at room temperature and therefore could not be considered as an impact-modifier additive.

Shrinkage should not be ignored during the curing of the UP resin. Strong polymerization shrinkage creates an internal stress passing in phase transition and finally forms microvoids in the additive phase (13). Microvoids could also occur around the fillers and fibers in the composites (49). The formation of microvoids helps to compensate for resin shrinkage and is the so-called *low-profile additive effect*.

However, most of the UP–additive systems (e.g., UP–PMMA, UP–PE, and UP–elastomer) are not miscible before reaction. After curing they show a continuous UP matrix and a dispersed additive phase. The morphology in the additive phase is the same as that of the UP–PVAc system. Thus, microvoid formation occurs in the dispersed additive phase. However, the microvoid content is smaller and the shrinkage compensation effect is worse with UP–additive systems than with PVAc. As a result, those additives are only used as low-shrinkage compounds.

Microvoid formation is an important phenomenon for the UP–additive system. In addition to compensating for resin shrinkage, microvoids also induce an intrinsic brittleness. Increased rubber content in the UP resin should produce higher impact strength; at the same time, if the microvoid content is also enhanced, impact strength is affected in the opposite way. This very intricate situation explains why UP resin impact strength cannot be improved easily by applying even a chemically modified elastomer. A good morphology of a significant toughening effect for the UP–additive system is still unknown.

Nevertheless, the dispersion of a CTBN-type complex elastomeric phase into a UP resin may reduce the number of catastrophic failures (i.e., the low-energy impact break). The size of the rubber domains must be controlled by using an emulsifying agent like ETBN or by using rubbers like isocyanate end-capped polybutadiene, which can react with the polyester hydroxyl chain ends and create, in situ, its own interfacial tension reducer. Those elastomer particles, well dispersed in the UP matrix, induce homogeneous distribution of the internal stresses due to the network formation (33, 36, 44, 45).

## Conclusion

The final morphology of the UP–additive systems depends on the competition of polymerization kinetics and thermodynamics. In most cases, the cured system shows a two-phase morphology. The strong shrinkage stress creates microvoids in the additive phase. The formation of microvoids can well compensate the resin shrinkage; however, it induces an intrinsic brittleness of the material and reduces impact strengh. The UP–elastomer system forms a dispersed rubber phase, but the rubber phase is still a two-phase region and the microvoids still exist.   When the rubber content in the UP resin is increased for higher impact strength, the microvoid content is simultaneously enhanced, which reduces the impact strength.

This very complicated situation explains why the impact strength of the UP resin cannot be improved easily by applying the elastomer. A good morphology of a UP–additive system with significant toughening effect is still unknown.   An attempt to toughen UP resin with elastomers is always a trade-off. However, the number of catastrophic failures (i.e., the low-energy-impact break) of the UP resin may be reduced by dispersion of a CTBN-type complex elastomeric phase. Finding an LPA that has a toughening effect on UP resin remains a real challenge for researchers.

## References

1.  Horie, K.; Mita, I.; Kambe, H. *J. Polym. Sci., Part* A-1, **1970**, *8*, 2839.
2.  Kubota, H. *J. Appl. Polym. Sci.* **1975**, *19*, 2279.
3.  Kamal, M. R.; Sourour, S.; Ryan, M. *SPE Antec Tech. Papers* **1973**, *19*, 187.
4.  Lee, L. J. *Polym. Eng. Sci.* **1981**, *21*, 483.
5.  Stevenson, J. F. *Polym. Proc. Eng.* **1983**, *1*, 201.
6.  Han, C. D.; Lem, C. W. *J. Appl. Polym. Sci.* **1983**, *28*, 3155, 3185, 3207.
7.  Kuo, J. F.; Chen, C. Y.; Chen, C. W.; Pan, T. C. *Polym. Eng. Sci.* **1984**, *24*, 22.
8.  Yang, Y. S.; Lee, L. J. *J. Appl. Polym. Sci.* **1988**, *36*, 1325.
9.  Yang, Y. S.; Lee, L. J.; Tom, S. K.; Menardi, P. C. *J. Appl. Polym. Sci.* **1989**, *37*, 2313.
10. Suspene, L., PhD Thesis, Institut National des Science Appliqúees de Lyon, France, 1989.
11. Kroekel, C. H., Paper presented at SAE Automotive Engineering Congress, Detroit, MI.; Society of Automotive Engineers: Warrendale, PA, January 1968.
12. Atkins, K. E. *Polymer Blends*; Paul, D. R.; Newman, S., Eds.; Academic Press: Orlando, FL, 1978; Vol. 2, Chapter 23, p. 391.
13. Suspene, L.; Fourquier, D.; Yang, Y. S. *Polymer* **1991**, *32*, 1593.
14. Pattison, V. A.; Hindersinn, R. R.; Shwartz, W. T. *J. Appl. Polym. Sci.* **1974**, *18*, 2763.
15. Pattison, V. A.; Hindersinn, R. R.; Shwartz, W. T. *J. Appl. Polym. Sci.* **1975**, *19*, 3045.
16. Kiaee, L.; Yang, Y. S.; Lee, L. J. *AIChE Symp.* **1988**, *84(260)*, 52.
17. Ross, L. R.; Hardebeck, S. P.; Backman, M. A. *43rd Ann. Conf., Comp. Inst.*; Society of Plastic Industry, Inc.: Washington, DC, 1988; p 17–C.
18. Tomasana, M.; Hidehiko, S.; Kiyoshi, H.; Gwilym, E. O. *44th Ann. Conf.*,

*Comp. Inst.*; Society of Plastic Industry, Inc.: Washington, DC, 1989; p 12-F.
19. *Rubber Thermoset Resins*; Riew, C. K.; Gillham, J. K. Eds.; Advances in Chemistry 208; American Chemical Society: Washington, DC, 1984.
20. Montarnal, S.; Pascault, J. P.; Sautereau, H. In *Rubber Toughened Plastics*; Riew, C. K.; Gillham, J. K., Eds.; Advances in Chemistry 222; American Chemical Society: Washington, DC, 1989; p 193.
21. Verchere, D.; Sauterau, H. Pascault, J. P.; Moschiar, S. M.; Riccardi, E. C.; Williams, R. J. J. *Polymer* **1989**, *30*, 107–115.
22. Tseng, H. S.; Lloyd, D. R.; Burns, C. M. *J. Appl. Polym. Sci., Part B* **1987**, *25*, 325.
23. Suspene, L.; Pascault, J. P. *SPE Antec Papers* **1989**, *35*, 604.
24. Suspene, L.; Fourquier, D.; Yang, Y. S. *45th Annual Conference, Composite Inst.*; Society of Plastic Industry, Inc.: Washington, DC, 1990; p 11–F.
25. Rosso, J. C.; Guieu, R.; Carbonnel, L. *Bull. Soc. Chim.* **1970**, *489*, 2849.
26. Rosso, J. C.; Guieu, R.; Carbonnel, L. *Bull. Soc. Chim.* **1970**, *490*, 2855.
27. Suspene, L.; Pascault, J. P. *J. Appl. Polym. Sci.* **1990**, *41*, 2665.
28. Koningsveld, R.; Staverman, A. J. *J. Polym. Sci., Part A2* **1968**, *6*, 349
29. Flory, P. J. *Principles of Polymer Chemistry*; Cornell University Press.: Ithaca, New York, 1953.
30. Van Krevelen, D. W.; Hoftyser, P. J. *Properties of Polymers*; Elsevier: Amsterdam, 1976; 2nd Ed.
31. Tseng, H. S.; Lloyd, D. R.; Burns, C. M. *J. Appl. Polym. Sci.* **1979**, *23*, 749.
32. Tompa, H. *Polymer Solutions*; Butterworth Scientific Publications: London, 1956.
33. Crosbie, G. A.; Phillips, M. G. *J. Mater. Sci.* **1985**, *20*, 5634.
34  Rowe, E. H. *34th Ann. Techn. Conf.*; Society of Plastics Industry, Inc.: Washington, DC, 1979; paper 23B.
35. Bucknall, C. B. *Toughened Plastics*; Applied Science Publishers: London, 1977.
36. Crosbie, G. A.; Phillips, M. G. *J. Mater. Sci.* **1985**, *20*, 182.
37. Scott-Bader Ltd., US Patent 4 530 962, July 23, 1985.
38. Yang, Y. S.; Lee, L. J. *Polym. Process. Eng.* **1988**, *5*, 327.
39. Kiaee, L.; Yang, Y. S.; Lee, L. J. *AIChe Symp.* **1988**, *84(260)*, 52.
40. Dusek, K.; Galina, H.; Mikes, J. *Polym. Bull.* **1980**, *3*, 19.
41. Yang, Y. S.; Lee, L. J. *Polymer* **1988**, *29*, 1793.
42. Kirk Patrick, J. P.; Gonzales, W. H.; Morgan, M. J. *45th Ann. Conf., Comp. Inst.*; Society of Plastics Industry, Inc.: Washington, DC, 1990; paper 6–C.
43. Mc Garry, F. J.; Rowe, E. H.; Riew, C. K. *Polym. Eng. Sci.* **1978**, *18(2)*.
44. Malinconico, M.; Martuscelli, E.; Volpe, M. G. *Inter. J. Polym. Mat.* **1987**, *11*, 295.
45. Malinconico, M.; Martuscelli, E.; Ragosta, G.; Volpe, M. G. *Inter. J. Polym. Mat.* **1987**, *11*, 317.
46. Rowe, E. H.; Howard, F. H. *33rd Ann. Conf., Comp. Inst., SPI, Reinf. Plastics/Comp. Inst.*; Society of Plastics Industry, Inc.: Washington, DC, 1978; paper 19–B.
47. Coleman, M. M.; Painter, P. C. *Appl. Spec. Rev.* **1984**, *20(3, 4)*, 255.
48. Lam, P. W. K. *Polym. Eng. Sci.* **1989**, *29*, 690.
49. Bucknall, C. B.; Partridge, I. K.; Phillips, M. J. *Polymer*, **1991**, *32*, 786.

RECEIVED for review March 6, 1991. ACCEPTED revised manuscript August 12, 1991.

# Particle–Matrix Interfacial Bonding

## Effect on the Fracture Properties of Rubber-Modified Epoxy Polymers

Y. Huang[1], Anthony J. Kinloch[1]*, R. J. Bertsch[2], and A. R. Siebert[2]

[1] Department of Mechanical Engineering, Imperial College of Science, Technology and Medicine, Exhibition Road, London SW7 2BX, United Kingdom
[2] BFGoodrich Company, R and D Center, Brecksville, OH 44141

*This study employed various butadiene–acrylonitrile rubbers and showed that both the functionality of the end groups and the acrylonitrile content have a strong influence on the microstructure and the interfacial bonding that are observed in the resulting rubber-toughened epoxy. For the rubbers examined, significant toughening is recorded only when the rubber forms a separate phase in the epoxy matrix with a particle size on the order of micrometers. These microstructural features are affected by both the functionality of the end groups and the acrylonitrile content of the rubber employed. However, once phase separation of the rubber has been achieved to give particles on the order of micrometers in size, then the interfacial bonding between the rubber particles and the epoxy matrix has only a small effect on the fracture properties of the rubber-toughened epoxy polymers.*

EPOXY POLYMERS ARE INCREASINGLY USED as adhesives and as matrices for fiber-reinforced composites because of their outstanding mechanical and thermal properties such as high strength, high modulus, and high glass-transition temperature ($T_g$). All these properties may be attributed to their highly cross-linked microstructure. However, the same microstructure leads to one very undesirable property, namely, a very low resistance to crack

* Corresponding author

0065–2393/93/0233–0189$06.50/0

initiation and propagation. It is important to increase the toughness of these materials without causing any major losses in the other desirable properties.

The most successful toughening method has been found to be the incorporation of a second rubbery phase into the glassy epoxy matrix through in situ phase separation (1, 2). In this process the rubber is initially miscible with the resin and curing agent. When the reaction starts, the rubber, which often first forms a copolymer with the resin, then phase-separates. The cured thermosetting polymer possesses, therefore, a dispersed rubbery phase.

Typically, with a rubbery volume fraction of 0.1 to 0.2, the fracture energy, $G_{Ic}$, may be increased by 10 to 20 times. The main toughening mechanisms initiated by the presence of such particles are localized shear deformation in the form of shear bands running between rubber particles (2–6) and internal cavitation, or debonding, of rubber particles (2–6) with subsequent plastic growth of voids in the epoxy matrix (7, 8). These mechanisms occur in the process zone ahead of the crack tip. In addition, the bridging of the surfaces of the cracks by stretched rubber particles, as proposed by Kunz-Douglass et al. (9), may also make a small contribution to the overall increase in fracture energy (10).

However, in nearly all the previous studies the interfacial bonding between the rubber particles and the epoxy matrix involved chemical grafting, with chemical reactions occurring between the reactive end groups of the rubber and epoxy matrix. Where poor interfacial bonding was engineered (11, 12) by using a nonreactive rubber, the increase in the measured fracture energy was lower than that achieved by using reactive rubbers. However, the microstructures attained from using nonreactive rubbers were not examined in detail to ensure that the extent of interfacial grafting was the only parameter that was being varied.

In one study (12) an increase in the fracture energy, compared to the unmodified epoxy, of about 20% was attained when using a nonreactive rubber. In contrast, when a reactive rubber was employed and chemical grafting occurred to give high adhesion, an increase of 70% was recorded. The authors concluded from these observations that strong interfacial bonding between the rubber particles and epoxy matrix was necessary to achieve the greatest increases in toughening. However, these authors also found that the rubbery particles formed from using the nonreactive rubber were substantially larger than those formed from using reactive rubbers. Hence, a direct comparison of the effects of interfacial bonding could not strictly be undertaken.

The present study involves an attempt to resolve the effects of the role of the particle–matrix adhesion on fracture properties. The interfacial bonding between the rubber and the epoxy phases was systematically varied by using rubbers with various types of end groups, including bifunctional, monofunctional, and nonfunctional reactive end groups. The relationships between the interfacial bonding and the fracture properties were established by studying both the microstructures and the fracture properties of these multiphase polymers.

## *Experimental Procedure*

**Liquid-Toughening Rubbers.**   The rubbers employed in the present study are described in Table I. The monofunctional and nonfunctional rubbers, research products of the BFGoodrich Company, contained reactive functional (–COOH) or unreactive end groups. The notations used in Table I indicate the acrylonitrile contents and functional groups contained in the respective rubbers. The terms "BF", "MF", and "NF" stand for bifunctional, monofunctional, and nonfunctional end groups, respectively; "17AN" and "26AN" indicate a 17 or 26% acrylonitrile content, respectively. The MF rubber should contain on a statistical basis 50% monofunctional, 25% bifunctional, and 25% nonfunctional end groups. It has been denoted as monofunctional rubber because, statistically, it could be regarded as having 100% monofunctional end groups. The BF/17AN and BF/26AN rubbers in the present study have usually been referred to previously by the codes "CTBN 1300 × 8" and "CTBN 1300 × 13", respectively.

Obviously, by changing the reactivities of the selected rubbers it is possible to prepare rubber-modified epoxy polymers that possess different levels of particle–matrix interfacial bonding. Use of rubbers with –COOH end groups would be expected to lead to rubber particles with strong chemical (primary) interfacial bonds across the particle–matrix interface; rubbers with nonfunctional end groups would be expected to form only relatively weak van der Waals (secondary) interfacial bonds. However, the functionality also affects other properties. For example, Table I shows that the viscosity of the rubber decreases with an increase in the percentage of nonfunctional end groups it contains. Further, the degree of functionality will

**Table I. Descriptions of the Various Rubbers**

| | End Groups[a] (%) | | | Acrylonitrile | Viscosity[b] |
|---|---|---|---|---|---|
| *Rubber* | *Monofunctional* | *Bifunctional* | *Nonfunctional* | *Content (%)* | *(MPa)* |
| BF/17AN | 0 | 100 | 0 | 17 | 132,000 |
| MF/17AN | 50 | 25 | 25 | 17 | 82,200 |
| NF/17AN | 0 | 0 | 100 | 17 | 36,700 |
| BF/26AN | 0 | 100 | 0 | 26 | 433,000 |
| MF/26AN | 50 | 25 | 25 | 26 | 229,200 |
| NF/26AN | 0 | 0 | 100 | 26 | 91,000 |

[a] The reactive functional end group is –COOH.
[b] Viscosity was measured at 27 °C.

influence the adducting of the rubber with the epoxy polymer and so affect the extent of phase separation and particle size achieved by the rubbery phase.

**Formulations and Preparation of Specimens.**  The formulations adopted in the present study were prepared following procedures outlined earlier. All formulations contained 100 phr of epoxy, 15 phr of rubber, and 5 phr of piperidine except the control, which contained no rubber (phr is parts per hundred parts of resin). The epoxy resin mainly consisted of the diglycidyl ether of bisphenol A, with an epoxy equivalent weight of approximately 190 g/mol (13). Piperidine was used as the hardener, and curing took place at 120 °C for 16 h. The MIX/26AN formulation contained 7.5 phr of the BF/26AN rubber and 7.5 phr of the NF/26AN rubber. Statistically, it employed the same amount of functional end groups as the formulation that contained 15 phr of the MF/26AN rubber.

The preparation of the rubber-modified epoxy is described in detail elsewhere (3). Essentially, the first step involved gently mixing together the liquid rubber and epoxy resin. The mixture was then heated to 65 ± 5 °C in a water bath before it was degassed in a vacuum oven at 65 °C. After degassing, the hardener was added and the mixture was cast into a mold and cured by using the appropriate conditions to produce a sheet of the material. The unmodified epoxy was prepared in exactly the same manner, but without the addition of the rubber. The prepared sheets of both the unmodified epoxy and the BF/26AN rubber-toughened epoxy were transparent. All the other materials were opaque brown.

**Test Procedures.**  A small, carefully polished specimen of the material was examined by using a scanning electron microscope (SEM). The micrographs were then processed by using an image analyzer to determine the parameters that characterize the microstructure: the volume fraction, size, and size distribution of rubber particles. Samples were also stained with osmium tetroxide and then microtoned and examined in the transmission electron microscope (TEM). The values of the glass-transition temperature, $T_g$, of the materials were measured by using Mettler differential scanning calorimetry (DSC) equipment at a heating rate of 20 °C/min.

Molded sheets of the epoxy polymers were cut into compact tension specimens, having dimensions of $10.0 \times 40.0 \times 38.4$ mm. The fracture toughness, or stress intensity at the onset of crack growth, $K_{Ic}$, was measured according to procedures outlined in the latest testing protocol (14), and the fracture surfaces were examined with the scanning electron microscope. Plane-strain compression tests (15) were performed to measure the yield stress, $\sigma_y$, and fracture strain, $\gamma_f$, of these materials. The elastic moduli, $E$, were measured by using the three-point bending flexural test, according to the appropriate American Society for Testing and Materials (ASTM) standard (16). The values of the fracture energy, $G_{Ic}$, were deduced from the relationship $K_{Ic}^2 = EG_{Ic}/(1 - \nu^2)$, where $\nu$ is the Poisson's ratio of the polymer and was taken to be 0.35.

## Characterization of Materials

A small piece of each material was sectioned and carefully polished before being examined in the SEM. No particles were observed by this SEM method in the materials modified with MF/17AN, MF/26AN, and BF/26AN rubbers. However, all the other rubbers were observed to phase-separate in the matrix during cure to form particles that were approximately spherical.

The specimens were then stained with osmium tetroxide ($OsO_4$) solution to enhance the contrast. After being stained the specimens were re-examined in the SEM. No significant changes were detected on the surfaces of the materials that had shown a multiphase structure before staining. However, rubber particles could now be seen in the multiphase polymers prepared by using the two monofunctional rubbers, because of the $OsO_4$ solution's selective attack on the rubbery phase. Figure 1 shows the microstructures of the MF/17AN and NF/26AN rubber-toughened epoxies. The poor bonding between the particles and the matrices is evident from the circumferential cracks that are apparent around the particles, particularly for the NF/26AN formulation. Such cracks are not seen when the bifunctional rubber (BF/17AN) is employed, when strong interfacial chemical bonding would be expected. Of course, the resolution of the electron micrographs is insufficient to preclude the possibility that failure has occurred in a boundary region, close to the interface, in the particle, or in the matrix. However, the direct correlation of the cracking phenomenon to the presence or absence of reactive end groups on the rubber clearly points to the interfacial particle–matrix adhesion being the important feature, with the locus of failure, such as that shown in Figure 1, at the interface rather than in a boundary region.

The BF/26AN rubber-modified material still did not show any phase structure in the scanning electron micrographs, even at a magnification as high as 5000 × . This observation is consistent with the work of Pearson and Yee (5), in which they concluded that no phase separation occurred when this rubber was used. However, phase separation was indeed observed when this rubber was examined with TEM, but very small particles, about 20–30 nm in diameter, were detected. The stained TEM of the BF/26AN rubber-modified epoxy is shown in Figure 2. Previous transmission electron microscopy (17) revealed that such small particles are not usually observed and that the BF/26AN materials appear to be somewhat unique in this respect.

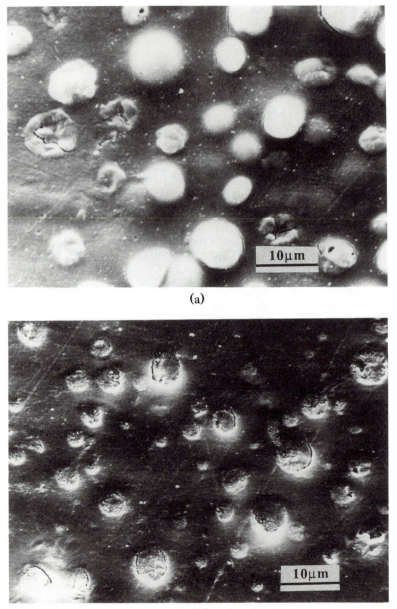

(a)

(b)

Figure 1. Scanning electron micrographs of the microstructure of the rubber-toughened epoxies. Key: a, MF/17AN rubber-toughened epoxy; the specimen was stained with OsO$_4$ solution before examination; b, NF/26AN rubber-toughened epoxy.

The SEM micrographs were then used with the image analyzer to deduce the size (particle diameter, $d_p$) and volume fraction ($V_f$) of rubber particles. The results are tabulated in Table II. As previously found (*12*), the size of the rubber particle and distribution of particle size increased as the

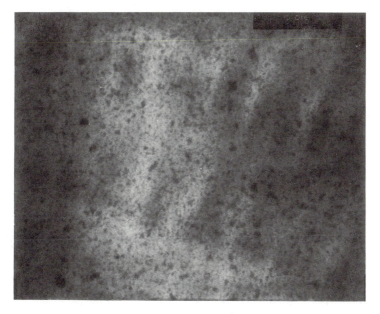

*Figure 2. Stained transmission electron micrograph of the microstructure of the BF26/AN rubber-modified epoxy (30,000 × ).*

**Table II. Microstructural Characterization of the Rubber-Modified Epoxies**

| CTBN Rubber | $V_f$ (%) | $d_p{}^a$ ($\mu m$) | $T_g{}^b$ (°C) |
|---|---|---|---|
| Control | — | — | 89 |
| BF/17AN | 18 | 1.8 | 90 |
| MF/17AN | 23 | 5.7 | 90 |
| NF/17AN | ~20 | ~30 | 88 |
| BF/26AN | nd[c] | ~0.025 | 87 |
| MF/26AN | 21 | 2.3 | 87 |
| NF/26AN | 17 | 3.7 | 85 |
| MIX/26AN | 11 | 2.1 | 88 |

[a] Value of $d_p$ for BF/26AN is from TEM studies.
[b] All glass-transition temperatures are from DSC measurements at a heating rate of 200 °C/min.
[c] Not determined.

concentration of the reactive functional end groups of the rubbery copolymer was decreased. The size of the rubber particles also increased with decreasing acrylonitrile content, again in agreement with previous work (18).

The volume fractions for the different rubber-toughened epoxies were in a range between 17 and 23%. Two exceptions were the MIX/26AN formulation, where a rubber volume fraction of 11% was obtained, and possibly the BF/26AN rubber, where very small particles were detected with TEM and the volume fraction could not be ascertained accurately. Because the scatter from the image analysis work was typically about ±2%, it is obvious that, apart from the MIX/26AN and possibly the BF/26AN rubbers, there were only minor differences in the volume fractions for the different rubber-modified epoxies.

Table II also indicates that the glass-transition temperatures are, within the experimental error of ±2 °C, all equivalent in value. This agreement indicates that complete phase separation has occurred for all the rubber-modified epoxies and that none, or only very little, of the added rubber remains in solution.

## Mechanical Properties

Table III presents the mechanical properties for the rubber-modified epoxies and the unmodified epoxy. The flexural moduli, $E$, of all these toughened materials were about the same in value; all decreased from 2.97 GPa for the unmodified epoxy to between 2.62 and 2.42 GPa for the rubber-toughened materials. This agreement suggests that the nature of the rubber–epoxy interface does not affect the modulus of the multiphase rubber-modified epoxies.

For all the formulations the yield stresses, $\sigma_y$, which were measured by using the plane-strain compression test, decreased with the addition of the

### Table III. Mechanical Properties of the Rubber-Modified Epoxies

| Rubber | $K_{Ic}$[a] $(MNm^{-3/2})$ | $E$ (flexural) (GPa) | $G_{Ic}$ $(kJ/m^2)$ | $\sigma_y$ (MPa) | $\gamma$ |
|---|---|---|---|---|---|
| Control | 0.86 | 2.97 | 0.22 | 97.7 | 0.46 |
| BF/17AN | 2.57 | 2.48 | 2.34 | 79.3 | 0.53 |
| MF/17AN | 2.30 | 2.58 | 1.80 | 77.3 | 0.43 |
| NF/17AN | 2.29 | 2.50 | 1.84 | nd[b] | nd |
| BF/26AN | 1.57 | 2.58 | 0.84 | 89.6 | 0.56 |
| MF/26AN | 2.15 | 2.52 | 1.61 | 77.9 | 0.53 |
| NF/26AN | 2.49 | 2.42 | 2.25 | 79.5 | 0.46 |
| MIX/26AN | 2.37 | 2.62 | 1.88 | 86.6 | 0.52 |

[a] Poisson's ratio was assumed to be 0.35 for all materials.
[b] Not determined.

rubbers. For the BF/26AN rubber-modified epoxy, the yield stress decreased from 97.7 MPa for the unmodified epoxy to 89.6 MPa, which represents a decrease of 8.3%. By contrast, the yield stresses for all the other rubber-toughened materials were in the range between 77.3 and 79.5 MPa, a decrease of about 20%. For the MIX/26AN rubber-toughened epoxy, the value of the yield stress was between these two cases, a result reflecting the combined effects on the yield stress from both the BF/26AN and the NF/26AN rubbers. The observation that the epoxies modified with the BF/17AN or BF/26AN rubbers have very different values of $\sigma_y$ suggests that the important factor influencing the value of $\sigma_y$ in these measurements is the the bifunctionality of the rubber. Rather, the very small particle size and possibly different volume fraction associated with the BF/26AN material are important.

The fracture strain, $\gamma_f$, was also measured by using the plane-strain compression test. If we take into account the experimental scatter of $\pm 0.03$, then the fracture strains are similar in value for all the rubber-modified formulations, and only marginally greater than that of the unmodified epoxy. However, the values of the relatively large strains that have been measured by using the plane-strain compression test must be considered as only approximate, because cracking of the specimens occurred around the ends of the compressed region. Nevertheless, clearly within the experimental limitations of the test, the interfacial bonding between the rubber particles and the epoxy matrix does not significantly affect the fracture strain of the rubber-modified epoxies.

## *Fracture Behavior*

The measured values of fracture toughness and fracture energy for the different formulations of materials are summarized in Table III. The surfaces of the fractured specimens were also studied by using the scanning electron microscope. The unmodified epoxy fractured in a very brittle manner, with a low fracture-toughness value of 0.86 MPam$^{1/2}$ (giving a fracture energy of 0.22 kJ/m$^2$). The fracture surface of the unmodified epoxy was mainly featureless, as shown in Figure 3, except for a few river lines in the crack-tip area. The main deformation micromechanism in the fracture of the unmodified epoxy polymers, according to previous reports (2–6, 10, 19), is very localized shear yielding in the crack-tip region.

The fracture properties and the fractographic studies on the rubber-modified epoxies will be discussed separately in two groups, with respect to the acrylonitrile contents of the respective rubbers.

**The 17AN-Rubber-Toughened Materials.** For BF/17AN, MF/17AN, and NF/17AN rubbers, the fracture-energy values, $G_{Ic}$, were

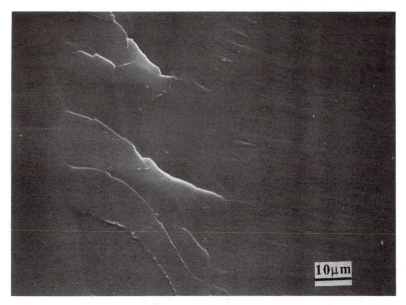

*Figure 3. Scanning electron micrograph of the fracture surface of the unmodified epoxy.*

2.34, 1.80, and 1.84 kJ/m$^2$, respectively, as summarized in Table III. Obviously, the addition of these three different rubbers significantly increased the fracture resistance of the brittle epoxy polymer, which initially had a fracture energy value of only 0.22 kJ/m$^2$. The maximum increase in the fracture toughness was achieved by using the bifunctional rubber (BF/17AN). In contrast, the epoxies modified with the monofunctional (MF/17AN) and the nonfunctional (NF/17AN) rubbers caused the overall toughness improvement to be only marginally lower by about 0.5 kJ/m$^2$. Hence, the present results clearly suggest that the interfacial reaction between the rubber and the epoxy phases only plays a minor role in the determination of the overall toughness.

The bifunctional rubber-modified epoxy fractured in a ductile stable manner with extensive stress whitening throughout the fracture surface; it is "type A", according to the classification suggested by Kinloch et al. (3). The other two rubber-modified materials exhibited a transition from ductile stable to brittle crack propagation (i.e., type A → B, where "type B" is a fast, unstable, and more brittle crack-propagation mode). Thus, the fracture surfaces of the MF/17AN and NF/17AN materials possessed two zones: a slow-growth zone near the crack tip and a fast-growth zone occupying the rest of the fracture surface. Figure 4 compares the fracture surfaces in the slow-growth region for the BF/17AN, MF/17AN, and NF/17AN rubber-toughened epoxy materials.

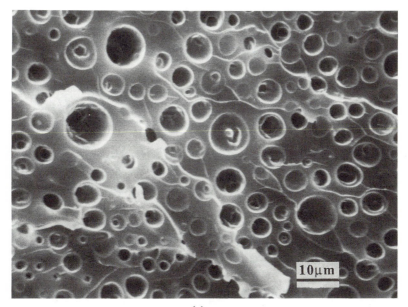

(a)

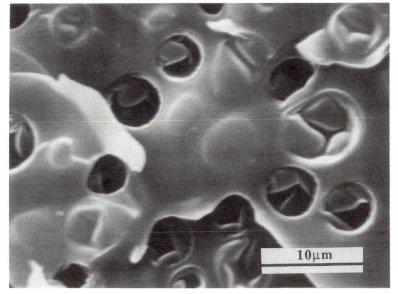

(b)

*Figure 4. Scanning electron micrographs of the fracture surfaces of the 17% acrylonitrile (17AN) series rubber-toughened epoxies, which show the respective slow-growth zones near the crack tip. Key: a, BF/17AN rubber-toughened epoxy; b, MF/17AN rubber-toughened epoxy; and c, NF/17AN rubber-toughened epoxy. Continued on next page.*

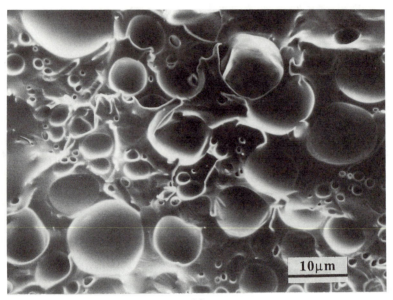

10μm

(c)

*Figure 4. Continued*

Apart from the plastic ridges, which are indications of plastic deformation, the fracture surface of the conventional bifunctional rubber-toughened epoxy (BF/17AN) shown in Figure 4a has many voids on the surface. These voids are caused by the internal cavitation of the rubber particles (3, 5) and the subsequent growth of the voids via plastic hole growth in the epoxy matrix. This latter deformation mechanism occurs because the voids may continue to enlarge in the epoxy matrix after the cavitation of the rubber particles. This irreversible plastic deformation process in the epoxy matrix causes the size of the voids to be appreciably larger than that of the original rubber particles, as has also been observed in previous investigations (3, 5). Plastic energy will obviously be dissipated during this plastic void growth process. Indeed, the contribution of plastic void growth during fracture has recently been quantitatively assessed (7, 8, 19) and may often represent a major contribution to the overall toughness.

These plastic deformations may well be reversible if the material is heated and maintained for some time above the $T_g$ of the matrix. However, at room temperature and within the time scale of the present studies the deformations were observed to be irreversible. Hence, they are referred to as plastic in nature. Finally, it may be seen that the rubber is left inside the particles, as was demonstrated previously (20) using a solvent swelling method.

The fracture surface of the monofunctional rubber-modified epoxy (MF/17AN, Figure 4b) appears rather different. The region of slow crack growth shows voids with rubbery "flaps" attached, as well as rubber particles. The mechanism that is suggested to account for the presence of these flaps is schematically depicted in Figure 5. The basic proposal is that in a given particle some of the rubber molecules are strongly bonded to the matrix because of the reaction between the reactive functional rubber that is present

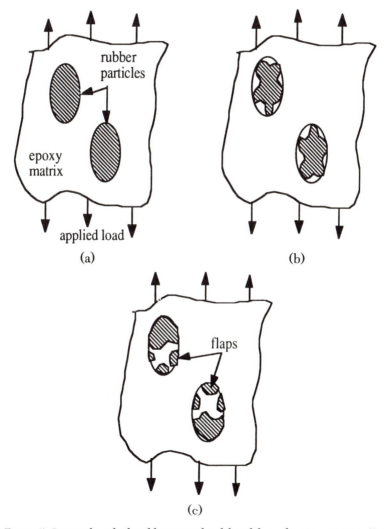

*Figure 5. Process by which rubber particles debond from the epoxy matrix. Key: a, particles stretched upon load application; b, initiation of particle tearing and debonding; and c, after debonding.*

and the epoxy matrix. However, other rubber molecules are only weakly bonded to the matrix through secondary (van der Waals) bonds associated with the nonfunctional rubbers present in the MF/17AN rubber (*see* Table I). Under the high triaxial stresses at the advancing crack tip the rubbery particle fails by a mixture of cohesive failure (internal cavitation) where the interfacial bonding is strong and interfacial failure where the bonding is relatively weak. Thus, on the fracture surface, the particles may stay half-bonded on either side. This mixed failure process leads to the formation of the flaps, as shown in Figure 5. This internal cavitation–debonding mechanism again enables further void growth to occur in the epoxy matrix; hence, even more extensive plastic deformation of the epoxy may be induced. As a result, the voids are considerably larger than the original rubber particles (cf. Figures 1 and 4b). Finally, because of debonding, the rubber-bridging micromechanism is no longer operative. This aspect will be considered in more detail.

The fracture surface of the nonfunctional (NF/17AN) rubber-modified epoxy contains many voids in the slow-growth zone (Figure 4c), although no voids were observed in the fast-growth zone. In the region of slow crack growth the voids exhibit a very smooth surface; this feature is a clear sign that debonding of the rubber particles has occurred. Because the rubber had nonfunctional end groups, no chemical reaction took place during cure. Bonding between the particle and the matrix was relatively very weak, arising solely from secondary (van der Waals) forces. Consequently, when high triaxial stresses were applied on these particles in front of the crack tip, they cleanly debonded from the matrix. Nevertheless, it appears that the toughening mechanisms were still largely operative, because the value of $G_{Ic}$ has a relatively high value of 1.84 kJ/m$^2$. However, as a result of this early debonding upon being loaded, the rubber particles in MF17/AN rubber-modified epoxies could no longer bear any load behind the crack tip. Thus, the rubber-bridging mechanism was no longer operative. The absence of this micromechanism may account for the slight shortfall in the $K_{Ic}$ and $G_{Ic}$ values, compared to those for the bifunctional (BF/17AN) rubber-modified epoxy.

The existence of the rubber-bridging mechanism in the bifunctional rubber-toughened epoxies was proposed by Kunz-Douglass et al. (9), and experimental evidence was provided by other researchers (21, 22). The absence of the rubber-bridging mechanism may account for the slightly lower fracture-energy values for the MF/17AN and NF/17AN rubber-toughened epoxies, compared to those for the BF/17AN rubber-toughened epoxy. A quantitative estimation may be made about the magnitude of toughening from the rubber-bridging mechanism via the following equation (8).

$$\Delta G_{Ic} = 4\Gamma_t V_f \qquad (1)$$

where $\Delta G_{\mathrm{Ic}}$ is the increase in fracture energy attributable to this mechanism, $\Gamma_t$ is the tearing energy of the rubber, and $V_f$ is the rubbery volume fraction. The tearing energies for the rubbers were measured by Kunz and Beaumont (23) at around 460 J/m$^2$ at room temperature. Assuming a rubber volume fraction of 0.2, the contribution of the rubber-bridging mechanism to the fracture energy may be calculated as $G_{\mathrm{Ic}} = 368$ J/m$^2$. The magnitude of the difference in the fracture energies between the bifunctional (BF/17AN), monofunctional (MF/17AN), and nonfunctional (NF/17AN) rubber-toughened epoxies was about 500 J/m$^2$, which is in approximate agreement with this calculation.

In the only other published report where poor interfacial bonding was studied in detail (12) the authors reported a 20% increase in fracture energy, compared with 70% when there was good adhesion. The relatively small difference in the fracture-energy data in these two cases approximately agrees with the magnitude of toughening offered by the rubber-bridging mechanism. The far smaller overall increase in toughness reported in this previous work compared to the present studies arises from the far higher $T_g$ of the epoxy matrix employed in the previous work. High values of $T_g$ for the matrix lead typically to a lower degree of plastic deformability of the matrix and, hence, to a lower degree of toughness being observed.

**The 26AN Rubber-Toughened Materials.**   This series contained the materials toughened by the bifunctional, monofunctional, and nonfunctional rubbers that possessed an acrylonitrile content of 26%. The behavior of the MIX/26AN formulation (i.e., 50:50 BF/26AN:NF/26AN blend; *see* Tables I and II) is also discussed in this section.

As shown in Table III, the addition of the BF/26AN rubber into the epoxy resulted in only a moderate increase in the fracture energy, compared to the unmodified material (i.e., an increase from 0.22 to 0.84 kJ/m$^2$). In contrast, the monofunctional (MF/26AN) and nonfunctional (NF/26AN) rubbers increased the fracture energy to 1.61 and 2.25 kJ/m$^2$, respectively. The poor toughening that results from using the BF/26AN rubber would appear to result from the formation of very small particles. The volume fraction could not be accurately assessed, but both the transmission electron micrographs and the $T_g$ measurements suggest that complete phase separation had occurred with this rubber.

Small particles did not undergo cavitation or debonding, which then enables further growth of the void by plastic hole growth of the epoxy matrix, as discussed in detail by Huang and Kinloch (8, 19). Thus, the important toughening mechanism of plastic void growth in the matrix did not appear to be operative when the rubber particles were about only 20–30 nm large. The MIX/26AN rubber-toughened epoxy possessed a fracture energy of 1.88 kJ/m$^2$, with a relatively smaller rubber volume fraction of 11%. However, the given values of $V_f$ and $d_p$ only consider the larger particles that were observed from the scanning electron micrographs. This formulation might well also contain smaller particles because it employs 7.5 phr of the BF/26AN rubber.

Crack propagation in all the materials in this series was type A → B (3) (i.e., there was a transition from a ductile stable propagation to a brittle unstable propagation). Thus, the fracture surface consisted of two zones, a slow-growth zone in the crack-tip area and a subsequent fast-growth zone. Studies using the SEM technique revealed that the fast-growth zones were typically flat and featureless, reflecting a rather brittle type of fracture. The slow-growth zones on the fracture surfaces of all the materials in this series are compared in Figure 6.

No rubber particles were observed on the fracture surface of the BF/26AN rubber-modified epoxy, which is shown in Figure 6a, even at magnifications as high as $5000 \times$. This observation confirms the studies that revealed that phase separation with this rubber resulted in particles of about 20–30 nm in diameter.

Figure 6b shows the fracture surface of the MF/26AN rubber-modified epoxy. Some of the voids have "flaps" attached to them, as in the MF/17AN rubber-toughened epoxy. These flaps were caused by the tearing and debonding of the rubber particles, as shown schematically in Figure 5. However, the flaps do not appear around all the voids, and this condition indicates that not all the rubbery particles have failed in this manner. The MF/26AN rubber actually contained 50% monofunctional, 25% bifunctional, and 25% nonfunctional end groups. The interfacial bonding strength for an individual rubber particle depends on its exact composition of the various end groups. Consequently, there may be a competition between cavitation and debonding for an individual rubber particle, depending on the strength of the particle–matrix interfacial bonding. If the interfacial bonding was strong enough, the critical cavitational stress for the particle would be reached prior to the critical debonding force, and the particle would then internally cavitate. If the interface was relatively weak, the critical debonding stress would be attained first, and the particle would debond from the epoxy matrix. Figure 6b suggests that about half of the particles voided by internal cavitation and the reminder via debonding.

The fracture surface of the NF/26AN rubber-modified epoxy (Figure 6c) is rather similar to that of the MF/26AN rubber-toughened epoxy (Figure 6b) because flaps are observed. This similarity suggests that the degree of interfacial bonding for this NF/26AN-toughened material is somewhat higher than for the equivalent 17% acrylonitrile (NF/17AN) material, where no flaps were observed (cf. Figures 4c and 6c). Because there are no functional groups on the NF/26AN rubber, this bonding must be attributed to the higher acrylonitrile content that promotes somewhat stronger secondary interfacial bonds, or a greater degree of interfacial mixing, with the epoxy matrix.

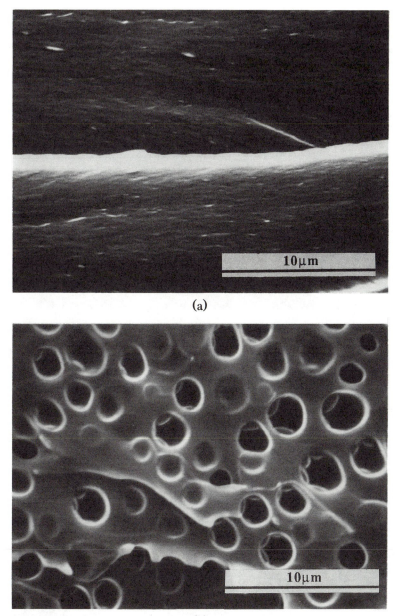

(a)

(b)

*Figure 6. Scanning electron micrographs of the fracture surfaces of the 26% acrylonitrile (26AN) series rubber-toughened epoxies, which show the respective slow-growth zones near the crack tip. Key: a, BF/26AN rubber-toughened epoxy; b, MF/26AN rubber-toughened epoxy. Continued on next page.*

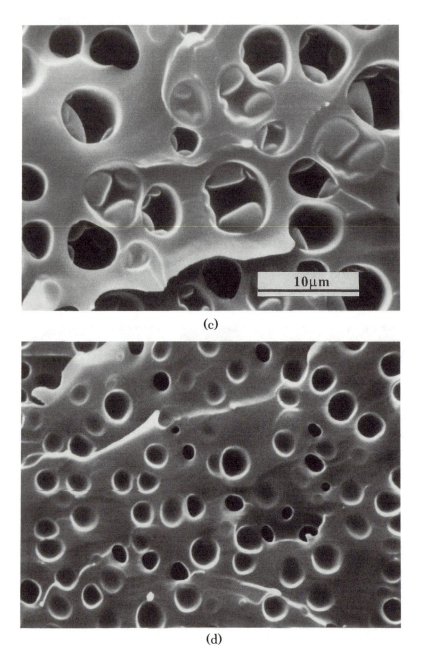

Figure 6. Continued. c, NF/26AN rubber-toughened epoxy; and d, MIX/26AN rubber-toughened epoxy.

Figure 6d shows the fracture surface for the MIX/26AN rubber-modified epoxy (i.e., the epoxy modified by the mixture of the BF/26AN and the NF/26AN rubbers). No flaps are visible around the voids. This observation again suggests that the 50:50 mixture of the BF/26AN and NF/26AN rubbers does not simply act in an additive manner; some form of synergism occurs.

### Comparison of the 17 and 26% Acrylonitrile Rubber Series.

Both the volume fraction and the size of the rubber particles decreased with increasing acrylonitrile content in the rubber, as discussed earlier (Tables II and III). Rubbers possessing an acrylonitrile content of either 17 or 26% may toughen the epoxy polymer if phase separation occurs to give micrometer-sized particles.

Among these rubber-toughened epoxies, BF/17AN and NF/26AN gave the highest fracture-energy values. The somewhat higher toughness of the BF/17AN rubber-toughened epoxy is considered to arise from the extra possibility of the rubber-bridging mechanism to be effective with this well-bonded rubbery phase. In the NF/26AN rubber-toughened epoxy, however, the particle–matrix interfacial bonding was relatively poor, as demonstrated by Figure 1b. Consequently, the rubber-bridging mechanism could no longer operate during the fracture of this material.

The rubbery volume fraction, the elastic modulus, and the $T_g$ for this material are significantly lower than those of the other materials in the 26AN series. This comparison suggests that a small amount of rubber has been trapped and remains in solution in the matrix, and that the matrix has consequently been plasticized. Thus, the matrix epoxy has become more ductile. This enhanced matrix ductility is reflected in the overall toughness of the multiphase material. Despite the similar fracture energies, the crack always propagated in a ductile stable manner (i.e., type A) in the BF/17AN rubber-toughened epoxy and crack propagation in the NF/26AN rubber-toughened epoxy showed a transition from a ductile stable manner to a brittle unstable one (i.e., type A → B).

A comparison of the fracture surface of the NF/17AN rubber-modified epoxy (Figure 4c) to that of the NF/26AN rubber-toughened epoxy (Figure 6c) illustrates that the particle–matrix interfacial bonding in NF/26AN was stronger. Rubber particles in NF/17AN debonded completely from the matrix, although some rubber particles in NF/26AN appeared to remain bonded. This contrast suggests that interfacial adhesion was improved by employing a rubber with a higher acrylonitrile content. This conclusion is supported by a comparison between the fracture surfaces of the MF/17AN (Figure 4b) and the MF/26AN (Figure 6b) rubber-toughened epoxies. In MF/17AN all the rubbery particles deformed and voided by debonding. Only about half of the MF/26AN particles interfacially debonded and the other half internally cavitated, a result indicating stronger interfacial bonding for the MF/26AN material.

**Interfacial Bonding and Toughening Micromechanisms.**   Our study revealed that when the microstructure is comparable, particle–matrix interfacial bonding is not of primary importance in determining the mechanical properties of rubber-modified epoxy polymers. The only influence it appears to have is upon the secondary rubber-bridging mechanism. Thus, there is only a small difference between the fracture energies of rubber-modified epoxy polymers with and without strong particle–matrix interfacial bonding. The two main toughening mechanisms are shear yielding and plastic void growth in epoxy matrix, the latter being catalyzed by the internal cavitation or the interfacial debonding of the rubbery particles. Obviously, these experimental results suggest that the strength of the interfacial particle–matrix bonding does not affect these two toughening mechanisms. This conclusion is supported by the following simple theoretical considerations.

Shear yielding in the matrix is governed by the stress distribution around the rubber particles and may be quantified by employing the techniques of numerical stress analysis. Previous (24–27) as well as the latest analyses (7, 19) revealed that, providing the rubbery particle has a tensile modulus of 1 to 2 MPa or lower (which is a very reasonable assumption), there are no significant differences in the magnitude and distribution of the stresses in the glassy matrix. This rule applies whether they are adjacent to a rubbery particle or a void, which is equivalent to a cavitated or debonded rubbery particle. Thus, the question of whether the rubbery particles deform and fail via internal cavitation or via interfacial debonding from the epoxy matrix would not theoretically be expected to affect the extent of plastic shear banding that occurs in the epoxy matrix.

A rubber particle with a Poisson's ratio, $v$, approaching 0.5 possesses a relatively high bulk modulus, $K$, having about the same value as the epoxy matrix, as shown by eq 2.

$$K = \frac{E}{3(1 - 2v)} \tag{2}$$

where $E$ is Young's modulus and is, of course, very much lower for the rubber particle than for the matrix. This equation suggests that a pure rubber particle, like the matrix, is a very rigid elastic body when subjected to triaxial stresses and will be highly resistant to any volumetric deformation. (Even when the rubbery particles form copolymers with the epoxy resin, the rubbery copolymer particles are still rubbery in character. For example, they still have very low values of $T_g$, on the order of –55 °C (20). Hence their Poisson's ratio will approach 0.5 in value and lead to a high, albeit not infinite, value of bulk modulus, $K$, for the rubbery particles.) Thus, the rubbery phase will not encourage any large-scale plastic dilatation in the matrix, unless the rubbery particles can either internally cavitate or interfacially debond from the epoxy matrix. Obviously, the occurrence of either event could equally enable the onset of plastic void growth in the epoxy matrix, which is the other main toughening mechanism (7, 8, 19, 28). However, the ability of rubberlike materials to cavitate relatively easily makes them very efficient toughening particles. Thus, unless the rubber particles are unable to cavitate, the plastic void growth mechanism is not expected to be affected by the degree of interfacial bonding in the rubber-modified epoxy polymer. It is theoretically likely that very small particles might not readily cavitate. Hence, in such a material the degree of interfacial bonding might play a more important role than is seen in the present work, with relatively weak interfacial bonding being an advantage. This is obviously an interesting area for further study.

## Conclusions

Our study revealed that when various butadiene–acrylonitrile rubbers are used, both the functionality of the end groups and the acrylonitrile content have a strong influence on the microstructure and the interfacial bonding observed in the resulting rubber-toughened epoxy. Significant toughening is achieved only when the rubber can form a separate phase inside the epoxy matrix with a particle size on the order of micrometers. This ability is affected by both the functionality of the end groups and the acrylonitrile content of the rubber employed. However, once phase separation has been achieved, the interfacial bonding between the rubber particles and the epoxy matrix has only a minor effect on the fracture properties of the rubber-toughened epoxy polymers.

During fracture there is competition between internal cavitation of the rubber particle and debonding of the rubber particle–epoxy matrix interface. When there is good interfacial bonding, the rubber particles internally cavitate during fracture of the rubber-toughened epoxy. However, when the interfacial bonding is poor, the rubber particles interfacially debond from the matrix. The former micromechanism is observed when a functional end group that can react chemically with the epoxy matrix is present on the rubber. The latter is observed when nonfunctional end groups are used, and only relatively weak, secondary, interfacial bonding results. The occurrence of either internal cavitation of the rubber particle or debonding of the rubber particle–epoxy matrix interface in the process zone ahead of the advancing crack tip is a necessary condition for further plastic deformation in the matrix via a plastic void growth micromechanism.

## Acknowledgments

The authors are grateful to the British Council for providing a studentship to Y. Huang. They are also grateful to The BFGoodrich Company for the supply of materials.

## References

1. Sultan, J. N.; Laible, R. C.; McGarry, F. J. *J. Appl. Polym. Sci.* **1971**, *13*, 29.
2. Kinloch, A. J. In *Rubber-Toughened Plastics*; Riew, C. K., Ed.; Advances in Chemistry 222, American Chemical Society: Washington, DC, 1989; p 67.
3. Kinloch, A. J.; Shaw, S. J.; Tod, D. A.; Hunston, D. L. *Polymer* **1983**, *24*, 1341.
4. Kinloch, A. J.; Shaw, S. J.; Hunston, D. L. *Polymer* **1983**, *24*, 1355.
5. Yee, A. F.; Pearson, R. A. *J. Mater. Sci.* **1986**, *21*, 2462.6.
6. Pearson, R. A.; Yee, A. F. *J. Mater. Sci.* **1986**, *21*, 2475.
7. Huang, Y., PhD Thesis, Univeristy of London, 1991.
8. Huang, Y.; Kinloch, A. J. *J. Mater. Sci. Lett.* **1992**, *11*.
9. Kunz-Douglass, S.; Beaumont, P. W. R.; Ashby, M. F. *J. Mater. Sci.* **1980**, *15*, 1109.
10. Garg, A. C.; Mai, Y.-W. *Compos. Sci. Tech.* **1988**, *31*, 79.
11. Riew, C. K.; Rowe, E. H.; Siebert, A. R. In *Toughness and Brittleness of Plastics*; Deanin, R. D.; Crugnola, A. M., Eds.; Advances in Chemistry 154; American Chemical Society: Washington, DC, 1976; p 326.
12. Chan, L. C.; Gillham, J. K.; Kinloch, A. J.; Shaw, S. J. In *Rubber-Modified Thermoset Resins*; Riew, C. K.; Gillham, J. K., Eds.; Advances in Chemistry 208; American Chemical Society: Washington, DC, 1984; p 261.
13. Savla, M. *Handbook of Adhesives*; Skeist, I., Ed.; Van Nostrand: New York, 1862; p 434.
14. European Group on Fracture Task Group, *Testing Protocol*; Williams, J. G., Ed.; Imperial College: London, 1990.
15. Williams, J. G.; Ford, H. *J. Mech. Eng. Sci.* **1964**, *6*, 7.
16. *ASTM Annual Book*; American Society for Testing and Materials: Philadelphia, PA, 1986; ASTM Standard D 790M−86.
17. Kinloch, A. J.; Hunston, D. L. *J. Mater. Sci. Lett.* **1987**, *6*, 131.
18. Rowe, E. A.; Siebert A. R.; Drake, R. S. *Modern Plastics* **1970**, *47*, 110.
19. Huang, Y.; Kinloch, A. J. *J. Mater. Sci.* **1992**, *27*.
20. *Structural Adhesives: Developments in Resins and Primers*; Kinloch, A. J., Ed.; Elsevier Applied Science Publishers: London, 1983; p 127.
21. Bandyopadhyay, S.; Silva, V. M.; Wrobel, H.; Nicholls, J. *Proc. Sixth Int. Conf. on Deformation, Yield and Fracture of Polymers*; Churchill College: Cambridge, England; April 1985.
22. Bascom, W. D.; Hunston, D. L. *Proc. Int. Conf. on Toughening of Plastics*; Plastics and Rubber Inst.: London, 1978.
23. Kunz, S. C.; Beaumont, P. W. R. *J. Mater. Sci.* **1981**, *16*, 3141.
24. Broutman, L. J.; Panizza, G. *Int. J. Polym. Mater.* **1971**, *1*, 95.
25. Agarwal, B. D.; Broutman, L. J. *Fibre Sci. Technol.* **1974**, *7*, 63.
26. Guild, F. J.; Young, R. J. *J. Mater. Sci.* **1989** *24*, 2454.
27. Finch, C. A.; Hashemi, S.; Kinloch, A. J. *Polym. Commun.* **1987**, *28*, 322.
28. Huang, Y.; Kinloch, A. J. *Polym. Commun.* **1992**, *33*, 1330.

RECEIVED for review August 13, 1991. ACCEPTED revised manuscript December 10, 1991.

# Macroscopic Fracture Behavior

## Correlation with Microscopic Aspects of Deformation in Toughened Epoxies

**S. Bandyopadhyay**[1]

**Defence Science and Technology Organisation Materials Research Laboratory, Melbourne, Materials Division, P.O. Box 50, Ascot Vale, Melbourne, Victoria 3032, Australia**

*The deformation and failure processes involved in the fracture of unmodified epoxies are discussed in this chapter. A review of the fracture behavior of the carboxyl-terminated butadiene–acrylonitrile copolymer (CTBN) rubber-modified diglycidyl ether of bisphenol A (DGEBA) polymers with or without a rigid particulate–fiber phase is presented in relation to the microscopic aspects of localized deformation and their relationship to microscopic fracture behavior are illustrated. The degree of improvement in fracture properties in modified materials depends to a great extent on the unmodified epoxy. If the latter is capable of even small-scale deformation at the crack tip, this induces in the modified system a number of additional microscopic failure mechanisms such as cavitation of rubber particles, enhanced shear deformation of the matrix, debonding and tearing of rubber, crack pinning, and debonding and pull-out of fibers. The recent research trend in toughening of high-temperature-grade TGMDA (tetraglycidyl 4,4'-methylenedianiline) resin is also outlined.*

THE IMPROVEMENT IN FRACTURE PROPERTIES of brittle epoxies (critical strain-energy release rate, $G_{Ic} \sim 100$–$200$ J/m$^2$, and stress-intensity factor, $K_{Ic} \sim 0.6$–$0.8$ MPa m$^{1/2}$, typically for unmodified resins) is well documented in the literature. Specifically, their $G_{Ic}$ is improved by incorporation of a

[1] Current address: School of Materials Science and Engineering, University of New South Wales, P.O. Box 1, Kensington, New South Wales 2033, Australia

rubbery phase. Epoxies containing rubber and a rigid phase have also been developed. In these epoxies, in addition to improved $G_{Ic}$, stiffness is restored to (or higher than) the stiffness of the neat resin. Thus hybrid epoxies can exhibit a good combination of $G_{Ic}$ and $K_{Ic}$ (e.g., 3 kJ/m$^2$, 3.8 MPa m$^{1/2}$). Rubber-modified epoxies are less rigid (hence lower $K_{Ic}$) than would have been expected if the modulus had not decreased, although their $G_{Ic}$ may be as high as 3 kJ/m$^2$.

The degree of improvement in fracture properties of modified materials depends to a great extent on the unmodified epoxy. If the parent (neat) resin is capable of showing even small-scale plastic deformation at the crack tip, this ability induces a number of additional microscopic failure mechanisms that impart toughness in the modified system. This requirement restricts the range of epoxies that have been actually toughened. The literature on toughened epoxy is based almost entirely on DGEBA (diglycidyl ether of bisphenol A) epoxies cured with 5 parts of piperidine, modified by CTBN (carboxyl-terminated butadiene acrylonitrile rubber) added in the range of 5–20 phr (parts per hundred parts of resin). The hybrid toughened epoxies are developed by further incorporating glass spheres, zirconia particles, or alumina fibers. References to these studies may be obtained in reviews on the subject (1–3).

This chapter presents a review of the fracture properties of toughened epoxies in the light of their microscopic aspects of deformation, illustrated mostly from my work and that of my co-workers. Load-displacement behavior of notched–precracked specimens of toughened epoxies under quasistatic loading conditions is compared with evidence of crack-tip deformation obtained from in situ fracture experiments carried out in the chamber of a scanning electron microscope. Study of fracture surfaces will provide another necessary link to the proposed microscopic–macroscopic behavior correlation. Impact properties of toughened epoxies will also be examined. The bulk of the chapter will be devoted to fracture of rubber-modified and hybrid toughened epoxies based on DGEBA resins. Some improved tetrafunctional epoxies developed by Pearce and co-workers (4) will be mentioned briefly.

## *Macroscopic Fracture Behavior of Unmodified Epoxies*

To discuss the macroscopic fracture behavior of toughened epoxies, it is important to consider the fracture of neat resins (3). Various workers have used LEFM (linear elastic fracture mechanics) specimen geometries such as compact tension (5–10), double cantilever beam (11–14), or three-point bend configuration (15–17). However, more useful information regarding fracture behavior was obtained from studies using linear compliance specimens like tapered (contoured) double cantilever beam (TDCB) (18–24) and double torsion (DT) specimens (25–32). These two geometries are designed in such a way that a change in compliance ($C$) with respect to crack length ($a$) (i.e., $dC/da$) is constant over almost the entire crack length. This relationship gives an advantage in that the value of $G_{Ic}$ or $K_{Ic}$ can be calculated from the load ($P$), without a need to know the crack length ($G_{Ic} \propto P^2$ and $K_{Ic} \propto P$). This calculation is very useful because the crack tip is often difficult to define accurately during crack growth, particularly in environmental testing or when there is a change in the crack-tip material characteristics during crack growth.

In general, two types of crack propagation have been observed in fracture studies of epoxy resins using DT and TDCB specimens. These types are continuous propagation at a constant load after initiation and intermittent stick–slip propagation in a macroscopic sequence of initiation and propagation by crack jump and arrest. These two different types of crack growth are shown schematically in Figure 1; a hypothetical crack front has been included as an inset (3).

In the continuous-propagation mode, crack initiation takes place as the load reaches a particular value. This initiation is followed by continuous crack propagation at almost the same load; positions a, b, c, and d on the load-displacement curve correspond to positions a, b, c, and d of the crack front. For the stick–slip mode of crack advance, crack initiation and sudden propagation take place when the load reaches point a. The change corresponds to a move from point a to b on the crack front in Figure 1  This propagation results in a drop in load on the load-displacement trace, and the crack is arrested. At this point, the specimen has to be loaded again to an initiation value so that further crack propagation occurs. The process is repeated until separation is complete. The level to which the load rises after the first crack jump depends on the crack-tip material characteristics, as will be discussed later.

Continuous crack propagation (Figure 1) would generally suggest very little crack-tip plastic deformation during propagation. On the contrary, stick–slip behavior would indicate energy-absorbing deformation processes at the crack tip. Often the same material can show both types of crack propagation under different conditions (test temperature and testing speed, curing agents and schedule, filler content, and environment). Continuous crack growth usually occurs at high strain rates and low temperatures (i.e., in situations that promote brittle failure). Under these conditions the $K_I$ or $G_I$ required for crack initiation (and propagation) is comparatively low and almost independent of test temperature and speed of testing. On the other hand, stick–slip crack growth is favored at slower strain rates and higher test temperatures, as well as by other factors that favor plasticization (softening) such as moisture absorption and under-optimum curing. (In very tough systems continuous crack growth can also be observed with extensive crack-tip plastic deformation. This situation is commonly referred to as *ductile tearing*.)

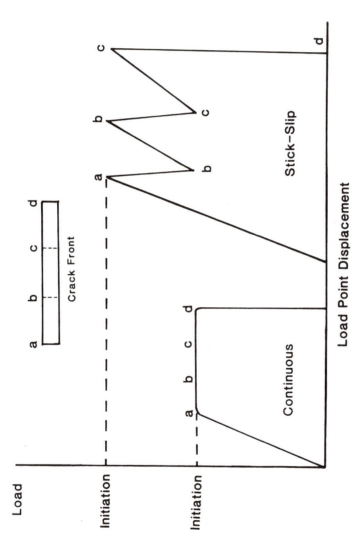
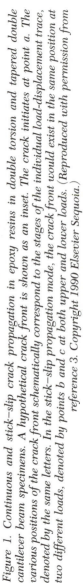

Figure 1. Continuous and stick–slip crack propagation in epoxy resins in double torsion and tapered double cantilever beam specimens. A hypothetical crack front is shown as an inset. The crack initiates at point a. The various positions of the crack front schematically correspond to the stages of the individual load-displacement trace, denoted by the same letters. In the stick–slip propagation mode, the crack front would exist in the same position at two different loads, denoted by points b and c at both upper and lower loads. (Reproduced with permission from reference 3. Copyright 1990 Elsevier Sequoia.)

As yield stress of epoxy resins generally decreases with decreasing strain rates and increasing temperature, stick–slip crack growth is attributed to plastic deformation at the crack tip prior to crack initiation (33). As a result, the higher initiation $K_I$ may result from crack-tip blunting. When the crack eventually propagates, the energy release rate is much greater than that required for continuous crack growth. Hence, the crack rapidly accelerates. Arrest occurs when the crack tip becomes depleted of the energy necessary for further crack propagation. After examination of the crack-propagation data of a wide range of epoxies, including DGEBA–piperidine systems and rubber-toughened resins obtained from various test conditions, Kinloch and Williams (33) showed a direct correlation between yield stress and mode of failure in epoxy resins.

At room temperature and under slow rates of testing, DT specimens of DGEBA–piperidine epoxies exhibit stick–slip behavior. In contrast, a TGMDA resin (tetraglycidyl 4,4'-methylenedianiline, such as Ciba-Geigy MY 720) cured with DDS (4,4'-diaminodiphenylsulfone) shows continuous crack propagation. These aspects will be discussed in the following sections.

## DGEBA–Piperidine–CTBN Rubber Materials

**Quasistatic Testing.** *Macroscopic Fracture Aspects.*   The fracture behavior of DGEBA–piperidine resins (Shell 828) reinforced with 5, 10, and 15 phr of CTBN rubber and with an alternate composition of 5 phr of CTBN plus 24 phr of bisphenol A will be examined. The distribution of rubber particles in the epoxy matrix are shown in Figure 2 (3), and the accompanying changes in mechanical properties and glass-transition temperature ($T_g$) as a function of rubber content are given in Figure 3.

The load-displacement characteristics of the materials obtained with DT specimens are described in Figure 4. Both the control specimen (zero rubber) (Figure 4a) and the 5-phr CTBN specimens (Figure 4b) show stick–slip behavior. However, there is a distinct difference between the two materials. In the control specimen the load for crack initiation ($P_i$) after the first arrest is close to the arrest load ($P_a$), whereas for the 5-phr rubber specimen the difference between $P_i$ and $P_a$ is considerably greater.

After the first crack jump (i.e., at $P_a$) there is a "natural" crack in the specimen. The load displacement traces for these two specimens suggest that the control material is capable of showing only small-scale crack-tip plastic deformation, as observed by other workers (21, 33). Furthermore, the degree of plastic deformation increases considerably with the addition of rubber. The $G_{Ic}$ of the control specimen was estimated to be between 100 and 230 J/m$^2$, corresponding to $P_a$ and $P_i$, respectively. However, the $G_{Ic}$ for the 5-phr CTBN material was estimated to be between 400 (at $P_a$) and 2400 (at $P_i$) J/m$^2$. This is a very wide range; the discussion with reference to Figure 1 indicates that $G_{Ic}$ should actually be closer to the lower value.

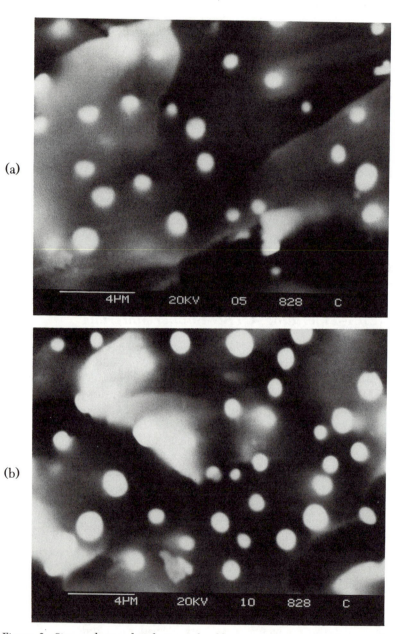

*Figure 2. Size and size distribution of rubber particles seen in slow crack growth fracture surface of DGEBA–epoxy resins cured with 5 phr of piperidine. Key: a, 5 phr of CTBN rubber; b, 10 phr of CTBN rubber; c, 15 phr of CTBN rubber; and d, 5 phr of CTBN rubber and 24 phr of bisphenol A. The rubber particles are highlighted by osmium staining and observed in the SEM with back-scattered electrons. (Reproduced with permission from reference 3. Copyright 1990 Elsevier Sequoia.)*

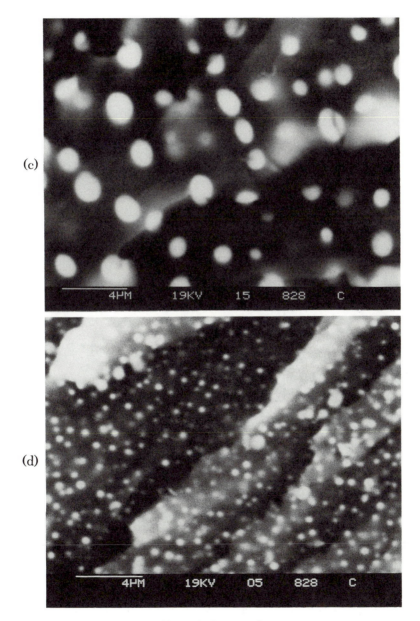

*Figure 2. Continued*

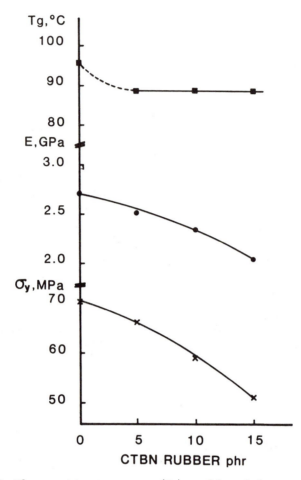

*Figure 3. Glass-transition temperature $(T_g)$, modulus of elasticity $(E)$, and tensile strength $(\sigma)$ of rubber-toughened epoxies of Figures 2a–2c as a function of rubber content.*

With increasing rubber additions (i.e., in 10 and 15 phr, as well as in the specimen containing bisphenol A (BPA)), load-displacement behavior showed a different nature (Figure 4c). Continuous change in the compliance prior to final failure, reminiscent of slow crack advance rather than crack jumping, was observed by Lee (*34*) in DT specimens of composite materials. The evidence of slow continuous crack growth in the rubber-modified specimens with increasing load was verified by optical microscopy. This behavior suggests that the $G_{Ic}$ of the materials increases with slow crack growth, presumably as a result of progressive energy-absorbing deformation processes taking place at the crack tip. The minimum and maximum values of $G_{Ic}$ for rubber

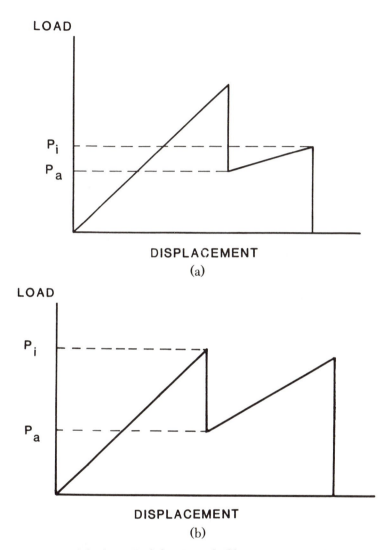

*Figure 4. Load-displacement behavior in double torsion testing. Key: a, 828–5 piperidine–0 rubber; b, 828–5 piperidine–5 rubber; and c, 828–5 piperidine–10 rubber. Diagrams 4a and 4b show stick–slip behavior; 4c shows $G_{Ic}$ with crack advance. Point X is a deviation from linearity, ($G_{Ic}$ min); point Y is where the stiffness is reduced by 5%; and point Z is the final fracture load ($G_{Ic}$ max). (Reproduced with permission from reference 3. Copyright 1990 Elsevier Sequoia.) Continued on next page.*

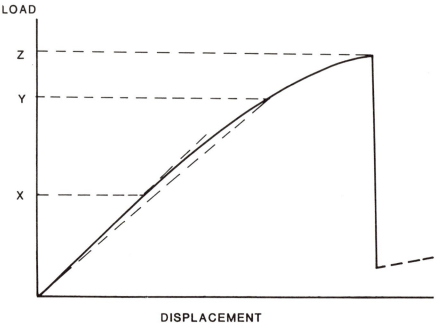

**DISPLACEMENT**

(c)

*Figure 4. Continued*

epoxies were 1300–3000 J/m$^2$ (10 phr), 1400–3000 J/m$^2$ (15 phr), and 3300–7000 J/m$^2$ for the BPA composition. Reported $G_{Ic}$ values for similar materials but with other specimen geometries fall within this range (*9, 11, 13, 16, 21*).

***Crack-Tip Deformation.***   In situ observation of crack-tip deformation has been reported in the literature from fracture experiments using the scanning electron microscope (SEM) with coated or uncoated specimens of unmodified epoxy (*1, 3, 35–37*), rubber-modified resins (*1, 3, 38, 39*), and glass-filled polymers (*1, 35*). In addition, in situ optical microscope observations have been made of crack-tip features in uncoated specimens of unmodified epoxies (*40*).

The range of information obtained from such experiments includes direct measurement of critical crack-tip opening displacement (CTOD), $\delta_c$, (*1, 3, 35–39*), crack-tip plastic zone (*1, 3, 37, 38*), and observation of features such as microcrack networks (*3, 37, 38, 40*) or void formation (*1, 3, 38*) near or ahead of the advancing crack. This information, which is difficult to obtain by other conventional means, can prove valuable in understanding deformation behavior, as shown in Figures 5–8 for the unmodified epoxy and the three

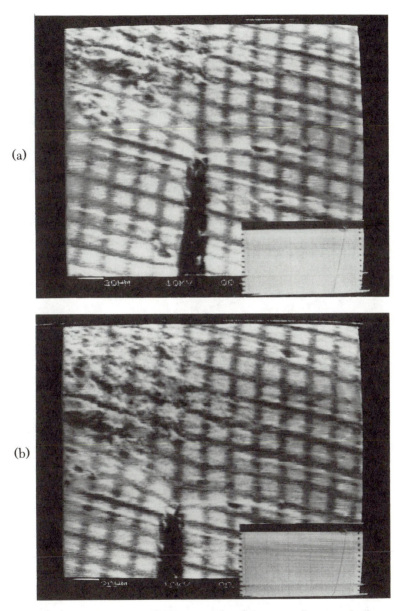

*Figure 5. Crack-tip photographs reproduced from a video record of in situ crack propagation in an unmodified 828–5 phr piperidine single-edge-notched specimen in the SEM. The discrete grid around the crack tip is obtained by application of vapor-deposited gold through a 17-μm copper mesh, which was then removed. Key: a, crack-tip opening; b, initiation of a sharp natural crack in the specimen upon loading. (Reproduced with permission from reference 3. Copyright 1990 Elsevier Sequoia.) Continued on next page.*

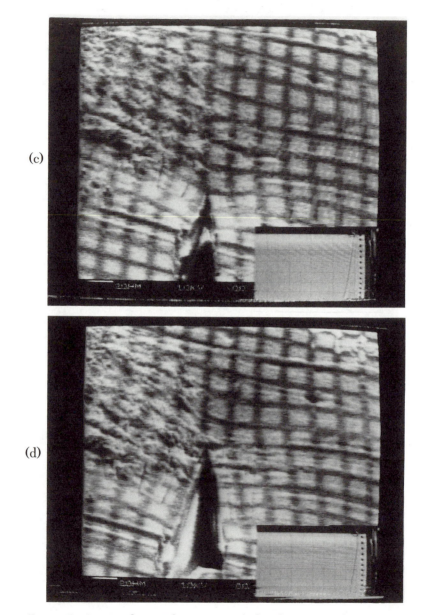

(c)

(d)

*Figure 5. Continued. c, Advancement of the sharp natural crack, with microcracks appearing near the crack tip; and d, catastrophic failure of the specimen. There was an increasing load at each of these stages. (Reproduced with permission from reference 3. Copyright 1990 Elsevier Sequoia.)*

rubber modifications of Figure 2. These results were obtained from experiments conducted in the SEM (*3, 37, 38, 41*) with single-edge-notched specimens approximately 1.5 mm thick and 10 mm wide. Each specimen, which contained a sharp starter crack made by a razor blade, was then pulled in a displacement-controlled tensile stage at a cross-head speed of 1 mm/min in the chamber of a Cambridge S–250 Mk II SEM. A discrete grid was created near the crack-tip region by depositing gold through a 17-μm copper mesh with 12-μm square apertures, then removing the grid. Thus, in Figures 5–8 (*3, 39*) the white square regions represent the gold-coated areas, which are separated from one another by uncoated regions appearing as dark lines.

Figure 5 shows crack-tip pictures reproduced from the video record for unmodified DGEBA–piperidine resin. Figure 5a shows the stage of crack-tip opening–blunting under load from the sharp starter cut. In Figure 5b a natural crack has started in the specimen as the load is increased. This crack soon starts growing further in Figure 5c, where microcracks can also be seen. Almost immediately the advancing crack reaches the catastrophic stage, as seen in Figure 5d. Once a natural crack has been initiated (Figure 5b), it is not possible to stop or slow down the crack growth in unmodified resin (e.g., stopping the cross-head movement has no effect). This feature illustrates their microscopically brittle nature. Kinloch and Williams (*33*) proposed that in DGEBA–piperidine epoxies the crack-tip blunting radius would range between 0.5 and 2.0 × $\delta_c$. This estimate agrees with Figure 5a if surface effects are neglected. Other in situ studies reported $\delta_c$ values for unmodified epoxy resins of 5 and 6 μm (*36*), 3.1 μm (*35*), and 2.4 μm (*1*).

The addition of rubber to the epoxy gives considerable stability to the crack-tip deformation processes, as seen in Figures 6a–6d for the material with 5 phr of rubber added to the unmodified epoxy. Figure 6a shows the sharp starter crack, and Figures 6b–6d show in sequence the crack-tip opening stage, crack initiation, and slow crack growth. All of these pictures were taken in the still mode by stopping the cross-head movement.

Figure 7 shows the stage of slow crack advancement in a 10-phr rubber specimen. Figure 8 shows the crack-tip opening and crack advance in a 15-phr rubber–epoxy material, again taken in the still mode.

The crack-tip pictures of Figures 5–8 reveal some interesting features. First, in the presence of a sharp crack, unmodified resin is capable of undergoing shear yielding as shown by the distortion of the grid lines in Figure 5. This ability is also seen in the three rubber-modified compositions (Figures 6–8). Second, the maximum shear strain in the unmodified epoxy is about 0.4, which is about the same as the maximum shear strain for the three materials containing rubber. Third, the value of $\delta_c$ on the surface of the specimens increases from approximately 15 μm (unmodified material) to 25 μm (5-phr material) and approximately 40 μm in the 10-phr material. These values agree with the proposed $\delta_c$ values for toughened epoxies, which range between 21 and 63 μm (*33*). Moloney et al. (*1*) reported a somewhat lower

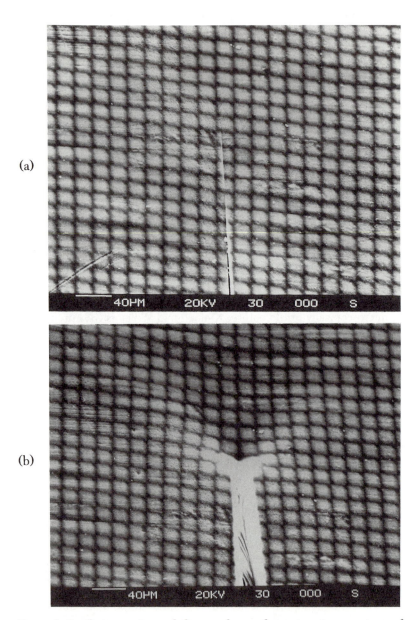

*Figure 6. Crack-tip opening and slow crack growth sequence in a specimen of DGEBA–5 piperidine–5 CTBN; in situ experiment in the SEM. Pictures were taken in the still mode by stopping the cross-head movement. Key: a, sharp starter crack; b, maximum crack-tip opening (crack-tip blunting); c, initiation of a crack in the specimen; and d, slow growth of the crack. (Reproduced with permission from reference 3. Copyright 1990 Elsevier Sequoia.)*

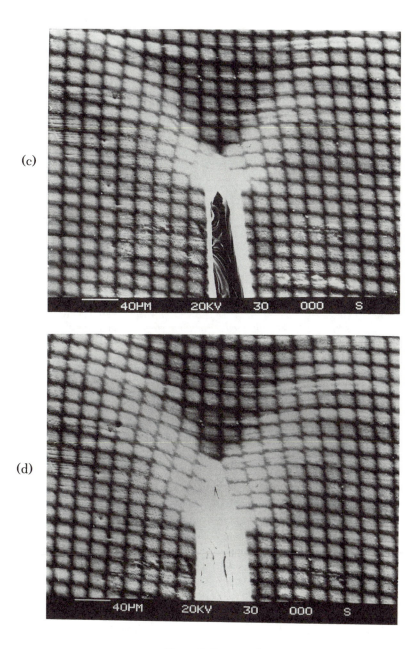

(c)

(d)

*Figure 6. Continued*

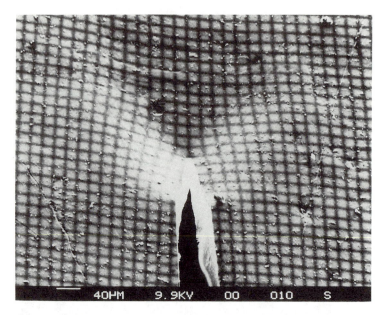

*Figure 7. Crack-tip picture in a DGEBA–5 piperidine–10 CTBN in situ specimen in the SEM in still mode, showing stage of slow crack growth. (Reproduced with permission from reference 3. Copyright 1990 Elsevier Sequoia.)*

$\delta_c$, ~ 10 μm for rubber-modified epoxies, but their plastic zone was "some hundred" micrometers.

To verify if rubber particles actually bridge crack faces, as proposed by Kunz-Douglass et al. (*42*) and Doyle and Goldthwait (*43*), a special in situ experiment was performed (*3, 39*). A tensile specimen with a center crack was pulled in an Instron machine to the stage of slow crack extension. At this stage the strained specimen was locked onto a clamping fixture and coated with gold for examination in the SEM. As shown in Figure 9a, deforming rubber particles can clearly be seen bridging the crack up to about 5 μm behind the crack tip. Figure 9b, a schematic drawing of Figure 9a, identifies six rubber particles. Figure 9 shows a remarkable similarity to the theoretical model of rubber stretching proposed by Doyle and Goldthwait (*43*).

The presence of a triaxial state of stress at the tip of a crack induces hydrostatic tension leading to dilatation in the matrix material and cavitation in rubber particles. Most published evidence is taken from fracture-surface studies in the SEM and observation of stress whitening on the fracture surface in the slow crack growth region (*7, 13, 17, 21*). Although no substantial evidence of voiding was observed in crack-tip deformation of rubber–epoxy (Figures 5–8), such evidence of surface void formation was observed in a rubber-toughened epoxy containing asbestos fibers, as demon-

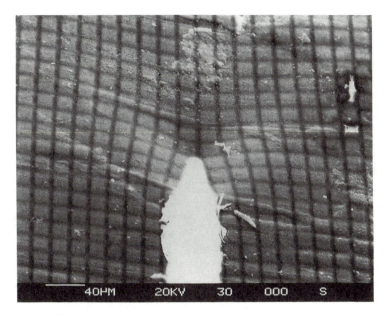

*Figure 8. Crack-tip picture in a DGEBA–5 piperidine–15 CTBN in situ specimen in the SEM in still mode during slow crack advancement. Distinct evidence of blunted crack tip can be seen compared to that shown in Figures 5c and 5d for the unmodified resin. (Reproduced with permission from reference 3. Copyright 1990 Elsevier Sequoia.)*

strated in Figure 10 (*3, 38*). Figure 10a presents the crack-tip region at no load, clearly showing no inherent voids in the material surrounding the crack tip. Upon loading, the crack tip opens up and eventually a slow crack advances. Noticeable void formation appears to take place during crack growth in Figure 10b (within the shear band in the left, for example), and some microcracks can be seen close to the crack tip. This specimen was subsequently slowly unloaded, then allowed to relax for 24 h, and examined again. The relaxed crack tip shown in Figure 10c proves that some voids have not closed down, although there has been a crack closure. The crack-tip picture of Figure 10c also shows a residual strain at the closed crack tip.

Void formation was observed in the surface of the specimen ahead of the advancing crack tip in rubber–epoxy containing alumina fiber. Final failure also appears to take place by void coalescence. Similar void formation was observed by Moloney et al. (*1*) in a hybrid toughened epoxy. Possibly the inorganic filler encourages more void formation in these materials than in systems containing only rubber.

Moloney et al. (*1*) mistakenly suggested that microcracks observed near the crack-tip region of epoxy resins in in situ experiments (*35, 37*) are

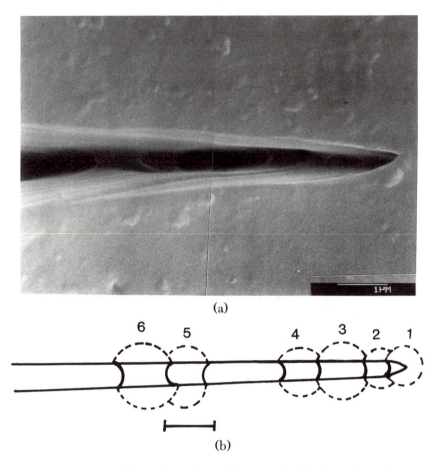

*Figure 9. Direct evidence of stretching rubber particles bridging the crack-tip region in a DGEBA–5 piperidine–10 CTBN specimen (a) and schematic sketch of six rubber particles (b) that are identified in part a: 1 is in front of 2; 2 is in front of 3; 3 is in front of 4; and 6 is in front of 5. (Reproduced with permission from reference 3. Copyright 1990 Elsevier Sequoia.)*

artifacts caused by cracking of the coating when the specimen is strained. Although such an argument can be applicable for specimens with a continuous coating, such as seen in Moloney's work (35), it is clearly not applicable and physically not viable for those specimens that are given discrete coating through a grid (Figures 5–8 and 10). In such specimens microcracks represent true specimen deformation characteristics. In fact, Lilley and Holloway (40) made optical microscopic observations of networks of microcracks perpendicular to the principal stress direction, in uncoated specimens of three unmodified epoxy resins. Their work provides indisputable evidence

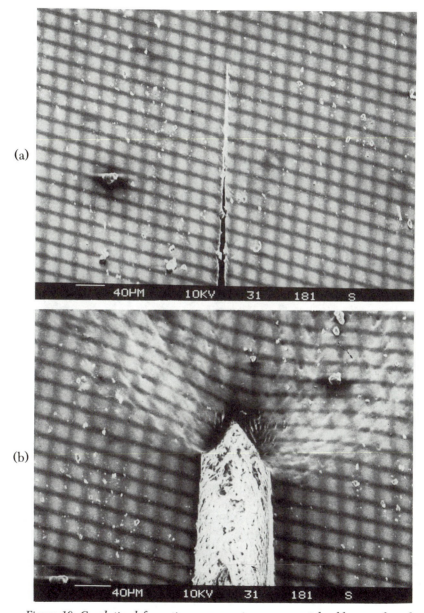

*Figure 10. Crack-tip deformation sequence in a commercial rubber-toughened epoxy containing 5 parts of asbestos filler; single-edge-notch specimen, in situ experiment in the SEM; pictured in still mode. Key: a, sharp crack tip prior to loading, no sign of voids; b, slow crack advance, formation of voids can be seen ahead of crack tip; asbestos fibers are visible across the crack edges; and c, relaxed crack tip upon unloading, showing crack-tip closure, residual deformation, and unclosed voids. (Reproduced with permission from reference 3. Copyright 1990 Elsevier Sequoia.) Continued on next page.*

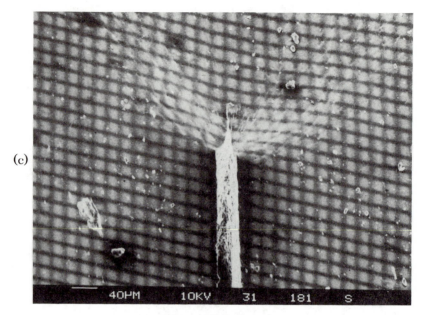

(c)

*Figure 10. Continued*

that microcracks can be realistic features in epoxies. Microcracks near the crack tip during fracture of epoxy resins have also been reported by other workers (*44*).

When the unmodified epoxy (Figure 5) is modified with rubber, which improves the deformability of the material, such microcracks are not observed (Figures 6–8) in specimens coated the same way as the unmodified epoxy. Walker et al. (*45*), who used a similar discrete grid to study in situ deformation in ductile glassy polymers such as polycarbonate, did not observe any microcracking in crack-propagation studies of that material. We, also, did not observe microcracks in in situ studies of crack-tip deformation in polycarbonate specimens (*37*).

The appearance of microcracks in the rubber-modified epoxy of Figure 10b may have been promoted by the presence of asbestos filler in this material, similarly to void formation observed in other hybrid toughened epoxies (*1*).

***Fracture Surfaces.*** Fracture surfaces of toughened epoxies have been studied extensively (*1, 3, 5, 7, 11, 17, 21, 39, 42, 46–52*). In general, two distinct types of appearance at slow and fast crack speed have been observed at room temperature. During crack initiation and slow crack propagation, the fracture appears to involve debonding of rubber particles and their eventual tearing, with stress whitening. At low magnification in a single-edge-notch specimen or compact-tension specimen, the fracture surface generally shows a transition from slow to fast fracture, as shown in Figure 11a, where crack propagation is from left to right (3). The length of the thumbnail-shaped slow growth in the specimen matches the calculated length of the stress-whitened zone (2 mm) in CTBN epoxy systems (2). High-magnification photographs of the fracture surface in slow growth and fast growth are shown in Figures 11b and 11c, respectively. In Figure 11b the fracture goes through the rubber particles with distinct evidence of cavitation in the rubber. An osmium-stained view of this fracture surface would appear similar to that shown in Figure 2. During fast growth, the crack front avoided the rubber particles to a great extent and proceeded through mainly the epoxy matrix. Osmium staining would reveal contours of rubber particles very close to the surface only (17). Figure 12 shows the transition from slow to fast crack propagation in a double-torsion specimen with 10 phr of CTBN rubber. In this figure the lower half represents slow crack advance, and the upper half represents catastrophic fracture.

**Fracture under Impact Conditions.** Addition of small amounts of CTBN rubber to epoxies results in improved impact properties (53). Limited studies on impact behavior of toughened epoxies have concentrated on using notched specimens, with a sharp starter crack (54–56) for evaluating $G_{Ic}$ under impact conditions and with blunt notches (55), following work of Plati and Williams (57), who used Izod or Charpy specimens. The results, however, do not show any general agreement.

Figure 13 shows the derivation of impact $G_{Ic}$ for an Izod specimen configuration in a 10-phr rubber–epoxy specimen. Figure 14 presents the impact $G_{Ic}$ (sharp-crack) and conventional impact resistance as a function of rubber content (3, 56). The results reveal two aspects. First, both the impact resistance (blunt-notch) and the impact $G_{Ic}$ (sharp-crack) increase with rubber content. However, they increase to a maximum of only 2–3 times that of the unmodified material, not an order of magnitude increase, as seen in slow tests. This limit may be attributed to the much higher strain rate attained in impact tests. Second, the blunt-notch impact resistance is about an order of magnitude higher than was measured for the sharp-crack impact $G_{Ic}$. Ting and Cottington (54), on the other hand, observed only approximately a factor of 2 difference in the $G_{Ic}$ under impact conditions and under slow testing. They did not report any blunt-notch data. Their systems incorporated a solid rubber in some compositions, in addition to liquid rubber, whereas the material of Figure 14 used only liquid CTBN rubber.

An explanation of the big difference in energy consumption between the sharp-crack and blunt-notch specimens in Figure 14 can be obtained from examination of the fracture surface for a 10-phr rubber specimen in Figure 15 (3). Apparently, in the blunt-notch impact specimen most of the energy is absorbed in the initiation of a large flaw and slow crack growth associated

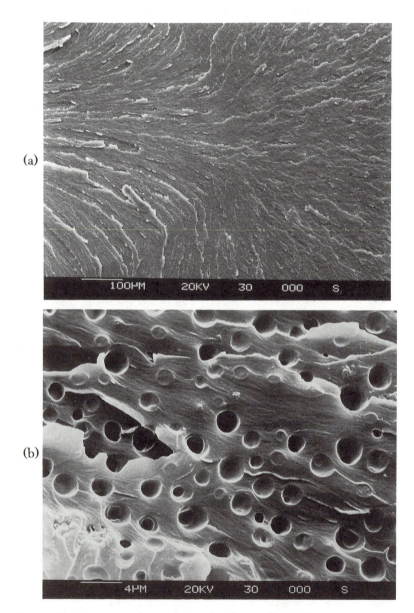

*Figure 11. Fracture surfaces of a single-edge-notched rubber-modified epoxy resin. Key: a, transition from slow-speed crack advance (left) to fast growth; b, high-magnification view of crack propagation through the rubber particles in the slow-growth region; and c, crack propagation mostly through the epoxy matrix, around the rubber particles. (Reproduced with permission from reference 3. Copyright 1990 Elsevier Sequoia.)*

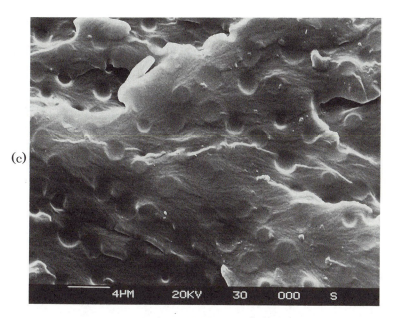

*Figure 11. Continued*

with debonding, tearing, and cavitation of rubber (Figures 15a and 15c). The large flaw in Figure 15a is very similar to that induced in an unnotched tensile specimen prior to failure (Figure 3a, ref. 3). In the sharp-crack impact-fracture specimen, the initiation and slow-growth region is small (Figures 15b and 15d). Consequently, far less energy is required for crack initiation, so $G_{Ic}$ mainly reflects the energy to propagate the crack.

Observation of the notch root region in transmission optical microscopy gives an indication of the degree of mutilation of the notch root material in impact testing (blunt-notch). This mutilation is shown in Figures 16a and 16b for an unmodified epoxy specimen before and after fracture (3). In contrast, a sharp-crack impact-fracture specimen of the same material as presented in Figures 17a and 17b shows little deformation in the region around the crack tip (3). The notch root deformation is best seen with unmodified epoxies, because these materials are transparent. Addition of rubber results in loss in transparency, and hence the effect seen in Figure 17 would not be appreciably visible in toughened materials. Figures 18a and 18b show fractured blunt-notch and sharp-crack specimens, respectively, in reflected light (3). The degree of damage in the surface of the blunt-notch specimen is quite obvious.

Low and Mai (55) reported much higher impact $G_{Ic}$ at room temperature with Charpy specimens of unmodified and rubber-modified materials

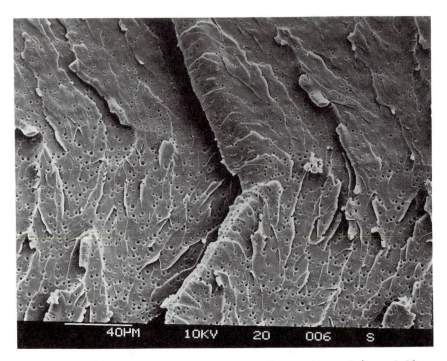

*Figure 12. Fracture surface appearance during slow crack growth (lower half) and fast fracture in a DT specimen containing 10 phr of CTBN rubber.*

(about 3 kJ/m$^2$). The high values were attributed to localized adiabatic heating at the crack tip. Although these magnitudes are not in agreement with those obtained from Izod specimens (*56*), they also do not agree with Low and Mai's subsequent derivation of impact $G_{Ic}$ (especially 0.2 kJ/m$^2$ for unmodified epoxy) extrapolated from blunt-notch specimen data (*55*, Figure 3). More work is necessary to understand deformation and failure behavior under impact conditions.

## Particulate-Filled and Hybrid Toughened Epoxies

The need to obtain higher stiffness or high $K_{Ic}$ resulted in research using rigid particulate filler such as alumina, silica, or glass spheres in epoxy resins (*1, 30, 58–62*) and in unsaturated polyester (*63, 64*). Use of filler in the form of short glass fibers has also been reported (*1*), and hybrid–particulate epoxies containing DGEBA epoxy, rubber, and glass spheres (*1, 65*), and a DGEBA epoxy–CTBN system reinforced with zirconia particles or short alumina fibers (*51, 66–68*) have been developed.

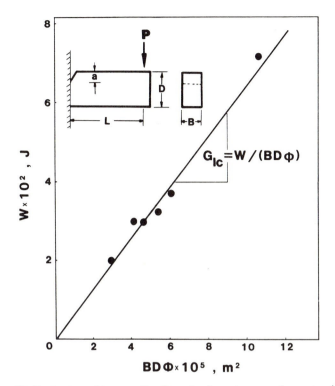

*Figure 13. Derivation of impact $G_{Ic}$ from Izod specimen configuration (shown as inset); d is a calibration factor. The graph shows the results for a specimen containing 10 phr of rubber. (Reproduced with permission from reference 3. Copyright 1990 Elsevier Sequoia.)*

In particulate-filled resins, a hard particle can improve fracture toughness by

- crack pinning (58, 63),
- deflecting the crack front by rigid particles (thereby generating larger crack-tip opening displacement (69),
- inducing plastic deformation in the matrix surrounding the particle, and
- fracture of the particle (2).

As hybrid epoxies contain both rubber and hard fillers, it is natural to assume that the resulting fracture mechanisms in hybrids will at least equal the sum of the individual contributions observed in the rubber-toughened epoxies and the particulate-filled resins. In addition, there may be synergistic effects. This

## IZOD IMPACT PROPERTIES

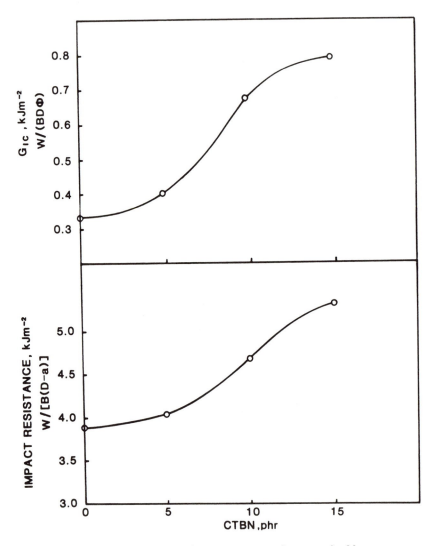

*Figure 14. Impact resistance and impact* $G_{Ic}$ *as a function of rubber content at room temperature. (Reproduced with permission from reference 3. Copyright 1990 Elsevier Sequoia.)*

variety is illustrated in the following section by comparing load-displacement behavior and fracture surfaces of the binary mixtures and the hybrid materials of epoxy–rubber–zirconia (ERZ) and epoxy–rubber–alumina fiber (ERF) systems obtained through collaborative research between the Materials Research Laboratory and the University of Sydney (66–68). The epoxy used in

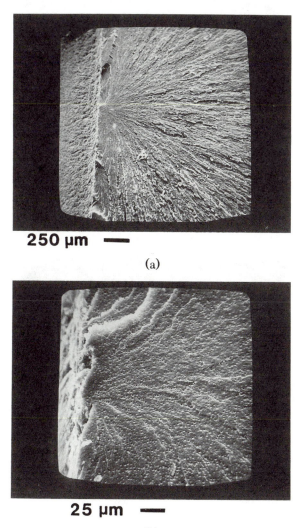

**250 μm** ▬▬

(a)

**25 μm** ▬▬

(b)

*Figure 15. Fracture surfaces of impact Izod and impact-fracture Izod specimens for 10-phr rubber epoxy material. Key: a, large flaw initiated in impact specimen, low magnification; b, high-magnification picture of flaw-initiation region in impact specimen shown in part a; the debonded–cavitated and separated rubber particular can be seen in this region; c, very small slow-growth initiation region in impact-fracture specimen, low magnification; and d, distinct features of slow growth seen in the initiation region of part c at higher magnification. (Reproduced with permission from reference 3. Copyright 1990 Elsevier Sequoia.) Continued on next page.*

250 µm ⸺

(c)

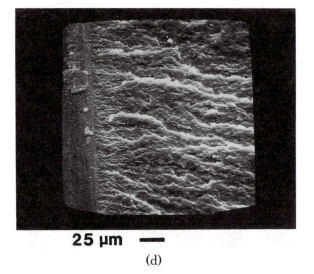

25 µm ⸺

(d)

*Figure 15. Continued*

this study was a DGEBA resin, cured with 5-phr piperidine. The rigid filler was alumina fibers or zirconia particles. Compact-tension specimens were made from cast sheets and then tested in an Instron machine at 1 mm/min to ascertain $K_{Ic}$ and crack-growth behavior (66).

The schematic load-displacement traces of the materials are shown in Figure 19 (68), and their $K_{Ic}$ values are given in Table I (66). Addition of rubber has a significant effect on energy absorption, and thus the hybrid

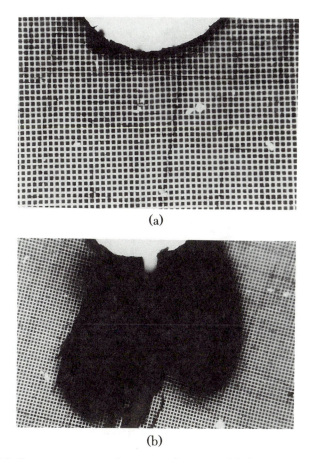

*Figure 16. Transmission optical pictures of an unmodified epoxy Izod impact specimen (blunt notch) before (a) and after (b) fracture. Gross material mutilation can be observed in part b. The grid is similar to that described for Figures 5–8. (Reproduced with permission from reference 3. Copyright 1990 Elsevier Sequoia.)*

specimens (ERZ and ERF) clearly have fracture properties superior to those of the nonhybrid epoxies. These properties are explained in terms of the fracture surfaces and crack-tip pictures as follows.

The featureless fracture surface of unmodified epoxy is shown in Figure 20a, and the increased deformability of the epoxy–rubber material (ER) is obvious in Figure 20b. Similar differences can be noted in the EZ and ERZ material (Figures 20c and 20d). High magnification of Figures 20c and 20d reveals crack pinning by the zirconia particles (Figure 20e and 20f). In the ERZ material, each pinning site also acts as a source for crack initiation by

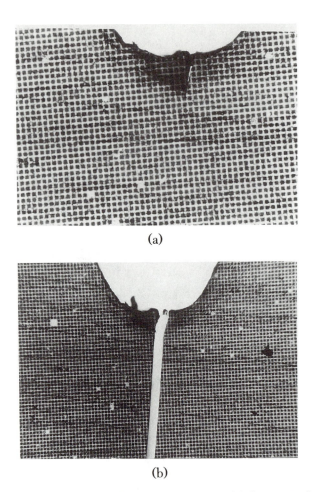

(a)

(b)

*Figure 17. Transmission optical pictures of an unmodified resin Izod impact-fracture specimen (sharp crack) before (a) and after (b) fracture. Apparently little energy is consumed in flaw initiation. (Reproduced with permission from reference 3. Copyright 1990 Elsevier Sequoia.)*

debonding, tearing, or cavitation of rubber particles, thereby giving synergistic energy-absorbing effects. Figure 20g shows fiber fracture and pinning in an EF specimen; Figure 20h presents the rough texture of an ERF specimen fracture surface at low magnification to reflect its higher toughness. Garg and Mai (2) consider that in low-temperature hybrid particulate epoxies, where crack pinning is the main toughening mechanism, fracture can be described by the critical crack opening displacement criterion (70). At high temperature, where plastic flow of matrix becomes a major parameter, the crack-tip blunting model (33) would better describe the fracture parameter. Even at

(a)

(b)

*Figure 18. Reflection optical photographs of the fractured specimens shown in Figures 16b and Figure 17b. Key: a, considerable material damage on or near the surface in blunt-notch specimen; and b, very little damage in sharp-crack specimen. (Reproduced with permission from reference 3. Copyright 1990 Elsevier Sequoia.)*

room temperature, sufficient crack-tip blunting is seen in the alumina hybrid material (Figure 20i)

Moloney et al. (*1*) showed evidence of fiber debonding and pullout rather than crack deflection, as suggested by Faber and Evans (*69*), in an epoxy–glass-fiber material. This debonding is not surprising because of the optimum bond that may exist in these materials for producing the pullout of fibers. Fiber pullout is seen in the epoxy–rubber–alumina hybrid in situ fractured specimen (Figure 21a), as well as on the fracture surface shown in Figure 21b (*3, 61, 67*). In addition to fiber pullout, a broken-fiber site in the

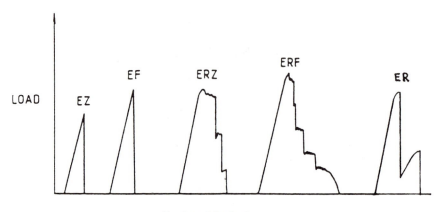

*Figure 19. Schematic load-displacement behavior of particulate and hybrid epoxy resins, with compact tension specimens. (For explanation of terms, see Table I). (Reproduced with permission from reference 68. Copyright 1990 Chapman & Hall.)*

### Table I. Fracture and Mechanical Properties of Particulate and Hybrid Epoxies

| Base Material | Weight Fraction of Fillers (phr) | | Fracture Toughness $(MPa\ m^{1/2})$ | | Flexural Modulus |
|---|---|---|---|---|---|
| | $ZrO_2$ (Z) | Fibers (F) | $K_{Ii}^a$ | $K_{Im}^b$ | (Gpa) |
| Epoxy resin– | 0 | 0 | 0.80 | — | 2.78 |
| 5 phr piperidine | 5 | — | 0.88 | — | 2.87 |
| (E) | — | 6 | 2.30 | — | 3.12 |
| | 25 | — | 0.96 | — | 3.17 |
| | — | 19 | 2.55 | — | 3.62 |
| Epoxy resin– | 0 | 0 | — | 2.18 | 2.58 |
| 5 phr piperidine– | 5 | — | — | 2.54 | 2.59 |
| 15 phr CTBN | | | | | |
| rubber | — | 6 | — | 2.61 | 3.05 |
| (ER) | 25 | — | — | 2.84 | 2.75 |
| | — | 19 | — | 3.81 | 3.80 |
| | — | 30 | — | 3.35 | 4.95 |

[a] Initiation fracture toughness (particulate epoxies).
[b] Maximum fracture toughness; hybrid materials show increasing crack resistance with increasing crack length.
SOURCE: Adapted with permission from reference 66. Copyright 1987 Chapman & Hall.

matrix can act as a site for stress-triaxiality and thus induce crack initiation by debonding, tearing, and cavitation failure of innumerable rubber particles (Figures 21c and 21d). This process progressively extends the slow crack growth region. This particular material develops a very large stress-whitening zone that gives rise to stable crack growth in compact tension test geometry, and hence to a rising crack-resistance curve $K_R$, as shown in Figure 22 (*3, 66, 67*).

Low (*68*) presented an interpretation of the fracture micromechanics of particulate and hybrid toughened epoxies, based on an examination of residual stresses that can develop at the interface between the matrix and the dispersed phase because of thermal expansion mismatch ($\Delta\alpha$). Low demonstrated that in EZ and EF materials compressive radial stresses of about 16.7 and 17 MPa, respectively, are induced at the filler–matrix interface on cooling from the curing temperature of 120 °C. These stresses serve to enhance the intrinsic bond developed at the interface. Consequently, Low argued that processes for premature debonding at the interface of an advancing crack tip are suppressed. The stress fields around the zirconia particles or alumina fibers caused the crack front to be pinned and bowed before breaking away to produce tail ends, which are characteristic features associated with crack pinning (Figures 20e and 20g).

Residual stress–strain distributions become more complicated when both rubbery and rigid fillers are dispersed in the epoxy matrix. This complication arises because the rubbery phase has a larger thermal expansion coefficient than that of epoxy. The converse is true for the rigid phase.

According to Low (*68*), $\Delta\alpha$ is negative in the rubber-modified epoxy system (ER), and tensile radial stresses of about 6 MPa are induced at the rubber–matrix interface. The addition of a third filler with low thermal

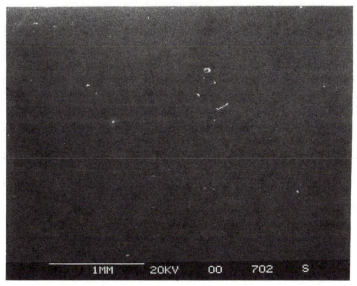

(a)

*Figure 20. Part a: Featureless fracture surface of unmodified epoxy.*

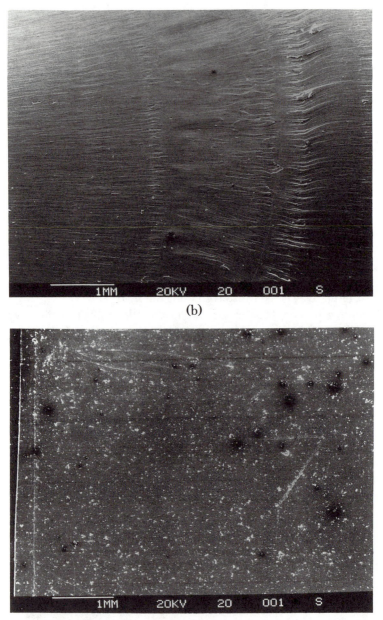

(b)

(c)

*Figure 20. Part b: low-magnification fracture surface of DGEBA epoxy cured with piperidine and containing 15 phr of CTBN rubber. Addition of rubber results in improved deformability; Part c: low-magnification fracture surface of DGEBA epoxy cured with piperidine and containing zirconia particles. Little evidence of crack pinning is seen at this magnification. (Reproduced with permission from reference 3. Copyright 1990 Elsevier Sequoia.)*

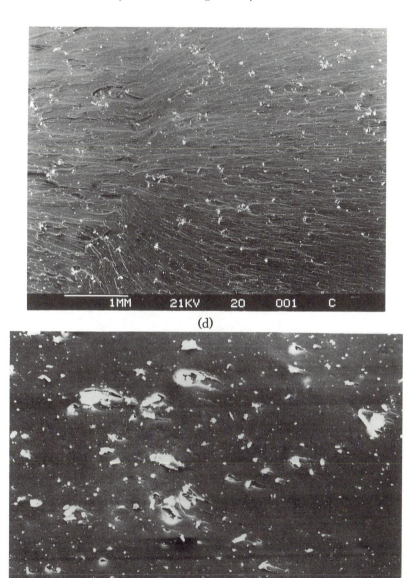

(d)

(e)

*Figure 20. Part d: low-magnification fracture surface of a hybrid epoxy containing both rubber and zirconia particles showing crack pinning, crack bowing, and matrix deformation; Part e: high-magnification view of the zirconia particles of Figure 20c, showing "tailing" of zirconia, thus indicating crack pinning by the particles. (Part d is reproduced with permission from reference 3. Copyright 1990 Elsevier Sequoia.)*

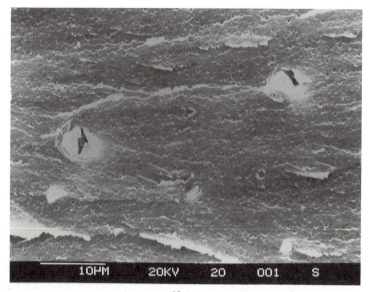

(f)

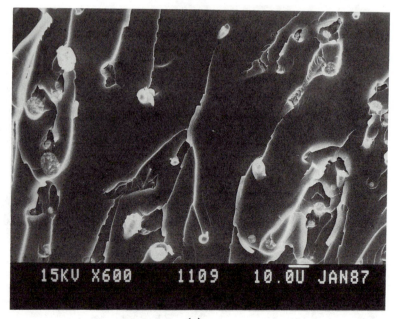

(g)

*Figure 20. Part f: high-magnification view of the fracture surface of the epoxy–rubber–zirconia hybrid material, showing that the pinned zirconia particles also act as a source for slow crack initiation by debonding, tearing, and cavitation of rubber particles; fracture of the large zirconia particles can also be seen; Part g: fiber fracture and crack pinning seen in an EF specimen. (Reproduced with permission from reference 3. Copyright 1990 Elsevier Sequoia.)*

(h)

(i)

*Figure 20. Part h: low-magnification view of epoxy–alumina hybrid specimen fracture surface showing very rough texture; Part i: in situ crack-tip picture of an alumina–epoxy hybrid specimen showing crack-tip opening, crack-tip deformation, and blunting. (Part h is reproduced with permission from reference 3. Copyright 1990 Elsevier Sequoia.)*

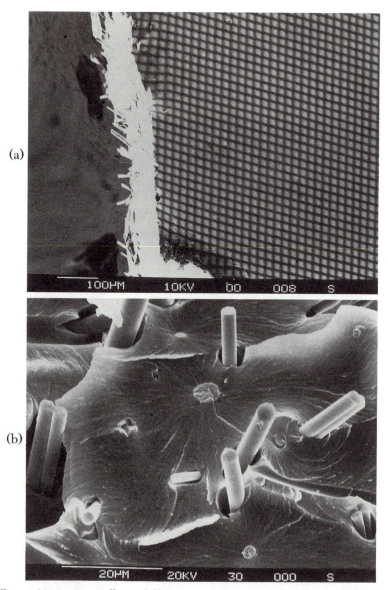

*Figure 21. Part a: Pullout of fiber seen in an in situ fractured specimen of alumina–hybrid epoxies; Part b: debonding and pullout of alumina fiber, as well as evidence of fiber fracture in epoxy–rubber–alumina hybrid materials; Part c: fractured fiber site acts as a source of triaxiality, inducing localized slow crack growth in the surrounding matrix that involves the debonding, tearing, and cavitation of innumerable rubber particles; an Al X-ray scan of the large inclusion confirms that it is an alumina fiber; Part d: osmium-stained back-scattered electron view of rubber particles in ERF specimen. (Sections b and c are reproduced with permission from reference 3. Copyright 1990 Elsevier Sequoia.)*

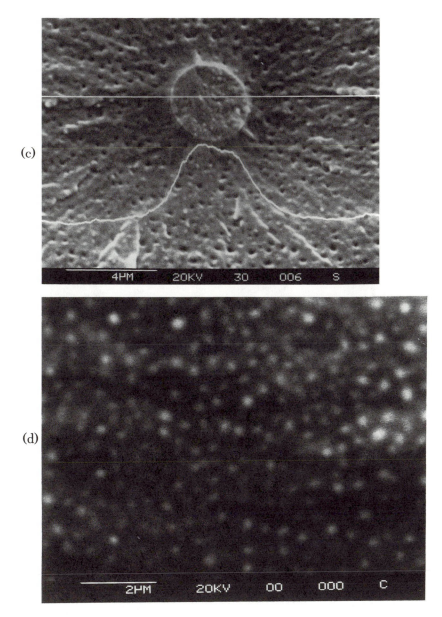

*Figure 21. Continued*

expansion to the epoxy matrix complicates this mechanism. The thermal expansion mismatch effects of the two fillers will be diametrically opposed to each other. In the ERF system, tensile stresses of about 6 MPa are induced at the rubber–epoxy interface with the concomitant generation of compressive tangential stress at the surrounding matrix. On the other hand, compres-

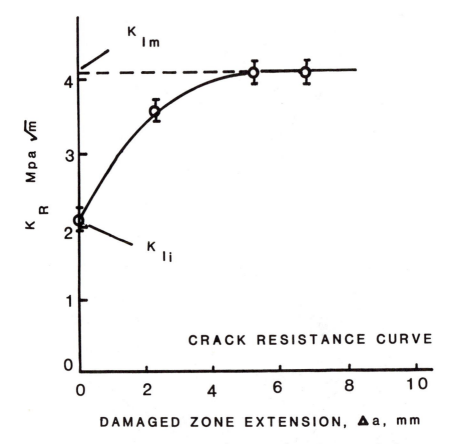

*Figure 22. $K_R$ curve of a fiber hybrid (30-phr fiber) tested at a cross-head rate of 0.05 mm/min. (Reproduced with permission from reference 3. Copyright 1990 Elsevier Sequoia.)*

sive radial stresses of about 17 MPa are induced at the fiber–matrix interface in concert with the production of tensile tangential stresses in the matrix. Because of the interference from the rubbery particles, the interface bonding between the fiber and the matrix is somewhat reduced by the presence of tensile radial stresses. Under these circumstances, the tendency for premature debonding at the interface is enhanced. This tendency results in a poor stress transfer by shear; only a moderate level of stresses, insufficient to cause fiber fracture, may be built up in the fibers.

Hence, fibers shorter than the critical transfer length will be pulled out. The formation of an extensive stress-whitened zone in this system is largely due to the concurrent display of debonding at the fiber interface and the cavitation of rubber particles. In this case the crack follows a fiber-avoidance path, resulting in a highly tortuous fracture surface. Avoidance of fibers by the advancing crack tip is accomplished by the initial tilting at the debonded interface and subsequent twisting of the crack front between fibers. Invariably, these sequential processes of fiber debonding, crack bridging, and fiber pullout will result in the stabilization of crack growth (Figure 19) and a display of an R-curve effect, as was shown in Figure 22 (*3, 66*).

Similar failure processes, according to Low (*68*), are anticipated for the ERZ materials. Again, the presence of tensile radial stresses at the rubber–matrix interface will substantially reduce the bonding strength at the $ZrO_2$–matrix interface. Consequently, premature debonding at the $ZrO_2$–matrix interface will be greatly enhanced and will lead to a substantial loss in the crack-pinning capability of these $ZrO_2$ particles.

## *Tetraglycidyl 4,4'-Methylenedianiline Resins*

Tetraglycidyl 4,4'-methylenedianiline (TGMDA) resins cured with DDS have much poorer fracture properties at room temperature (typically $G_{Ic} \sim 40$ $J/m^2$), although these materials have higher $T_g$ ($\sim 200$ °C) and are hence more suitable for high-temperature use. With DT specimen configuration, these materials show continuous crack propagation. Attempts to improve their toughness with CTBN rubber resulted in only 3–4-fold increase in $G_{Ic}$ (*4, 71, 72*).

Pearce and co-workers (*4*) developed toughened TGMDA resins by using piperidine and piperidine adducts as the curing agent, then incorporating CTBN rubber. Some of their results are presented in Table II, where it can be seen that the TGMDA–piperidine combination was inherently 4 times tougher than the control TGMDA–DDS. Addition of the rubber provided further improvement. Thus, the addition of 20 phr of CTBN gave a $G_{Ic}$ 14 times that of the basic DDS-cured resin. Some improvement in the $K_{Ic}$ and $G_{Ic}$ was observed with TGMDA–piperidine adduct (Table II), and the crack propagation changed to stick–slip behavior. Further improvement was observed when using phenyl isocyanate piperidine adduct (PIPIP), with a 24-fold gain in $G_{Ic}$ after addition of 20 parts of rubber. In these materials as well, the crack-propagation mechanism changed from continuous to stick–slip type. The improved ductility of the TGMDA–CTBN–piperidine materials is reflected in the crack-tip pictures shown for one system (Figure 23) in which it was possible to stop the cross-head movement during the slow crack growth and to photograph the crack-tip deformation. This procedure was impossible in the TGMDA–DDS materials, where crack propagation became catastrophic almost immediately once crack initiation started.

The $T_g$ of the toughened TGMDA epoxies is in the range of 170–180 °C. This range is close to that of the control TGMDA–DDS resin (*4*).

## Table II. Toughened TGMDA Epoxies

| Materials | Composition | Cure Schedule[a] | $GdI_c$ $(J/m^2)$ | $K_{Ic}$ $(MPa\ m^{1/2})$ |
|---|---|---|---|---|
| TGMDA | 100:0:28 | A | $38^b$ | $0.44^b$ |
| CTBN X–13 | 100:10:28 | | $72^b$ | $0.58^b$ |
| DDS | 100:20:28 | | $121^b$ | $0.66^b$ |
| TGMDA | 100:0:5 | B | $152^b$ | $0.76^b$ |
| CTBN X–13 | 100:10:5 | | $310^b$ | $0.99^b$ |
| Piperidine | 100:20:5 | | $525^b$ | $1.08^b$ |
| TGMDA | 100:0:11 | B | $84–123^c$ | $0.60–0.73^c$ |
| CTBN X–13 | 100:10:11 | | $81–266^c$ | $0.75–0.97^c$ |
| TGMDA–piperidine adduct | 100:20:11 | | $60–314^c$ | $0.85–0.94^c$ |
| TGMDA | 100:0:12 | B | $209^b$ | $0.68^b$ |
| CTBN X–13 | 100:10:12 | | $166–264^c$ | $0.72–0.91^c$ |
| Phenyl isocyanate–piperidine adduct (PIPIP) | 100:20:12 | | $699–907^c$ | $1.30–1.52^c$ |

[a] Cure cycles: A, 1.5 h at 135 °C plus 2 h at 175 °C; B, 1 h at 150 °C plus 1 h at 170 °C.
[b] Continuous crack propagation mode (double torsion specimens).
[c] Stick–slip crack propagation mode (double torsion specimens).
SOURCE: Adapted from reference 4.

## Summary

**Correlation between Failure Behavior in Unmodified Epoxies and Toughness in Modified Resins.**  Understanding of the localized failure processes in the crack-tip region of unmodified epoxies has been gained mainly from the load-displacement behavior data obtained by using tapered double cantilever beam and double torsion geometries. By using these constant-compliance specimens, $G_{Ic}$ or $K_{Ic}$ can be determined without knowledge of the position of the advancing crack front. Two types of crack-propagation behavior have been identified. The first, continuous propagation, involves negligible crack-tip plastic deformation. It takes place with a critical CTOD on the order of 0.5–2 μm and a crack-tip plastic zone of about the same size. Such crack propagation is favored under conditions that promote brittle failure, such as low temperature and high strain rate. The second type of crack propagation occurs when fracture proceeds in the stick–slip manner. This crack propagation takes place under conditions favoring crack-tip yielding, such as higher test temperature, slower testing speeds that reduce the yield strength, or the presence of plasticizing environment at the crack tip. The same material can thus show both types of fracture behavior, depending on the test conditions. For stick–slip behavior, the unmodified epoxies may exhibit a critical CTOD in the 3–15-μm range and a plastic zone of comparable size at the crack tip. The load at arrest would give a more realistic reflection of the $G_{Ic}$ or $K_{Ic}$ in this mode of failure.

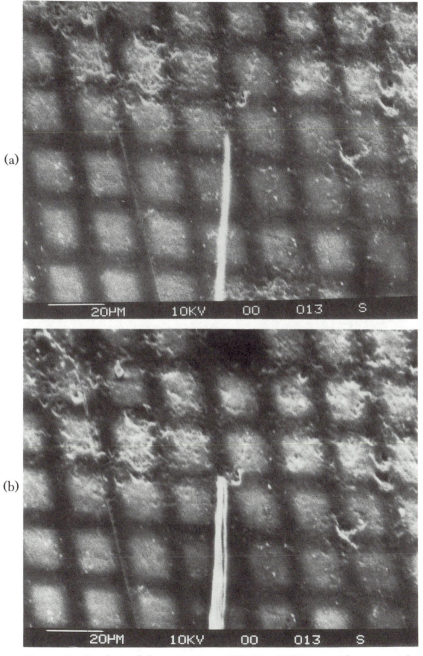

*Figure 23. Crack-tip deformation in a TGMDA–CTBN–piperidine material system from in-situ SEM experiment, taken in the SEM still mode. Key: a, original crack tip; b, crack-tip opening; and c, slow crack advance. Continued on next page.*

*Figure 23. Continued*

This understanding of the fracture behavior in unmodified epoxies provides the basic background for development of modified epoxy systems. For a particular application, an epoxy system that would exhibit stick–slip behavior at the temperature of use and perhaps at room temperature should be considered. For example, DGEBA–piperidine resin shows stick–slip behavior, and this unmodified material is also capable of shear yielding. When this material is reinforced with a dispersion of the rubbery phase, the shear deformation ability of the matrix produces a number of effects, such as deformation of the rubber–matrix interface, dilatation of the matrix, cavitation in rubber, and debonding and stretching of rubber particles.

This deformation brings into operation a number of mechanisms that could impart toughness to the epoxy individually or jointly. Even in the absence of rubber, such as in a particulate-filled material, the presence of a large number of hard particles and interfaces would induce shear deformation through crack pinning and crack bowing. Thus hybrid epoxies bring about the benefit of both the binary systems, as discussed earlier. This discussion is based mainly on the mechanical properties of the material; nevertheless, its chemical structure, cure chemistry, morphology, and other characteristics such as thermal and thermomechanical stability would be important.

In contrast, TGMDA resins cured with DDS show continuous crack propagation in double torsion testing at room temperature and have little crack-tip deformation ability. When this material is modified by rubber, those toughening mechanisms that are synergistic because of matrix shear yielding are absent, and only marginal improvement in fracture energy at room temperature is reported when rubber is added. The approach taken by Pearce et al. (*4*), induction of crack-tip plastic deformation by using the piperidine system as a curing agent, seems to be the appropriate direction in such materials.

**Relative Contribution of Various Mechanisms to Improvement of Fracture Toughness in Modified Epoxies.**   This aspect has not been addressed to any significant extent in the literature. The difficulties obviously lie in the fact that a number of localized deformation mechanisms can be operative at the crack-tip regions. Structural parameters such as

- chemistry of the resin,

- resin–second-phase interfacial bonding,

- size, size distribution, and orientation (for fibers of the added phase), and

- mechanical and fracture behavior of the unmodified resins

have significant effects. In addition, external factors like temperature, strain rate, loading conditions, humidity, and plasticizer, which affect the yield strength of the material, can change the failure from ductile to brittle mode and vice versa. Although the $G_{Ic}$ of a system may be increased by one of these factors, the modulus of elasticity, $E$, could well decrease. The relative change in these two properties will determine the change in fracture toughness, $K_{Ic}$.

Evans et al. (*73*) provided a quantitative basis for assessing trends in rubber-toughened polymers from a consideration of the rubber stretching and plastic dilatation of the matrix that result from debonding or cavitation of rubber particles. However, comparison with experimental data is still in an early stage. The addition of a third phase in hybrid epoxies must be accommodated in the theoretical analysis.

Notwithstanding, it is quite clear that epoxy resins, whether unmodified or modified, have as a single class of material generated tremendous interest in and made significant contribution to the understanding of fracture mechanisms of the entire area of toughened polymers.

## Acknowledgments

A number of my colleagues have been involved in fracture studies of epoxy resins over the past 8 years: Carolyn Morris, Peter Pearce, Sal Mestan,

Veronica Silva, Brian Ennis, Don Pinkerton, Evan Gellert, Ivan Grabovac, Tony Camilleri, Yiu Wing Mai, and It Meng Low. Dr. Low provided the SEM picture of Figure 20g. Technical assistance by Jim Nicholls and Harald Wrobel is also acknowledged.

## References

1. Moloney, A. C.; Kausch, H. H.; Kaiser, T.; Beer, H. R. *J. Mater. Sci.* **1987**, *22*, 381.
2. Garg, A. C.; Mai, Y. W. *Composites Science and Technology* **1988**, *31*, 179.
3. Bandyopadhyay, S. *Materials Science and Engineering* **1990**, *A125*, 157.
4. Pearce, P. J.; Morris, C. E. M.; Ennis, B. C. *Polymer Communications* **1988**, *29*, 93.
5. Kinloch, A. J.; Shaw, S. J. *J. Adhesion* **1981**, *12*, 59.
6. Ting, R. Y.; Cottington, R. L. *Polymer Bulletin* **1980**, *2*, 211.
7. Kinloch, A. J.; Shaw, S. J.; Tod, D. A.; Hunston, D. L. *Polymer* **1983**, *24*, 1341.
8. Kinloch, A. J.; Shaw, S. J.; Hunston, D. L. *Polymer* **1983**, *24*, 1355.
9. Kinloch, A. J.; Hunston, D. L. *J. Mater. Sci. Lett.* **1987**, *6*, 137.
10. Clarke, J. A. *Polymer Communications* **1985**, *25*, 113.
11. Sultan, J. N.; Liable, R. C.; McGarry, F. J. *J. Appl. Polym. Sci, Applied Polym. Symp.* **1971**, *16*, 127.
12. Cherry, B. W.; Thomson, K. W. *J. Mater. Sci.* **1981**, *16*, 1913.
13. Bucknall, C. B.; Yoshii, T. *The British Polym. Journal* **1978**, *10*, 53
14. Mizutani, K.; Iwatsu, T. *J. Appl. Polym. Sci.* **1981**, *26*, 3447.
15. Narisawa, I.; Murayama, T.; Ogawa, H. *Polymer* **1982**, *23*, 291.
16. Yee, A. F.; Pearson, R. A. *J. Mater. Sci.* **1986**, *21*, 2462.
17. Pearson, R. A.; Yee, A. F. *J. Mater. Sci.* **1986**, *21*, 2475.
18. Mostovoy, S.; Crosley, P. B.; Ripling, E. J. *ASTM Special Tech. Publ.* **1976**, *601*, 234.
19. Mijovic, J. S.; Koutsky, J. A. *J. Appl. Polym. Sci.* **1979**, *23*, 1037.
20. Mijovic, J.; Koutsky, J. A. *Polymer* **1979**, *20*, 1095.
21. Bascom, W. D.; Cottington, R. L.; Jones, R. L.; Peyser, P. *J. Appl. Polym. Sci.* **1975**, *19*, 2545.
22. Gledhill, R. A.; Kinloch, A. J. *Polym. Eng. Sci.* **1979**, *19*, 82.
23. Scott, J. M.; Phillips, D. G. *J. Mater. Sci.* **1979**, *10*, 551.
24. Bandyopadhyay, S.; Morris, C. E. M. *Micron* **1982**, *13*, 269.
25. Yamini, S.; Young, R. J. *J. Mater. Sci.* **1980**, *15*, 1814.
26. Yamini, S.; Young, R. J. *Polymer* **1977**, *18*, 1075.
27. Phillips, D. G.; Scott, J. M.; Jones, M. *J. Mater. Sci.* **1978**, *13*, 311.
28. Yamini, S.; Young, R. J. *J. Mater. Sci.* **1980**, *28*, 1823.
29. Gledhill, R. A.; Kinloch, A. J.; Yamini, S.; Young, R. J. *Polymer* **1978**, *19*, 574.
30. Young, R. J.; Beaumont, P. W. R. *J. Mater. Sci.* **1975**, *10*, 1343.
31. Bandyopadhyay, S.; Mestan, S. A.; Pearce, P. J.; Morris, C. E. M. *Proc. Polymer 85, the International Symposium on Characterisation of Polymers*; Polymer Division, Royal Australian Chemical Institute: Melbourne, Australia, 1985; pp 140–142.
32. Yamini, S.; Young, R. J. *J. Mater. Sci.* **1979**, *14*, 1609.
33. Kinloch, A. J.; Williams, J. G. *J. Mater. Sci.* **1980**, *15*, 987
34. Lee, S. M. *J. Mater. Sci. Lett.* **1982**, *1*, 511.
35. Moloney, A. C.; Kausch, H. H. *J. Mater. Sci. Lett.* **1985**, *4*, 289.

36. Mao, T. H.; Beaumont, P. W. R.; Nixon, W. C. *J. Mater. Sci. Lett.* **1983**, *2*, 613.

37. Bandyopadhyay, S.; Silva, V. M.; Wrobel, H.; Nicholls, J. *Proc., 6th Churchill Coil. Conf. on Deformation, Yield and Fracture of Polymers*; Plastics and Rubber Institute: London, 1985; paper 42.

38. Bandyopadhyay, S.; Silva, V. M. *Proc. 6th Inter. Conf. Fracture (ICF-6);* Valluri, S. R., Ed.; Pergamon Press: Oxford, England, 1984; p 2971.

39. Bandyopadhyay, S.; Pearce, P. J.; Mestan, S. A. *Crack Tip Micromechanics and Fracture Properties of Rubber Toughened Epoxy Resins, Proc. the 6th Churchill College Conf. on Deformation, Yield and Fracture of Polymers*; Plastics and Rubber Institute: London, 1985; paper 18.

40. Lilley, J.; Holloway, D. G. *Phil Mag.* **1973**, *28*, 215.

41. Bandyopadhyay, S. *J. Mater. Sci. Lett.* **1984**, *3*, 39.

42. Kunz-Douglass, S.; Beaumont, P. W. R.; Ashby, M. F. *J. Mater. Sci.* **1980**, *15*, 1109.

43. Doyle, M. J.; Goldthwait, R. G. *ANTEC 83, 41st Annual Technical Conference*; Society of Plastics Engineers: Chicago, IL, 1983; pp 781–784.

44. Leksovskii, A. M.; Baskin, B. L.; Ye Gorenberg, A.; Usmanov, G. KL.; Regel, V. R. *Fiz Tverd Tela (Lenin)* **1983**, *25*, 1096.

45. Walker, N.; Haward, R. N.; Hay, J. N. *J. Mater. Sci.* **1981**, *16*, 817.

46. Rowe, E. H.; Riew, C. K. *Plastics Engineering* **March 1975**, *45*.

47. Daly, J.; Pethric, R. A.; Fuller, P.; Cluniffe, A. V.; Datta, P. K. *Polymer* **1981**, *22*, 32.

48. Yorkgitis, E. M.; Tran, C.; Eiss, N. S.; Hu, T. Y.; Yilgor, I.; Wilkes, G. L.; McGrath, J. E.; In *Rubber Modified Thermoset Resins*; Riew, C. K.; Gillham, J. K., Eds.; ACS Advances in Chemistry 208; American Chemical Society: Washington, DC, 1984; p 137.

49. Trostyanskaya, Ye B.; Bahayevskii, P. G.; Kulik, S. G. *Polym. Science USSR* **1986**, *21*, 1450.

50. Hunston, D. L.; Bitner, J. L.; Rushford, J. L.; Oroshnik, J.; Rose, W. S. *J. Elastomers and Plastics* **July 1980**, *12*, 133.

51. Low, I. M.; Mai, Y. W. *Composites Science and Technology* **1988**, *33*, 191.

52. Manzione, L. T.; Gillham, J. K. *J. Appl. Polym. Sci.* **1981**, *26*, 889.

53. Bowerman, H. H.; McCarthy W. C. *Proc. 28th Annual Technical Conf.* Reinforced Plastics/Composites Institute, The Society of the Plastics Industry, Inc, 1973; Section 9-A, pp 1–6.

54. Ting, R. Y.; Cottington, R. L. *J. Appl. Polym. Sci.* **1980**, *25*, 1815.

55. Low, I. M.; Mai, Y. W. *J. Mater. Sci.* **1989**, *24*, 1634.

56. Mestan, S. A.; Pearce, P. J.; Morris, C. E. M.; Bandyopadhyay, S. presented at the Inter. Symp. on Deformation, Failure and Strengthening of Polymers, Monash University, Australia, May 23–25, 1985, organised jointly by Monash University and Royal Australian Chemical Institute, Polymer Division.

57. Plati, E.; Williams, J. G. *Polym. Eng. Sci.* **1975**, *15*, 470.

58. Lange, F. F.; Radford, K. C. *J. Mater. Sci.* **1971**, *6*, 1197.

59. Young, R. J.; Beaumont, P. W. R. *J. Mater. Sci.* **1977**, *12*, 684.

60. Moloney, A. C.; Kausch, H. H.; Stieger, E. R. *J. Mater. Sci.* **1983**, *18*, 208.

61. Spanoudakis, J.; Young, R. J. *J. Mater. Sci.* **1984**, *19*, 473.

62. Spanoudakis, J.; Young, R. J. *J. Mater. Sci.* **1985**, *20*, 87.

63. Brown, S. K. *British Polym. J.* **March 1980**, *12*, 24.

64. Brown, S. E. *British Polym. J.* **March 1982**, *14*, 1.

65. Kinloch, A. J.; Maxwell, D.; Young, R. J. *J. Mater. Sci.* **1985**, *20*, 4169.

66. Low, I. M.; Mai, Y. W.; Bandyopadhyay, S.; Silva, V. M. *Materials Forum* **1987**, *10(4)*, 241.
67. Bandyopadhyay, S.; Silva, V. M.; Low, I. M.; Mai, Y. W. *Plastics and Rubbers Processing and Applications* **1988**, *10*, 193.
68. Low, I. M. *J. Mater. Sci.* **1990**, *25*, 2144.
69. Faber, K. T.; Evans, A. G. *Acta Metall.* **1983**, *31(4)*, 565.
70. Williams, J. G. *Int. J. Fract. Mech.* **1972**, *8*, 393.
71. Lee, B. H.; Lizak, C. M.; Riew, C. K., paper presented at the 12th National SAMPE Technical Conference, Seattle, Washington, 1980.
72. Diamant, J.; Moulton, R. J. *SAMPE Qrly.* **October 1984**, 13.
73. Evans, A. G.; Ahmad, Z. B.; Gilbert, D. F.; Beaumont, F. W. R. *Acta Metall.* **1986** *34*, 79.

RECEIVED for review March 6, 1991. ACCEPTED revised manuscript September 9, 1991.

# Optimization of Mode-I Fracture Toughness of High-Performance Epoxies by Using Designed Core-Shell Rubber Particles

H.-J. Sue[1], E. I. Garcia-Meitin[1], D. M. Pickelman[2], and P. C. Yang[2]

[1]Analytical and Engineering Sciences Department, Texas Polymer Center, Dow Chemical U.S.A., Freeport, TX 77541
[2]Advanced Polymeric Systems and Advanced Composites Laboratories, Central Research, The Dow Chemical Company, Midland, MI 48674

*The fracture behavior of high-performance epoxies modified with seven types of designed core-shell rubber particles is examined by using various microscopic techniques. The mode-I plane-strain fracture-toughness values ($K_{Ic}$) of the designed rubber-modified epoxy systems that have highly cross-linked epoxy vary with the architecture of the interface between the matrix and the rubber particles. Accordingly, the toughening mechanisms observed among these systems are very different: rubber-particle cavitation and matrix shear yielding are found in systems that exhibit higher $K_{Ic}$ values; only crack deflection is found in systems that possess lower $K_{Ic}$ values. Synergistic toughening can be obtained if both the matrix shear-yielding and the crack-deflection mechanisms are operative upon fracture. An approach for toughening highly cross-linked epoxies is addressed.*

TOUGHNESS IS AN IMPORTANT AND SOMETIMES DECISIVE FACTOR for material selection, especially for structural applications. For polymeric materials, many brittle or notch-sensitive polymers can be effectively toughened by elastomeric inclusions (*1–24*). Certain types of rubber are more effective in toughening the brittle and notch-sensitive polymers than others (*15–17*).

0065–2393/93/0233–0259$09.25/0

However, there is still controversy over the exact role(s) of rubber particles in toughening (*1–3, 10–13, 25, 26*). To optimize the toughening effect for a rubber-modified polymer system, the failure mechanisms must be fully understood and the exact role(s) the rubber plays in the toughening process must be identified.

Many studies have been conducted on specific aspects of the effects of the rubber particles in epoxy toughening. The effect of rubber type in epoxy toughening was addressed by Pearson and Yee (*27*) and by Sue (*28*). The effect of rubber particle size in polymer toughening was investigated by Sultan and McGarry (*10*), Pearson and Yee (*27*), Bascom et al. (*9*), and Borggreve and co-workers (*15*). However, understanding the role(s) played by the interfacial architecture between the rubber particle and the matrix in toughness optimization is still lacking. The effects of shell chemistry of the core-shell rubber in toughening a polycyanate (PCN) matrix recently were studied by Yang et al. (*29*), who suggest that the chemical bonding at the interface between the rubber particle and the matrix is not essential for effective toughening of PCN. The fracture-toughness results are, however, affected by the change of the shell chemistry of the core-shell rubber particle. Yang et al. (*29*) recognize that the underlying toughening mechanism(s) in these systems may be different. In other words, the changes of shell chemistry in the core-shell rubber may alter or affect the toughening process, which, in turn, is reflected in the variation of the fracture-toughness results. To clearly determine how the fracture-toughness results are altered by the changes of the core-shell rubber architecture systematic micromechanism(s) investigations are needed.

Knowledge about the effect of interfacial property changes in polymer toughening is largely limited to characterization and studies of chemical or physical interaction at the molecular level at the interface between the matrix and the toughener phase (*12, 30–34*). The links between these chemical or physical interactions with the mechanical performance are frequently inexplicit. This type of correlation provides no direct evidence of the true relationship between the toughness and the interfacial property. Interfacial strength between the matrix and the toughener phase can be affected by thermal stresses, mechanical interlocking, chemical bonding, hydrogen bonding, and physical molecular interactions (*34*). Furthermore, as pointed out by Bucknall (*8*), toughening effects due to interfacial adhesion and particle size are not always separable. When these two parameters are separable (i.e., with the use of preformed particles), the interfacial adhesion and the degree of particle dispersion may be interdependent. Consequently, misleading results and erroneous conclusions may be made if only the effects at the interface due to chemical bonding or polarity of molecules are considered. Thus, care must be taken for a clear understanding of the effect of the interface on toughening.

The preformed core-shell rubber particles, also termed grafted rubber concentrate (GRC), are well suited for elucidating the effects of a core-shell rubber interfacial architecture on toughening highly cross-linked epoxy systems. These core-shell particles are produced by a two-stage latex emulsion polymerization technique (35). The core is a graftable elastomeric material that, preferably, is either cross-linked, insoluble in the hosting liquid (uncured) resin phase, or both. One example of both cross-linking and insolubility is cross-linked polybutadiene and a liquid epoxy resin based on polyhydric phenols. The extent of cross-linking of the core rubber can be estimated from a determination of percent gel (nonextractable polymer) and swell index (imbibition of solvent by test polymer). Typical percent gel ranges from ~ 50 to ~ 95% of the total polymer; the swell index ranges from ~ 3 to ~ 50. The cores are presized, and they maintain this morphology before, during, and after the resin cure process (28, 29, 35). The core diameter can range in size from ~ 30 nm to ~ 2 μm. For toughening applications, a high amount of elastomeric core relative to total particle is desirable. The useful average composition range of cores is from ~ 50 to ~ 90 wt % of the particles.

The shell component is grafted to the core to effectively stabilize the particle in the resin phase. Rubbery core particles having a low glass-transition temperature require a protective shell for colloidal stability in a nonaqueous environment even if the cores are cross-linked or insoluble in the resin phase. The shell phase typically has chemically attached polymer chains and nongrafted chains. To minimize the viscosity of the resin phase, the level of nongrafted chains should be at a minimum and the molecular weight of the shell chains should be in the 20,000 to 100,000 range. The shell composition should be compatible with the resin phase to achieve the optimum colloid stability necessary to survive mixing operations at elevated temperatures and homogeneous dilution. Mixed monomer systems that yield copolymer compositions for the shell can be designed to provide an interface between the rubbery core and the continuous resin phase that fosters compatibility. Some of the copolymer parameters are chain flexibility, polarity, and postcore activity with the curing resin phase.

To study property effects of a core-shell rubber additive in a thermoset resin system, a four-component monomer system is chosen for the shell. The shell monomers, styrene, methyl methacrylate, acrylonitrile (AN), and glycidyl methacrylate (GMA), are varied in composition, core–shell ratio, polarity, and core activity, and all use the same core rubber. The grafted-shell thickness is varied by simply changing the core–shell weight ratio from 84:16 to 75:25 and 65:35. The considerations for shell chemistry changes are polarity and latent core activity. To track the effects of polarity, the AN, ranging from 0 to 25 wt % of shell composition, is prepared; to track the effects of interfacial (resin–particle surface) coreactivity, the vinyl epoxide monomer GMA, ranging from 0 to 30 wt % of shell composition, is utilized (Table I).

**Table I. Composition of the Designed Core-Shell Particles**

| Rubber Particle | Core–Shell | Particle Diameter (nm) | | Shell Composition[a] | | | | $\delta^c$ |
|---|---|---|---|---|---|---|---|---|
| | | Experimental[b] | Theory | S | MMA | AN | GMA | |
| Core[d] | 100–0 | 119 | — | — | — | — | — | — |
| GRC-A | 84–16 | —[e] | 126 | 3.6 | 3.6 | 4.0 | 4.8 | 10.3 |
| GRC-B | 84–16 | 127 | 126 | 4.8 | 4.8 | 4.0 | 2.4 | 10.2 |
| GRC-C | 84–16 | 127 | 126 | 6.0 | 6.0 | 4.0 | 0.0 | 10.1 |
| GRC-D | 84–16 | — | 126 | 7.0 | 7.0 | 2.0 | 0.0 | 9.7 |
| GRC-E | 84–16 | — | 126 | 8.0 | 8.0 | 0.0 | 0.0 | 9.3 |
| GRC-F | 75–25 | 133 | 131 | 7.5 | 7.5 | 6.25 | 3.75 | 10.2 |
| GRC-G | 65–35 | 140 | 138 | 10.5 | 10.5 | 8.75 | 5.25 | 10.2 |
| Neat epoxy | | | | | | | | 10.3 |

[a] Parts of the shell are S, styrene; MMA, methyl methacrylate; AN, acrylonitrile; and GMA, glycidyl methacrylate.
[b] Universal light scattering photometer (Brice–Phoenix).
[c] Obtained by the small group method.
[d] Core composition is 7% styrene and 93% butadiene.
[e] Dash means not measured.

The present work focuses on gaining knowledge about whether the interfacial architecture between the toughener phase and the matrix can alter the toughening of high-performance epoxies. Seven different types of designed GRC particles are used to modify the diaminodiphenylsulfone (DDS) cured Dow D.E.R. 332 epoxy resin, which is a liquid diglycidyl ether of bisphenol A (DGEBA) epoxy (see Table I). The goal is to experimentally investigate how the toughening mechanisms can be altered by the changes of interfacial architecture between the rubber particle and the matrix. The role(s) of the designed rubber particles in the fracture process is also examined. The measured $K_{Ic}$ (mode-I plane-strain fracture toughness) results are related to the observed operative toughening mechanisms. Another intention is to determine whether the highly cross-linked high-performance epoxies can undergo shear yielding around the crack tip when the rubber particles are incorporated.

The double-notch four-point-bend (DN-4PB) technique (36), which is effective in probing the toughening mechanisms of rubber and rigid polymer-modified systems (5–7, 36–39), is used to generate a subcritically propagated crack. Transmitted optical microscopy (TOM), scanning electron microscopy (SEM), and transmission electron microscopy (TEM) techniques are used to study the fracture surface and the damage zone around the subcritically propagated crack of the DN-4PB rubber-modified epoxies.

## Toughening Principles

Garg and Mai (40) review the observed toughening mechanisms in rubber-modified epoxies and find that more than 14 toughening mechanisms potentially can operate around a propagating crack. Among the available toughening mechanisms, some may operate inclusively, exclusively, or sequentially, depending on the testing condition and the nature of the polymer (*5*, *7*). Experimentally, the shear-banding type of mechanism provides the highest toughening effect, followed by crazing, crack-bridging, microcracking, crack-bifurcation, crack-deflection, and crack-pinning mechanisms (*40*, *41*). The shear-banding mechanism potentially can provide an order of magnitude increase in toughness (*1–3*, *10*, *14*), whereas only a several-fold increase results from the crazing mechanism (*8*), and only a moderate increase is seen for crack bridging (*11*, *14*). For crack-bifurcation, crack-deflection, and crack-pinning mechanisms, only a fractional increase in toughness is reported (*28*, *40*, *41*).

The concept of toughening epoxy resins by generating a dispersed rubbery phase normally produces an order of magnitude increase in toughness for some rubber-toughened epoxy systems (*1–3*, *10*, *14*). However, epoxy systems that can be effectively toughened by the dispersed rubbery phase generally have relatively low cross-link density and, therefore, low glass-transition temperatures ($T_g$) in most cases (*1–3*). Low-$T_g$ epoxies in high-temperature, high-performance aerospace applications are of limited use. In high-$T_g$ brittle epoxy resins, such as DGEBA epoxy of low epoxy equivalent weight (172–176) cured by DDS, rubber modification is ineffective for toughening (*1–3*). The general consensus is as follows. Theoretically, because the molecular mobility in the highly cross-linked epoxies is so restricted, the plastic deformations, such as crazing and shear banding, are not possible. Experimentally, only a very limited increase in $K_{Ic}$ can be achieved (*1–3*, *24*). Only crack–particle bridging, crack deflection, and crack pinning can be operative for rubber-toughened epoxy systems of high cross-link density. (The crazing mechanism does not occur in highly cross-linked thermosets.) The addition of the rubbery phase does little to improve toughness, whereas it reduces the thermal and mechanical properties of the epoxy matrix. Thus, there has been little progress in the toughening of epoxy systems having high cross-link density.

To effectively toughen a brittle or notch-sensitive polymer requires an efficient mechanism for absorbing a large amount of strain energy, such as profuse shear banding around a growing crack (*1–3*). At the crack tip, under mode-I fracture, the material experiences a triaxial tensile stress state (*42*). The octahedral shear-stress component is suppressed, and microcracks (or crazes if the matrix polymer is not cross-linked), rather than the more energy-absorbing mechanism of shear deformation, tend to nucleate in front of the crack tip. As a result, the amount of energy absorbed prior to fracture is limited. For shear bands to form in front of a crack tip, the triaxial tension must be relieved by either cavitation or crazing of the toughener phase or by debonding at the interface between the two phases. Cavitation, crazing, or debonding alone does not guarantee shear deformation (*28*, *39*, *43*); the resultant octahedral shear-stress component must be sufficiently close to the yield stress of the matrix polymer to induce localized shear deformation (43). Therefore, selection of appropriate combinations of the matrix and the toughener phase as well as the interface is critical for obtaining an optimal toughening effect.

Experimental work conducted by Kinloch et al. (44) and Glad (45) shows that an epoxy system of high cross-link density can undergo strain softening and strain hardening under compression. These findings imply that epoxy of high cross-link density is capable of undergoing shear banding (46) when subjected to an appropriate stress state. This finding also suggests that it may be feasible to promote the shear-banding mechanism around the crack tip (under mode-I loading) of highly cross-linked epoxy systems by incorporating appropriate rubber particles into the matrix, provided the addition of the particles can alter the crack-tip stress field from a highly triaxial to a highly shear stress state (39, 43). The mechanical interaction between the matrix and the toughener phase before and after cavitation occurs must be significant enough to alter the crack-tip stress state from a state that causes brittle failure to a state that leads to a tougher material response, such as shear banding.

When the shear-banding mechanism cannot be activated, alternative toughening mechanisms, such as crack bridging (47), crack bifurcation (48, 49), crack pinning (50, 51), and crack deflection (52, 53), must be promoted. The requirements for achieving the optimal toughening effects of these mechanisms greatly differ from those of the shear-banding mechanism. Detailed descriptions as well as modeling of these alternative toughening mechanisms in rubber-modified epoxy are reviewed and discussed by Garg and Mai (40). The relevant experimental evidence will be presented and discussed in this chapter.

Moreover, to effect toughening, a certain degree of interfacial adhesion between the matrix and the dispersed phase is critical. The adhesion strength necessary between the matrix and dispersed phase to optimize the toughness of an alloy system is still unknown. Poor adhesion between the phases can cause debonding at the interface when a finite load is applied. Debonding can either nucleate cracking or crazing or initiate the onset of localized shear deformation, depending on the polymer. In a strongly adhering interface, both phases share and transmit the load, which leads to greater stiffness in the system. Cavitation still must occur to relieve high triaxial tension. Furthermore, depending on the operative toughening mechanisms, the interfacial adhesion requirement may differ. To promote the crazing mechanism, a good interfacial adhesion is beneficial for optimizing toughness (8). It is still unclear whether good interfacial adhesion is needed to induce shear banding. Wu (12) suggests that in the rubber-modified nylon-6,6 system, chemical bonding is unnecessary for Izod impact toughness improvement. From the mechanics point of view (38) as well as from the experimental work conducted by Chan et al. (54), depending on the intrinsic property of the matrix, poor interfacial adhesion may result in ineffective toughening of polymers. The need for interfacial adhesion between the toughener phase and the matrix also appears to be affected by the testing conditions (e.g., temperature and rate effects) (55). Depending on the viscoelastic or viscoplastic nature of the toughener phase, the matrix, and the interfacial adherend, the degree of interfacial adhesion needed for optimal toughening may also vary.

Finally, the surface free energy at the interface resulting from either chemical or physical means alters the degree of interfacial adhesion and affects the extent to which the toughener phase disperses in the matrix. These changes, in turn, may alter the operative toughening mechanism(s) in the system.

## *Experimental Details*

**Materials**.   The D.E.R. 332 epoxy resin (epoxide equivalent weight ~ 174) was mixed with a stoichiometric ratio of DDS at 130 °C and cured at 180 °C for 2 h, followed by postcuring at 220 °C for 2 h.

Seven types of designed GRC particles with varied shell chemistry and shell thickness of the core-shell rubber, but with core size and composition the same (Table I), were used to toughen the DDS-cured epoxy systems (Table II).

Preparation of the core-shell rubber involves grafting the shell monomers onto a monodispersed submicrometer rubber latex with an average particle diameter of 119 nm as measured by a universal light scattering photometer (Brice–Phoenix). The rubber latex particles are cross-linked styrene–butadiene copolymer (7% styrene, 93% butadiene) stabilized by 3% sodium dodecylbenzene sulfonate soap (based on polymer). The grafting procedure

**Table II. Summary of $K_{Ic}$, $G_{Ic}$, and $T_g$ of Designed GRC-Modified Epoxy Systems**

| D.E.R. 332 Epoxy Resin–DDS | $K_{Ic}$[a] $(MPa \cdot m^{1/2})$ | $G_{Ic}$[b] $(J/m^2)$ | $T_g$[c] $(°C)$ |
|---|---|---|---|
| Neat Resin | 0.83 ± 0.02 | 180 | 220 |
| GRC-A | 1.20 ± 0.07 | 490 | 222 |
| GRC-B | 1.20 ± 0.04 | 490 | 219 |
| GRC-C | 1.30 ± 0.09 | 580 | 223 |
| GRC-D | 1.07 ± 0.05 | 390 | 229 |
| GRC-E | 1.05 ± 0.03 | 380 | 224 |
| GRC-F | 1.33 ± 0.07 | 620 | 223 |
| GRC-G | 1.37 ± 0.06 | 640 | 220 |

[a] A 63% confidence interval is used.
[b] $G_{Ic} = K_{Ic}(1 - \nu2)/E$, where $E$ is Young's modulus and $\nu$ is Poisson's ratio.
[c] Second-heat midpoint $T_g$ is reported.

includes a two-stage monomer feed and temperature change before post-cooking and residual monomer removal.

In each of a series of runs, 3675 g of the previously described rubber latex containing 1176 g of rubber solids are charged into a 5-L reactor equipped with a mechanical stirrer, temperature control, cold-water-jacketed reflux condenser, and an inert atmosphere of nitrogen. To the stirred (150 rpm) reactor, 2.24 g of VAZO 64 Dupont; [2,2'-azobis(2-methylpropane-nitrile)] are added. The head space in the reactor is continuously flushed with nitrogen, and the contents are heated to a temperature of 70 °C.

For each run, 224 g of a mixture of monomers corresponding to one of the compositions listed as core-shell examples in Table I is added to the reactor at an initial feed rate of 20% of the total shell monomers over 150 min. During the remaining 80% feed addition over 90 min, the reactor temperature is increased to 80 °C and maintained for another 120 min.

With each 2.5 kg of the core-shell latexes, 2.5 kg of methyl ethyl ketone (MEK) is mixed to form a dispersion in MEK–water. To the dispersion, 1 L of deionized water is added without mixing. Approximately 1 h later, the bottom water phase is removed, additional water–MEK (77:23) is added, and the mixture is agitated for 5 min. The washing procedure is repeated until the desired ionic level is attained. The upper layer (i.e., the GRC particles and MEK mixture) is then mixed with the liquid epoxy resin, and in all cases, 10 wt % of each designed core-shell rubber is added to the epoxy resin on a solids basis. The volatile components are removed by distillation, leaving the rubber particles dispersed in the liquid resin.

The GRC particles are mixed with the resin, the DDS is added, and the mixture is degassed at 130 °C for about 10 min before curing. The curing schedule for all the designed GRC rubber-toughened D.E.R. 332 epoxy resin–DDS systems is as follows: 2 h at 180 °C and 2 h at 220 °C.

**Glass-Transition Measurements**. The glass-transition tempera-ture ($T_g$) of the neat and rubber-modified epoxy systems is measured by using differential scanning calorimetry (Mettler DC-30) ranging from 25–250 °C, with a heating rate of 10 °C/min. The midpoint $T_g$ of the second heat was recorded and is reported in Table II.

**Fracture-Toughness Measurements**. After the epoxy resin was cured and slowly cooled to room temperature (25 °C) in the oven, the 0.635-cm-thick epoxy plaque was machined into bars with dimensions of 12.7 cm × 1.27 cm × 0.635 cm for the DN-4PB experiments (*see* Figure 1 for the schematic of the DN-4PB geometry) and with dimensions of 6.35 cm × 1.27 cm × 0.635 cm for the single-edge-notch three-point-bend (SEN-3PB) experiments. These bars were then notched with a 250-μm radius notching cutter (TMI model 22-05) followed by razor blade tapping to wedge open a sharp crack with a parabolic crack front. The ratio of the final crack

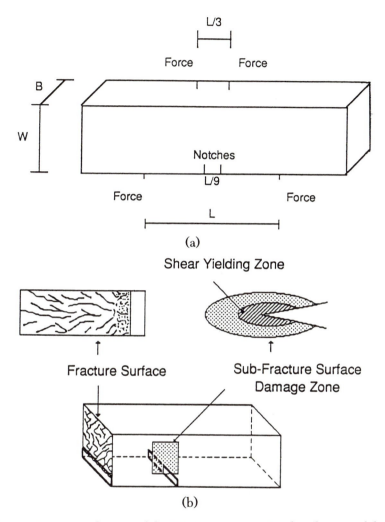

Figure 1. Part a: Schematic of the DN-4PB geometry. Part b: Schematic of the regions from the DN-4PB specimen for SEM, TOM, and TEM investigations.

length ($a$) and the specimen width ($W$) was held in the range between 0.3 and 0.7.

A screw-driven mechanical testing machine (Sintech-2) was used to conduct both the SEN-3PB and the DN-4PB experiments. A crosshead speed of 0.508 mm/min was utilized. When the DN-4PB experiment was performed, care was taken to ensure that the upper contact loading points were touching the specimen simultaneously.

**Investigation of Toughening Mechanisms**. The damage zone around the surviving DN-4PB crack was cut in half along the thickness direction using a diamond saw (Figure 1). The plane-strain core region was prepared for TOM and TEM examination. The fracture surface of the failed crack was coated with Au–Pd for SEM investigation.

In the TOM investigation, thin 40-$\mu$m sections of the designed GRC-toughened epoxies were obtained by polishing, following the procedure described by Holik et al. (56). The thin sections were taken from the midsection (plane-strain region) of the fractured DN-4PB specimens. The sections were made normal to the fracture surface but parallel to the cracking direction. The thin sections were then examined with a microscope (Olympus Vanox-S) under both bright field and cross-polarization.

In the TEM work, the core region of the damage zone was carefully trimmed to an appropriate size, that is, an area of $\approx 5$ mm $\times$ 5 mm, and embedded in D.E.R. 331 epoxy resin–diethylenetriamine (12:1 ratio by weight). The embedment was cured at 38 °C for 16 h. The cured block was then further trimmed to a size of $\approx 0.5$ mm $\times$ 0.5 mm, with the crack tip in the damage zone roughly at the center of the trimmed surface. A research microscope (Olympus Vanox) was used in the reflectance mode to observe the sample during the trimming process to ensure a good crack location on the thin section. A diamond knife was then utilized to face off the trimmed block.

The faced-off block of each of the GRC-modified epoxy specimens was placed in a vial containing 1 g of 99.9% pure osmium tetroxide ($OsO_4$) crystals and stained for 65 h. The $OsO_4$-stained block was microtomed to thin sections ranging from 60 to 80 nm thick by using a microtome (Reichert–Jung Ultracut E) with a diamond knife at ambient temperature. The thin sections were placed on 200-mesh Formvar-coated copper grids. The $OsO_4$ had penetrated to a depth of about 500 nm, so only six ultra-thin sections, ranging from 60 to 80 nm, were obtained for investigation. The thin sections were examined with an analytical transmission electron microscope (JEOL 2000FX) operated at an accelerating voltage of 100 kV.

## Results

In this work, direct observations on how different experimental GRC particles respond to a growing crack are made, instead of analyzing the chemistry at the interface between the GRC particle and the epoxy matrix. In all cases, the surviving cracks (i.e., the subcritically propagated crack) of the DN-4PB specimens of the designed GRC rubber-modified epoxy systems are prepared for various microscopic investigations of failure mechanisms. The relationship between the $K_{Ic}$ values and the corresponding toughening mechanisms for each designed GRC-modified system is ascertained.

**Fracture-Toughness Measurements.** To study the differences in toughening effect resulting from the variation of the shell architecture in the core-shell rubber, $K_{Ic}$ measurements are conducted.

To ensure a valid $K_{Ic}$ value, the peak load is plotted against $BW^{1/2}/Y$ ($B$ is thickness, $W$ is width, and $Y$ is correction factor) for each epoxy system. For each testing geometry and technique, the correction factor is different and can be found in published literature (57). The correction factor for the SEN-3PB geometry, which has a span–width ratio of 4 in this study, is given as

$$Y = 11.6(a/W)^{1/2} - 18.4(a/W)^{3/2} + 87.2(a/W)^{5/2}$$
$$- 150.4(a/W)^{7/2} + 154.8(a/W)^{9/2}$$

where $a$ is the crack length. The least-squares slope of the line drawn from a plot of the peak load against $BW^{1/2}/Y$ (for at least six specimens) is defined as $K_c$. If this line is straight and the $K_c$ value fulfills the conditions (58) $B$, $(W - a)$, and $a \geq 2.5(K_{Ic}/\sigma_{ys})^2$ where $\sigma_{ys}$ is the yield stress, then a valid $K_{Ic}$ value is attained. All of the tests in the current study fulfill the preceding requirements for a valid $K_{Ic}$. The $K_{Ic}$ values of the experimental GRC-modified epoxy systems are summarized in Table II. A 63% confidence interval is used for calculating the standard error. A typical plot of the peak load vs. $BW^{1/2}/Y$ is shown in Figure 2.

As shown in Table II, when the GMA content is altered from 30 wt % in the shell to 15 and 0 wt %, the $K_{Ic}$ values are effectively the same. (Even though the GRC-C modified system exhibits a slightly higher $K_{Ic}$ value, the standard error is also higher.) However, when the AN content is changed from 25 wt % in the shell (GRC-C) to 12.5 (GRC-D) and 0 wt % (GRC-E), significant differences in $K_{Ic}$ values are detected. Altering the shell thickness while retaining the core size and composition results in an observable increase in $K_{Ic}$ as the shell thickness is increased. To definitely understand how the $K_{Ic}$ values are affected by the change of the designed GRC particles, the toughening mechanisms of these GRC-modified highly cross-linked epoxy systems are investigated.

**Investigation of Toughening Mechanisms.** To determine how the $K_{Ic}$ values are altered among the designed GRC-modified epoxy systems (*see* Table II), the damage zone introduced by the DN-4PB method on each system is studied. Because of the brittleness of the epoxy matrix as well as the small size of the GRC particles, OM and SEM are impractical for the present study (5, 26). (Typical OM and SEM micrographs of the GRC-modified

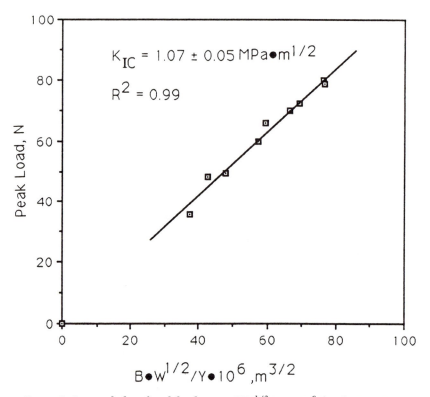

*Figure 2. A typical plot of peak load versus* $BW^{1/2}/Y \times 10^6$ *for the D.E.R. 332 epoxy resin–DDS–GRC-D system. The slope is defined as the* $K_{Ic}$ *value.* $R^2$ *is the coefficient of determination.*

epoxy are shown in Figures 3 and 4.) Consequently, TEM investigation is conducted.

To determine exactly what role the rubber particles play in toughening the highly cross-linked epoxy system, TEM investigations are conducted. TEM micrographs of the damage zone around the crack tip of the experimental GRC-modified epoxy systems (labeled from GRC-A to GRC-G) are shown in Figures 5–15. Except for the GRC-D and GRC-E systems, the major toughening mechanisms are rubber-particle cavitation and matrix shear yielding. The GRC-B modified epoxy, which exhibits a random dispersion of GRC particles, demonstrates the occurrence of the rubber-particle cavitation and matrix shear-yielding mechanisms observed in the majority of the designed GRC-modified epoxy systems.

The TEM micrograph taken at the damaged crack tip reveals that the originally spherical GRC-B particles are elongated as much as 60%

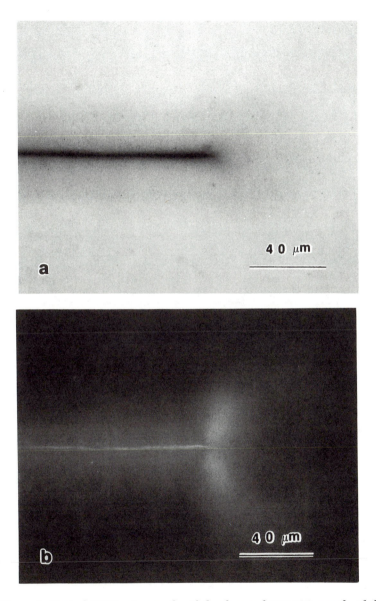

*Figure 3. Typical TOM micrographs of the damaged DN-4PB sample of the GRC-modified epoxy system taken under bright field (a) and cross-polars (b). As shown in Part a, a faint cavitation zone is formed around the crack tip. In Part b, the birefringent zone is found around the crack tip. The birefringent zone is partially affected by the residual stress field around the crack tip. As a result, the shear-yielded zone cannot be defined unequivocally. The crack propagates from left to right.*

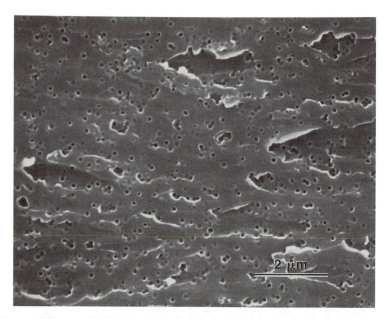

*Figure 4. A typical SEM micrograph taken on the stress-whitened zone of the designed GRC-modified epoxy systems. The crack propagates from left to right. (Reproduced with permission from reference 28. Copyright 1991 Society of Plastics Engineers.)*

immediately adjacent to the crack plane, as shown in Figure 6. This elongation implies that the matrix material around the rubber particles has plastically deformed to a similar degree, that is, 60% of plastic strain. Careful investigation around the damage zone, as shown in Figures 7 and 8, indicates that the rubber particles change from being elongated and possessing a larger cavity around the crack tip to being spherical and having a smaller cavity away from the crack tip. Outside the cavitation zone ($\approx 100$ μm above the crack tip), intact nondeformed rubber particles are found (Figure 9). For comparison, a TEM micrograph taken at the precracked crack-tip region before the DN-4PB test is shown in Figure 10. It is evident from the contrast between Figures 7 and 10 that the events of rubber-particle plastic dilatation and shearing (Figure 7) are the result of the DN-4PB test. No resolvable rubber-particle deformation is observed at the precracked crack-tip region (Figure 10).

Even though the major toughening mechanisms of most of the GRC-modified epoxy systems are the same, the $K_{Ic}$ values are quite different. To account for the variation of the $K_{Ic}$ values among the designed GRC-modified systems, it is necessary to correlate how the GMA content, the AN content, and the shell thickness affect the morphology or interfacial adhesion of the system and how the resultant morphology or interfacial adhesion affects the toughening mechanisms of the brittle epoxy systems.

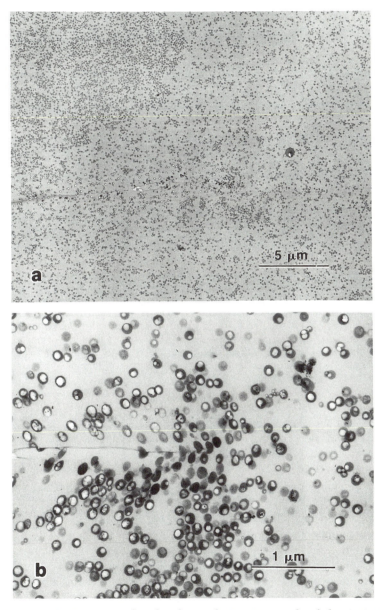

*Figure 5. TEM micrographs of a damaged DN-4PB sample of the GRC-A-modified epoxy system at a low (a) and at a high (b) magnification. A multimodal dispersion of the GRC particles is observed. The crack propagates from left to right.*

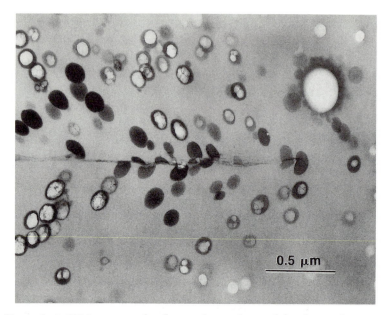

*Figure 6. A TEM micrograph taken at the crack tip of the damaged DN-4PB specimen of the GRC-B-modified epoxy system. A 60% elongation of the GRC rubber particles at the crack tip is found, which indicates that the highly cross-linked D.E.R. 332 epoxy resin is capable of undergoing large-scale plastic deformation. The crack propagates from left to right. (Reproduced with permission from reference 28. Copyright 1991 Society of Plastics Engineers.)*

**Variation of GMA Content.** When the GMA content is altered from 30 wt % in the shell (GRC-A) to 15 (GRC-B) and 0 wt % (GRC-C), the particle–resin morphology of the three systems changes from a dispersion with a multimodal distribution of nearest-neighbor distance (Figure 5) to a more random dispersion (Figures 6 and 7) and to a local clustering (Figure 11) of the GRC particles. These results indicate that a minor addition of GMA can help disperse the GRC particles more randomly. If the GMA content is too high, the distribution of the GRC particles in the epoxy becomes multimodal (Figure 5). This phenomenon may be due to a GMA concentration that is too high at the interface and that locally interacts with the surrounding GRC particles.

The major toughening mechanisms in these three systems are the same: cavitation of the GRC particles, followed by shear yielding of the matrix around the crack tip. These mechanisms are supported by the observation of GRC particle cavitation and shearing around the crack tip. In the GRC-C modified system, an additional crack-deflection mechanism is also operative (Figure 11), but the toughening effect due to this crack-deflection mechanism is overwhelmed by the scattering of the $K_{Ic}$ data, which may result from the local clustering of the GRC particles in this modified epoxy system.

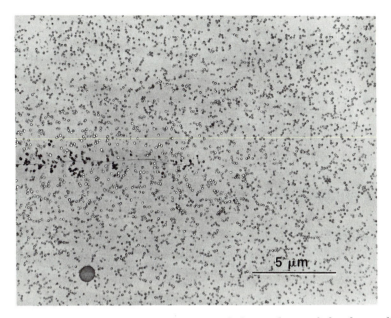

*Figure 7. A TEM micrograph taken around the crack tip of the damaged DN-4PB specimen of the GRC-B-modified epoxy sample. The highly localized plastic deformation of the material around the crack tip is apparent. The rubber particles have cavitated around the crack tip. The cavity size is found to be smaller outward from the crack tip. The crack propagates from left to right. (Reproduced with permission from reference 28. Copyright 1991 Society of Plastics Engineers.)*

Consequently, the GMA content (from 0 to 30% in the shell) plays, if any, a minor role in toughening highly cross-linked epoxies.

The study just described indicates that the GMA content is not critical for the toughening of D.E.R. 332 epoxy resin–DDS. Without the presence of GMA, the interfacial strength between the matrix and the GRC particle is still higher than the cavitational strength (cohesive strength) of the GRC particle. As a result, the interfacial adhesions in these systems are appropriate.

***Variation of AN Content.*** Because the GMA does not critically affect the toughening of highly cross-linked epoxies, it is omitted in the shell in studying the effect of AN content on toughening.

When the AN content varies from 25 wt % in the shell (GRC-C), to 12.5 (GRC-D), and to 0 wt % (GRC-E), clear changes of morphologies are observed. A noticeable decrease of $K_{Ic}$ values from 1.30 ± 0.09 to 1.07 ± 0.05 and 1.05 ± 0.03 MPa · $m^{1/2}$ is also found. The reduction of AN content induces enormous segregation of the GRC particles (Figures 12 and 13), in contrast to the GRC-C modified system shown in Figure 11. The enormous

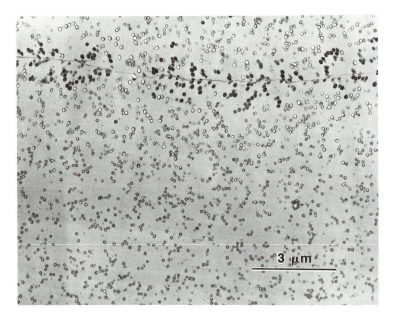

*Figure 8. A TEM micrograph taken at the damaged crack wake of the GRC-B modified epoxy sample. The degree of plastic deformation as well as the size of the rubber cavity is decreased away from the crack wake. The crack propagates from left to right. (Reproduced with permission from reference 28. Copyright 1991 Society of Plastics Engineers.)*

clustering, which forms irregular GRC particle domains and acts as a big rubber particle, prohibits cavitation of individual GRC particles. As a result, shear yielding of the matrix is suppressed by the plane-strain constraint, which should have been relieved by the cavitation of the GRC particles (28). Because the GRC-C modified system possesses a good global dispersion of GRC particles, the toughening mechanism is dominated by rubber-particle cavitation and shear yielding of the matrix (Figure 11). In the GRC-D and GRC-E modified systems, because of the poor dispersion of GRC particles, the major toughening mechanism is found to be crack deflection (Figures 12 and 13, respectively). The toughening effect is greatly reduced for the GRC-D and GRC-E modified systems.

This study suggests that without the presence of GMA, the AN content, which controls the solubility parameter in the shell of the GRC particles (Table I), plays a crucial role in dispersing the GRC particles in the epoxy matrix. Therefore, as demonstrated, the dispersion of GRC particles is extremely important in effective toughening of the epoxy matrix having high cross-link density.

***Variation of Shell Thickness.*** To study the effect of shell thickness in toughening, the shell and core compositions are kept the same. The size of

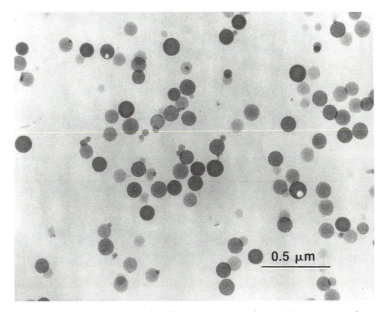

*Figure 9. A TEM micrograph taken at a region about 100 μm away from the crack tip of the GRC-B modified epoxy system. The GRC rubber particles appear spherical. No cavities are found inside the rubber particles. (Reproduced with permission from reference 28. Copyright 1991 Society of Plastics Engineers.)*

the core is also kept constant. As a result, when the shell is thicker, the actual GRC particle size is larger. However, the diameter increase is only ~11.5 nm as the shell–core weight ratio increases from 16:84 to 35:65. The GRC particle diameter differences among GRC-B, GRC-F, and GRC-G are effectively negligible (*see* Table I).

Comparison of the morphologies of GRC-B, GRC-F, and GRC-G modified epoxy systems indicates that the GRC particles are globally well dispersed, whereas the GRC-F and GRC-G modified systems show local clustering of the GRC particles (Figures 14 and 15, respectively). Difference in the dispersion of GRC particles between the GRC-F and GRC-G systems is not noticeable. When the dispersions of the GRC particles are altered from a random to a locally clustered dispersion, the $K_{Ic}$ values change from 1.20 ± 0.04 MPa · m$^{1/2}$ for the GRC-B modified epoxy to 1.33 ± 0.07 MPa · m$^{1/2}$ for the GRC-F modified epoxy and to 1.37 ± 0.06 MPa · m$^{1/2}$ for the GRC-G modified epoxy. The major toughening mechanisms in these systems are still cavitation of the GRC particles, followed by shear yielding of the matrix around the crack tip. For the GRC-F and GRC-G modified systems, an additional crack-deflection mechanism is triggered by the local clustering of the GRC particles (Figure 15). The crack-deflection mechanism, which

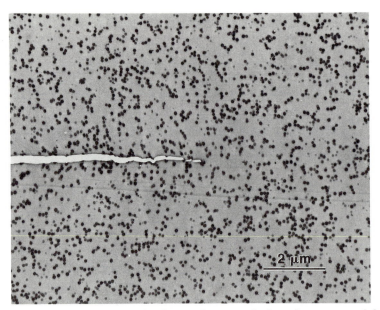

*Figure 10. A TEM micrograph taken at the precracked crack-tip region of the GRC-B modified epoxy system. No rubber-particle cavitation and shearing are observed around the crack tip. The crack propagates from left to right. (Reproduced with permission from reference 28. Copyright 1991 Society of Plastics Engineers.)*

operates in both the GRC-F and GRC-G modified systems, should be the main cause for the increase of the $K_{Ic}$ values over that of the GRC-B system.

## Discussion

The intent of the present work is to account for the fracture-toughness variations of the designed GRC-modified highly cross-linked epoxy systems with the corresponding toughening mechanisms. An approach for toughening brittle polymers is also discussed.

The current study confirms that the D.E.R. 332 epoxy resin–DDS, modified with designed core-shell particles, can undergo shear yielding under the mode-I loading condition. However, it is still not known if an order-of-magnitude increase in toughness can be achieved for such a highly cross-linked epoxy system. Compression test studies of the epoxy systems of high cross-link density, conducted by Kinloch et al. (44) and Glad (45), indicate that the highly cross-linked epoxy systems have limited ability to undergo large plastic deformation compared with the lowly cross-linked epoxy systems. Hence, it is important to understand the ability and the stress state for which the DGEBA epoxy–DDS system can plastically deform to achieve its optimal toughening

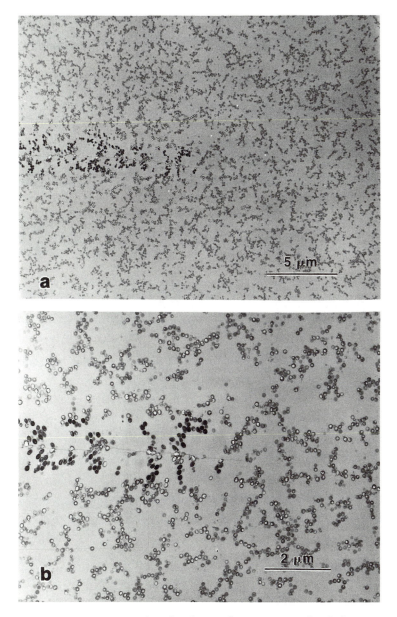

*Figure 11. TEM micrographs of a damaged DN-4PB sample of the GRC-C modified epoxy system taken at a low (a) and a high (b) magnification. The GRC particles are globally well dispersed and locally clustered. The crack propagates from left to right.*

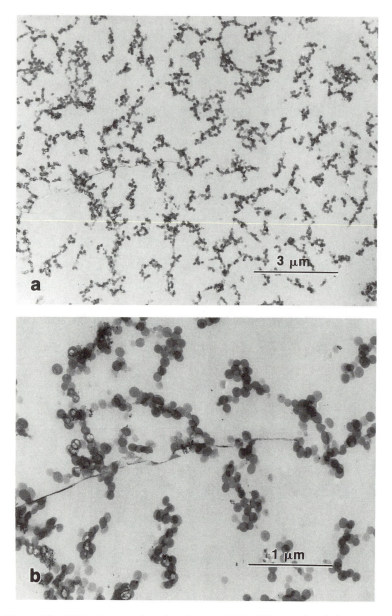

*Figure 12. TEM micrographs of a damaged DN-4PB sample of the GRC-D modified epoxy system taken at a low (a) and a high (b) magnification. Cavitation of GRC particles is highly suppressed by the poor dispersion of the GRC particles. The crack propagates from left to right.*

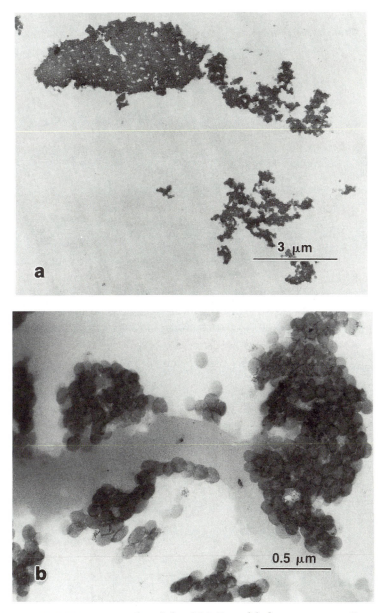

*Figure 13. TEM micrographs of the GRC-E modified epoxy system. Part a:
Without GMA and AN in the shell, the GRC particles severely agglomerate and
form irregular domains. Part b: At the damage zone, only crack deflection is
observed; rubber particle cavitation is suppressed. The crack propagates from
left to right.*

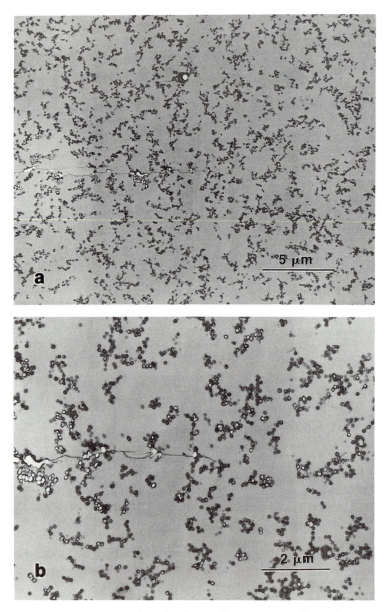

Figure 14. TEM micrographs of a damaged DN-4PB sample of the GRC-F modified epoxy system taken at a low (a) and a high (b) magnification. The GRC particles are globally well dispersed and locally clustered. The crack propagates from left to right.

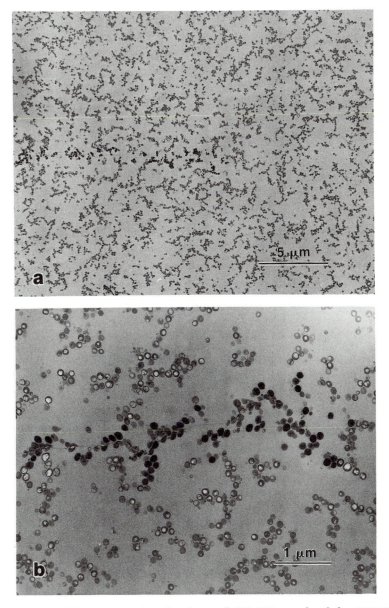

*Figure 15. TEM micrographs of a damaged DN-4PB sample of the GRC-G modified epoxy system taken at a low (a) and a high (b) magnification. The GRC particles are globally well dispersed and locally clustered. The crack propagates from left to right.*

efficiency. Knowledge of micromechanics of polymers and fundamental polymer physics has to be employed.

The improvement of $K_{Ic}$ for the designed GRC-modified systems (except for the GRC-D and GRC-E systems) with respect to the neat resin system can be attributed mainly to the cavitation and shear-yielding mechanisms. Because the rubber cavitation mechanism alone has little effect in absorbing fracture energy (7, 39), the main energy-absorbing mechanism is shear yielding. The observation of the highly elongated rubber particles (an indication of the shear-yielding mechanism) around the crack tip is somewhat surprising. The cavitation of the rubber particles must have considerably altered the highly triaxial crack-tip stress field to a stress state more prone to shear yielding. However, rubber-particle cavitation does not guarantee the formation of the shear-yielded zone. The stress level at which rubber particles cavitate must be sufficiently high to form the shear-yielded zone (7, 24, 39, 43). The sequence of events for rubber particle cavitation and matrix shear yielding in the GRC-modified epoxy system is summarized in Figure 16.

Among the designed GRC-modified systems, the dominant toughening mechanism in the GRC-D and GRC-E systems appears to be crack deflection. An improvement of $\approx 25\%$ in $K_{Ic}$ (or 110% in plane-strain strain

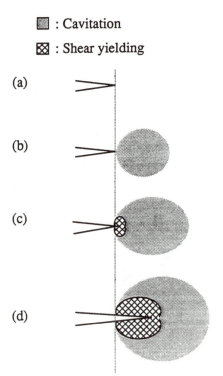

■ : Cavitation

⊠ : Shear yielding

(a)

(b)

(c)

(d)

*Figure 16. A sketched sequence of the toughening events for most of the designed GRC-modified epoxy systems. Key: a, the initial starter crack; b, formation of a cavitation zone in front of the crack tip when the specimen is initially loaded; c, formation of initial shear-yielded plastic zone around the crack tip when the hydrostatic tension is relieved due to the cavitation of the rubber particles; and d, once the build-up of shear-strain energy reaches a critical value, the material undergoes shear yielding, and the crack propagates, leaving a damage zone surrounding the propagating crack before the crack grows unstably. (The size of the plastic zone is not drawn to scale.) (Reproduced with permission from reference 28. Copyright 1991 Society of Plastics Engineers.)*

energy release rate $G_{Ic}$) is observed. For the GRC-A and GRC-B systems, the major toughening mechanisms are GRC particle cavitation and matrix shear yield, which help increase $K_{Ic}$ by $\approx 45\%$ (or by 170% in $G_{Ic}$). For the GRC-C, GRC-F, and GRC-G systems, in addition to the GRC particle cavitation and matrix shear yield, the crack-deflection mechanism also contributes to the toughening of the highly cross-linked epoxy. An increase of $\approx 65\%$ in $K_{Ic}$ (or over 250% increase in $G_{Ic}$) can be achieved. Clearly the GRC particle cavitation–matrix shear-yield mechanisms and the crack-deflection mechanism can produce a synergistic toughening effect. To optimize the mode-I fracture toughness of highly cross-linked epoxies (or any matrices at any given testing condition) that allow the coexistence of the particle cavitation–matrix shear-yield mechanisms and the crack-deflection mechanism, it may be preferable to disperse the toughener phase in such a way that it is globally randomly dispersed and locally clustered, rather than the intuitively preferred uniformly dispersed rubber particles.

Because the rubber-particle composition and size are kept constant, changes in the shell architecture of a core-shell rubber appear to affect only the dispersion of the rubber particles. No detectable changes of interfacial adhesion are found, that is, the interfacial adhesion due to physical means, such as surface free energy and molecular entanglement at the interface, may be sufficient (i.e., the interfacial strength is greater than the cohesive strength of the GRC particles) to prevent debonding at the interface. Consequently, dispersion of the GRC particles appears to be more critical than the degree of interfacial adhesion between the matrix and the GRC particles. The present finding agrees with observations made by Vollenberg (30) and Gent and Park (59). They both experimentally and theoretically demonstrate that small filler particle size requires excessive energy to debond the interface. Consequently, cohesive failure of either the matrix or the toughener phase takes place.

When the GRC particle cavitation and matrix shear-yield mechanisms are operative, the crack-bridging mechanism is unimportant in the designed GRC-modified epoxies. As shown in Figures 6–8, the rubber particles are broken apart as the crack propagates through them. In front of the crack tip (Figure 6), the rubber particles have already been highly stretched, so they have only reduced ability to span the two crack faces and to resist the crack opening when the crack grows. As the crack propagates through these highly stretched and cavitated rubber particles, the particles can no longer fully perform the assumed role for the crack–particle bridging mechanism. These observations are again in agreement with previous findings in the polycarbonate–polybutadiene core-shell rubber (7) and polyamide 6,6-polyphenylene oxide (37) systems, but contradict the conclusions of Kunz-Douglass et al. (11, 47) that show in rubber-toughened epoxy systems that the rubber particle forms a bridge between two opposite crack faces, resulting in the high toughness of the rubber-toughened epoxy systems. Experimentally,

Kunz-Douglass et al. present the existence of a rubber-particle bridging mechanism when the rubber particles are significantly larger than the crack size. When the rubber particles are small compared to the crack size and when the shear-yielding mechanism is operative around the crack tip, the crack-bridging mechanism has not been observed (5–7, 24, 37–39). The toughening model proposed by Evans et al. (13), which states that the crack-bridging and the shear-yielding mechanisms are multiplicative contributors to the overall toughness of a system, is likely to be invalid if the rubber-particle size is small and uniform.

The change of orientation of the deformed GRC particles from perpendicular to the crack face ahead of the crack tip to an angle along the crack wake and the mismatch between the two crack planes when the crack is closed because of unloading demonstrate the existence of a rather complicated stress state when the crack propagates (7, 37, 39, 60–63). To understand the complicated crack evolution phenomena, in situ observations of crack initiation and propagation processes are needed (63).

Some of the GRC rubber particles that are adjacent to the crack face do not show cavitation, mainly because the staining solution tends to penetrate through the crack path faster (i.e., due to the capillary force) than through the matrix. The rubber particles adjacent to the crack face are overstained, and the cavities of these rubber particles are likely to be covered by the stain. This situation has been observed and verified in other systems (28, 29, 39).

The present study only visually correlates the toughening effect with the dispersions of the GRC particles in the modified epoxy matrix. Quantitative image analysis using measurable morphological parameters is essential for design optimization purposes and is the subject of future investigations.

## *Approach for Toughening Brittle Epoxies*

The toughening principles for relatively ductile polymers are reviewed and discussed by Yee and co-workers (1–3, 5, 43) and Kinloch (24). The important mechanisms for toughening highly cross-linked epoxies are summarized by Garg and Mai (40). The present work, which is part of a larger effort aimed at utilizing both the materials science understanding and the fracture mechanics approach to study toughened epoxy systems of high cross-link density, focuses on the investigation of the toughening mechanism(s) in designed rubber-modified high-performance epoxy systems. The following discussion is based on the evidence obtained in the present study for one approach to toughening the D.E.R. 332 epoxy resin–DDS.

The investigation of the GRC-modified epoxy system shows that the highly cross-linked epoxy system can undergo shear yielding around the propagated crack tip, even when the crack experiences the plane-strain mode-I loading condition. The major toughening mechanisms for most of these systems are the result of cavitation of the rubber particles, followed by the formation of the shear-yielded zone around the propagated crack (*see* Figure 16). If this small-scale yielding can be extended further, then potentially an order-of-magnitude increase in fracture toughness can be achieved (*1–3*). Not all the rubbers can produce the same result as that of the GRC rubber particles. The carboxyl-terminated butadiene–acrylonitrile rubber (CTBN) used by Yee and Pearson (*1–3*) and Levita (*64*) and the dispersed acrylic rubber (DAR) (*65, 66*) and acrylonitrile–butadiene copolymer (Proteus 5025) rubbers studied previously (*28*) all show negative results for inducing extended shear yielding of the matrix. Even with the use of GRC particles, if the particle dispersion is poor (e.g., the GRC-D and GRC-E systems), the rubber-particle cavitation and matrix shear-yielding mechanisms are suppressed. Thus, the type of rubber or rubber-particle size and extent of particle dispersion in the matrix will induce an entirely different toughening mechanism(s) in modified epoxies. It is still not clear if an order-of-magnitude increase in fracture toughness can be attained in highly cross-linked epoxies. To clarify this uncertainty, the nature of shear yielding–shear banding in brittle epoxy systems as well as the quantitative estimation of the crack-tip stress field before and after the rubber particles cavitate must be understood.

Shear yielding can occur in epoxy systems of high cross-link density, and shear yielding and crack–particle bridging mechanisms, if they coexist, do not show synergistic toughening in a rubber-modified system of small and uniform particle size (*7, 29*). To test if the synergistic effect due to both shear yielding and crack–particle bridging exists, a bimodal size distribution of rubber particles was attempted by Pearson and Yee (*27*). Pearson and Yee incorporated both the 1-2-μm and 10-20-μm CTBN rubber particles into D.E.R. 331 epoxy resin cured with piperidine. The bimodal rubber particle modification did not further improve the fracture toughness of the system in comparison to that of the unimodal (1-2-μm particle size) distribution system.

If the shear yielding cannot be induced for other reasons, alternative mechanisms, such as crack–particle bridging (*11, 47*), microcracking (*67*), crack-deflection (*52, 53*), crack-pinning (*50, 51*), and crack-bifurcation (*48, 49*) mechanisms should be considered. These mechanisms appear to occur independently in the rubber-modified systems. For the crack–particle bridging mechanism to take place, the toughener phase must be drawable, have a large rubber-particle size (preferably three times the characteristic crack-tip radius), and have strong interfacial adhesion to the matrix. In this way, when the crack propagates inside the material, the large rubber particles begin to span the two crack faces behind the crack tip. As a result, the crack-tip stress intensity factor is reduced, and this change toughens the system.

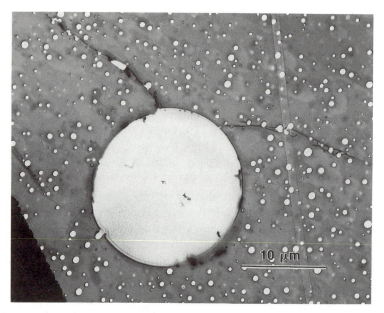

*Figure 17. A TEM micrograph taken at the damaged crack wake of the dispersed acrylic rubber-modified D.E.R. 332 epoxy resin–DDS. The crack is deflected by a void. The crack propagates from left to right. (Reproduced with permission from reference 28. Copyright 1991 Society of Plastics Engineers.)*

For the microcracking-type mechanism to occur, the interfacial adhesion between the matrix and the toughener phase need not be strong. Debonding at the interface, internal cavitation of the toughener phase, and microcracking and crazing of the inclusion phase can effectively shield the crack from growing. These dilatational processes can also result in matrix microcracking (28), depending on the physical nature of the matrix material. Consequently, a toughened system is obtained.

For the crack-deflection and crack-bifurcation mechanisms to occur, the toughener phase needs to generate a sufficiently different stress field in front of the crack tip. When the crack propagates, the crack path will be altered by the stress field perturbed by the crack-tip toughener particles. This alteration causes crack-deflection and crack-bifurcation mechanisms to occur. The rubber-particle size plays an important role in deflecting the crack. Comparison of Figures 7 and 8 with Figures 17 and 18 shows that a larger rubber particle (or a hole) can deflect the crack more effectively than a smaller particle (28). Furthermore, when the smaller particles cluster locally, they can be mechanically treated as a large particle (i.e., the GRC-C, GRC-F, and GRC-G systems) and consequently, the crack-deflection mechanism can be enhanced. Therefore, a larger rubber-particle size or a local clustering of small particles that maintains a good global particle dispersion should be used to promote the crack-deflection mechanism.

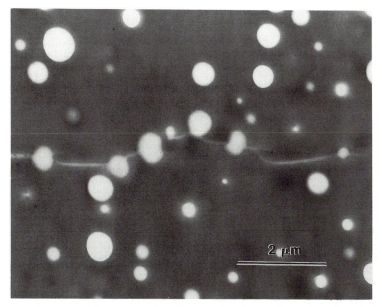

*Figure 18. A TEM micrograph taken at the damaged crack wake of the dispersed acrylic rubber-modified D.E.R. 332 epoxy resin–DDS. The crack is deflected by the DAR rubber particles in front of it. The crack propagates from left to right. (Reproduced with permission from reference 28. Copyright 1991 Society of Plastics Engineers.)*

For the crack-pinning mechanism to occur, the toughener phase must adhere strongly to the matrix. When the crack grows through the particles, the crack front is bowed and requires more energy to propagate the crack, and the system is, therefore, toughened.

## Conclusion

The toughening in the D.E.R. 332 epoxy resin–DDS system was mostly the consequence of rubber-particle cavitation, which relieved the triaxial tension in front of the crack tip, followed by the formation of shear bands. The crack-deflection mechanism was also observed in systems where the GRC particles cluster. The crack-bridging mechanism was not observed because of the small size of the rubber particles and the formation of an extended shear-yielded zone in the matrix.

## Acknowledgment

The authors thank A. F. Yee for his constant valuable discussions and R. S. Porter for permitting the use of work previously published in *Polymer*

*Engineering and Science.* Special thanks are given to R. E. Jones, D. L. Barron, C. C. Garrison, N. A. Orchard, C. E. Allen, C. J. Bott, L. M. Kroposki, T. E. Fisk, and E. P. Woo for their input, experimental assistance, and material supplied for this work.

## References

1. Yee, A. F.; Pearson, R. A. *J. Mater. Sci.* **1986**, *21*, 2462.
2. Pearson, R. A.; Yee, A. F. *J. Mater. Sci.* **1986**, *21*, 2475.
3. Pearson, R. A.; Yee, A. F. *J. Mater. Sci.* **1989**, *24*, 2571.
4. Yee, A. F.; Maxwell, M. A. *Polym. Eng. Sci.* **1981**, *21*, 205.
5. Yee, A. F.; Pearson, R. A.; Sue, H.-J. *7th International Conference on Fracture*; Pergamon Press: New York, 1989; Vol. 4, p 2739.
6. Sue, H.-J.; Pearson, R. A.; Yee, A. F. *Polym. Eng. Sci.* **1991**, *31*, 793.
7. Parker, D. S.; Sue, H.-J.; Huang, J.; Yee, A. F. *Polymer* **1990**, *31*, 2267.
8. Bucknall, C. B. *Toughened Plastics*, Applied Science Publishers Ltd.: London, 1977.
9. Bascom, W.; Ting, R.; Moulton, R. J.; Riew, C. K.; Siebert, A. R. *J. Mater. Sci.* **1981**, *16*, 2657.
10. Sultan, J. N.; McGarry, F. J. *Polym. Eng. Sci.* **1973**, *13*, 19.
11. Kunz-Douglass, S.; Beaumont, P. W. R.; Ashby, M. F. *J. Mater. Sci.* **1980**, *15*, 1109.
12. Wu, S. *Polymer* **1985**, *26*, 1855.
13. Evans, A. G.; Ahmad, Z. B.; Gilbert, D. G.; Beaumont, P. W. R. *Acta Metall.* **1986**, *34*, 79.
14. Kinloch, A. J.; Shaw, S. J.; Tod, D. A.; Hunston, D. L. *Polymer* **1983**, *24*, 1341.
15. Borggreve, R. J. M.; Gaymans, R. J.; Schuijer, J.; Ingen Housz, J. F. *Polymer* **1987**, *28*, 1489.
16. Argon, A. S.; Cohen, R. E.; Gebizlioglu, O. S.; Schwier, C. In *Advances in Polymer Science: Crazing in Polymers*; Kausch, H. H., Ed.; Springer: Berlin, Germany, 1983; Vol. 52/53, p 275.
17. Sjoerdsma, S. D. *Polym. Commun.* **1989**, *30*, 107.
18. Borggreve, R. J. M.; Gaymans, R. J. *7th International Conference on Deformation, Yield and Fracture of Polymers*; The Plastics and Rubber Institute: London, 1988; p 34.
19. Breuer, H.; Haaf, F.; Stabenow, J. *J. Macromol. Sci.-Phys.* **1977**, *B14(3)*, 387.
20. Boyce, M. E.; Argon, A. S.; Parks, D. M. *Polymer* **1987**, *28*, 1681.
21. Yee, A. F. *J. Mater. Sci.* **1977**, *12*, 757.
22. Low, I. M.; Mai, Y. W. *Composites Sci. Tech.* **1988**, *31*, 191.
23. Garg, A. C.; Mai, Y. W. *Composites Sci. Tech.* **1988**, *31*, 225.
24. Kinloch, A. J. In *Rubber-Toughened Plastics*; Riew, C. K., Ed.; Advances in Chemistry 222; American Chemical Society: Washington, DC, 1989; p 67.
25. Guild, F. J.; Young, R. J. *J. Mater. Sci.* **1989**, *24*, 2454.
26. Hobbs, S. Y.; Dekkers, M. E. J. *J. Mater. Sci.* **1989**, *24*, 1316.
27. Pearson, R. A.; Yee, A. F. *J. Mater. Sci.* **1991**, *26*, 3828.
28. Sue, H.-J. *Polym. Eng. Sci.* **1991**, *31*, 275.
29. Yang, P. C.; Woo, E. P.; Sue, H.-J.; Bishop, M. T.; Pickelman, D. M. *Polym. Mater. Sci. Eng.* **1990**, *63*, 315.
30. Vollenberg, P. H. Th. Ph.D. Thesis, Eindhoven University of Technology, 1987.
31. Angola, J. C.; Fujita, T.; Sakai, T.; Inoue, T. *J. Polym. Sci. B: Polym. Phys.* **1988**, *26*, 807.
32. Xanthos, M. *Polym. Eng. Sci.* **1988**, *28*, 1392.

33. Raghava, R. S. *J. Polym. Sci. B: Polym. Phys.* **1987**, *25*, 1017.
34. Wu, S. *Polymer Interface and Adhesion*; Marcel Dekker: New York, 1982.
35. Henton, D. E.; Pickelman, D. M.; Arends, C. B.; Meyer, V. E. U.S. Patent 4,778,851, 1988.
36. Sue, H.-J.; Pearson, R. A.; Parker, D. S.; Huang, J.; Yee, A. F. *Polym. Prepr.* (*Am. Chem. Soc. Div. Polym. Chem.*) **1988**, *29*, 147.
37. Sue, H.-J.; Yee, A. F. *J. Mater. Sci.* **1989**, *24*, 1447.
38. Sue, H.-J. Ph.D. Thesis, The University of Michigan, Ann Arbor, MI, 1988.
39. Sue, H.-J.; Garcia-Meitin, E. I.; Burton, B. L.; Garrison, C. C. *J. Polym. Sci. B: Polym. Phys.* **1991**, *29*, 1623.
40. Garg, A. C.; Mai, Y. W. *Composites Sci. Tech.* **1988**, *31*, 179.
41. Fu, Z.; Sun, Y. *Polym. Prepr.* (*Am. Chem. Soc. Div. Polym. Chem.*) **1988**, *29*, 177.
42. Knott, J. F. *Fundamentals of Fracture Mechanics*; Butterworth: London, 1976.
43. Yee, A. F. *Toughened Composites*; Johnston, N., Ed.; American Society for Testing and Materials: Philadelphia, PA, 1986; ASTM STP 937, p 377.
44. Kinloch, A. J.; Finch, C. A.; Hashemi, S. *Polym. Commun.* **1987**, *28*, 322.
45. Glad, M. D. Ph.D. Thesis, Cornell University, Ithaca, New York, 1986.
46. Bowden, P. B. In *The Physics of Glassy Polymers*; Haward, R., Ed.; Applied Science Publishers: London, 1973; Chap. 5.
47. Kunz, S. C. Ph.D. Thesis, Churchill College, Cambridge University, 1978.
48. Clark, A. B. J.; Irwin, G. R. *Exp. Mech.* **1966**, *6*, 321.
49. Ramulu, M.; Kobayashi, A. S. *Exp. Mech.* **1983**, *23*, 1.
50. Lange, F. F.; Radford, K. C. *J. Mater. Sci.* **1971**, *6*, 1199.
51. Lange, F. F. In *Composite Materials, Vol. 5: Fracture and Fatigue*; Broutman, L. J., Ed.; Academic: Orlando, FL, 1974; p 2.
52. Faber, K. T.; Evans, A. G. *Acta Metall.* **1983**, *31*, 565.
53. Faber, K. T.; Evans, A. G. *Acta Metall.* **1983**, *31*, 577.
54. Chan, L. C.; Gillham, J. K.; Kinloch, A. J.; Shaw, S. J. In *Rubber Modified Thermoset Resins*; Riew, C. K.; Gillham, J. K., Eds.; Advances in Chemistry 208; American Chemical Society: Washington, DC, 1984; p 261.
55. Sue, H.-J.; Yee, A. F. *J. Mater. Sci.* **1991**, *26*, 3449.
56. Holik, A. S.; Kambour, R. P.; Hobbs, S. Y.; Fink, D. G. *Microstruct. Sci.* **1979**, *7*, 357.
57. Towers, O. L. *Stress Intensity Factor, Compliance, and Elastic η Factors for Six Geometries*; The Welding Institute: Cambridge, England, 1981.
58. *ASTM Standard, E 399–81*; American Society for Testing and Materials: Philadelphia, PA, 1981.
59. Gent, A. N.; Park, B. *J. Mater. Sci.* **1984**, *19*, 1947.
60. Budiansky B.; Rice, J. R. *J. Appl. Mech.* **1973**, *40*.
61. Chudnovsky, A. Private communication, February 1990.
62. Chudnovsky, A. *Crack Layer Theory*; National Aeronautics and Space Administration: Washington, DC, 1984; Report No. CR–174634.
63. Yee, A. F. Private communication, April 1990.
64. Levita, G. In *Rubber-Toughened Plastics*; Riew, C. K., Ed.; Advances in Chemistry 222; American Chemical Society: Washington, DC, 1989; p 93.
65. Hoffman, D. K.; Arends, C. B. U.S. Patent 4,708,996, 1987.
66. Hoffman, D. K.; Arends, C. B. U.S. Patent 4,789,712, 1988.
67. Evans, A. G.; Faber, K. T. *J. Am. Ceram. Soc.* **1984**, *67*, 255.

RECEIVED for review March 6, 1991. ACCEPTED revised manuscript January 8, 1992.

# 11

# Toughening of Epoxy Resin Networks with Functionalized Engineering Thermoplastics

Jeffrey C. Hedrick[1], Niranjan M. Patel[1], and James E. McGrath*

**Department of Chemistry and National Science Foundation and Technology Center: High Performance Polymeric Adhesives and Composites, Virginia Polytechnic Institute and State University, Blacksburg, VA 24061**

*Epoxy resin thermosets are traditionally toughened by the incorporation of an elastomeric component that phase-separates during cure to form a multiphase network. The toughness enhancement in these systems is considerable. However, in most instances, toughness enhancement is achieved at the expense of stiffness and high-temperature performance. An alternative approach to increasing the toughness and impact strength involves the use of functionalized engineering thermoplastics as toughening agents. The improvement in toughness is accomplished in this case without significant sacrifice of properties at elevated temperatures. In this chapter, a brief review of the state of the art of thermoplastic-modified epoxy resin networks is presented. Emphasis is placed on the types of modifiers, morphological character, and the various mechanisms proposed to be responsible for the toughening behavior.*

$E$POXY RESIN NETWORKS ARE CURRENTLY USED as coatings, structural adhesives, and advanced composite matrices in many applications involving both the aerospace and electronics industries. In addition to their outstanding adhesive properties, these highly cross-linked networks possess excellent

---

[1] Current address: IBM Research Division, Thomas J. Watson Research Center, Yorktown Heights, NY 10598.
* Corresponding author.

0065–2393/93/0233–0293$06.00/0
© 1993 American Chemical Society

thermal and dimensional stability as well as high modulus and strength (1, 2). The widespread use of epoxies, however, is limited in many high-performance applications because of their inherent brittleness. Several methods have been proposed to improve on this attribute of epoxy networks. The most common method involves the addition to the system of a second polymeric component that phase-separates upon curing (3, 4). Traditional modifiers include functionalized rubbers such as the widely known carboxyl-terminated butadiene–acrylonitrile copolymers (CTBNs). A two-phase morphology consisting of relatively small (0.1–1μm) rubbery particles dispersed in and bonded to an epoxy resin network is produced by incorporating CTBNs into epoxy resins. The toughness of the modified networks is dependent on the properties of the original epoxy, the particle size, particle volume fraction, interfacial bonding, and the properties of the elastomeric component (5).

A major limitation to toughening epoxy resins with elastomers such as CTBNs is that increased toughness can be achieved only at the expense of high-temperature performance. Because of the low glass-transition temperature $(T_g)$ of the rubbery phase, rubber modification often lowers both the use temperature and the thermoset modulus. With the growing demands of the aerospace industry for materials that display high thermal stability as well as toughness, alternative methods of toughening epoxy resins based on the incorporation of functionalized thermoplastics have emerged. Thermoplastic modifiers are tough, ductile engineering polymers possessing high $T_g$s. Network systems based on this technology are toughened without negatively affecting their high-temperature performance. In this chapter, the aim is to provide a general overview of this relatively new class of material systems with a focus on the variations in the modifier chemistry, aspects of phase separation and the resulting morphology, and some of the mechanisms that have been proposed to explain the toughening behavior.

## Thermoplastic Modifiers

Initial reports in the literature concerning the use of thermoplastics to toughen epoxy resins appeared in the early 1980s. Raghava (6) and Bucknall and Partridge (7) were the first to publish the results of their investigations in which a commercial poly(ether sulfone) (Vitrex 100P, Chart 1), was employed as a toughening agent in epoxy resins. Although the existence of a two-phase structure was evident from both dynamic mechanical analysis (DMA) and scanning electron microscopy (SEM), Bucknall and Partridge (7) observed no toughness improvement in systems in which both chemistry and stoichiometry of the resin and curing agent were varied. The lack of toughening agreed with the findings of Raghava (6), who, in a later publication (8), attributed it to poor adhesion between the phase-separated components.

The importance of chemically prereacting a thermoplastic polymer into the epoxy resin to control the compatibility and interfacial adhesion of the phase-separated network was realized by McGrath and co-workers (9–11). In their approach, a phenolic hydroxyl-terminated bisphenol A polysulfone (PSF; Chart 1) was incorporated into an epoxy resin system based on diglycidyl ether of bisphenol A and diaminodiphenyl sulfone (DDS) as depicted in Scheme I. The PSF was prereacted with a large excess of epoxy resin using a quaternary ammonium catalyst to functionalize the PSF oligomers with epoxide groups prior to the addition of the DDS curing agent. The concentration and molecular weight of the PSF modifier were varied to investigate their effect on the properties of the resulting networks. Mechanical property results demonstrated that the fracture toughness ($K_{Ic}$) of the cured networks increased substantially (from 0.6 to 1.7 $N/m^{3/2}$) when both the concentration and the molecular weight of the PSF were increased. Moreover, the increase in fracture toughness was accomplished without significant reduction in the modulus, thermal stability, or solvent resistance (9–11).

Chu et al. (12) reported in patent literature similar results utilizing amino-functionalized thermoplastics. Specifically, they demonstrated that amine-terminated aromatic polyethers, polysulfones, and poly(ether sulfone) of molecular weights in the range of 2000–10,000 g/mol and $T_g$s between 125 and 250 °C produced multiphase morphologies and significant enhancements in the fracture toughness over the neat resin. In addition, Chu et al. (12) reported that the modified epoxy compositions resulted in stiff, tough thermoset composites with increased compression-after-impact (CAI) strengths. Other thermoplastics such as amino-functional poly(arylene ether ketone)s (PEKs) have been tested as toughening agents for epoxy resins. The PEK oligomers, however, appear to be less compatible with epoxy resins; inclusions of greater than 10–15 wt % (depending on the molecular weight) resulted in macrophase separation (13, 14).

Recently, Bucknall and Gilbert (15) demonstrated that simple physical blends of poly(ether imides) (PEI; Chart 1) also can be used for effective toughening of epoxy resin networks. They showed that fracture toughness increased linearly with PEI content up to 25 wt % with only modest reductions in Young's modulus. Whether the linear correlation between fracture toughness and modifier concentration is valid in the entire composition range has been of some interest. For instance, Recker et al. (16, 17) proposed that an optimum morphology exists at intermediate compositions at which a maximum in fracture toughness is achieved. The work of Recker et al. involved correlating the fracture toughness of modified epoxy networks to damage tolerance in composite laminates. Fracture toughness as measured by fracture energy ($G_{Ic}$) and the strain energy release rate under shear ($G_{IIc}$) was studied as a function of modifier molecular weight and concentration. The $G_{Ic}$ was directly proportional to the number-average molecular weight up to about 10,000 g/mol, beyond which there was little or no change. Additionally, the toughness appeared to have a sharp maximum at an inter-

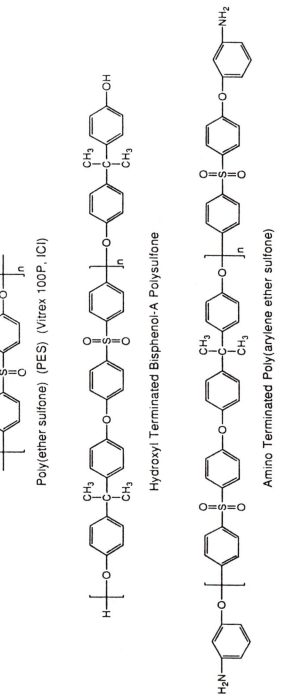

Poly(ether sulfone) (PES) (Vitrex 100P, ICI)

Hydroxyl Terminated Bisphenol-A Polysulfone

Amino Terminated Poly(arylene ether sulfone)

Amino Terminated Poly(arylene ether ketone)

Poly(ether imide) (PEI) (Ultem 100, General Electric)

*Chart 1. Engineering thermoplastics typically incorporated into epoxy resin networks to enhance toughness.*

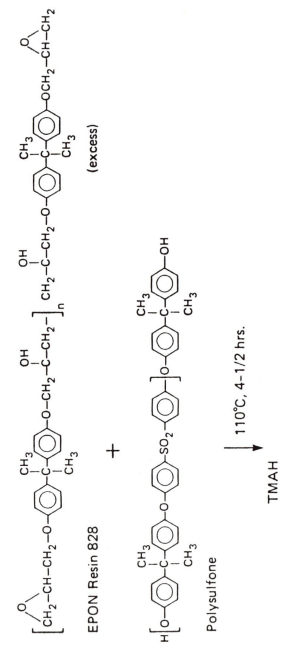

EPON Resin 828

(excess)

+

Polysulfone

TMAH

110°C, 4-1/2 hrs.

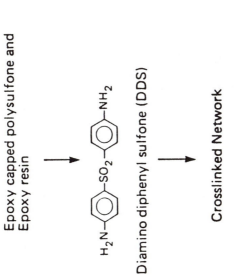

Epoxy capped polysulfone and
Epoxy resin

Diamino diphenyl sulfone (DDS)

Crosslinked Network

*Scheme I. Preparation of thermoplastic-modified epoxy networks.*

mediate modifier concentration (ca. 25 wt %). Similarly, Almen and co-workers (18) reported that fracture toughness achieves a maximum at intermediate compositions in thermoplastic-modified epoxies. Almen et al. correlated their results with morphological studies and proposed that a co-continuous morphology was responsible for the maximum and that this morphology better translated the increased toughness into composite structures. The observed co-continuous morphology was attributed to spinodal decomposition, the occurrence of which can be controlled by varying the backbone structure of the thermoplastic while holding other variables, such as cure schedule, molecular weight, and concentration of the modifier, constant (18). The foregoing studies have generated a great deal of interest in exploring the means to systematically control phase-separation behavior and, hence, the ensuing morphology in thermoplastic-toughened thermosets.

## Phase-Separation Behavior

Phase separation during thermoset cure in the presence of a modifier is controlled by the competing effects of thermodynamics, kinetics, and polymerization rate. Thermodynamic changes provide the primary impetus for the phase separation to occur; however, kinetic factors govern the extent to which the system follows the equilibrium path as the cure progresses. The rate of polymerization determines whether thermodynamics or kinetics dominates the final outcome of phase separation. Kinetics becomes the governing mechanism in systems with fast cure rates. Conversely, the kinetic restrictions become insignificant when cure rates are slower and phase separation is dictated by thermodynamics. The phenomenon of gelation introduces a further consideration: At gelation, the viscosity of the system approaches infinity, and the mobility of the phase-separating species diminishes drastically. As a result, further phase separation is halted, and the morphology is essentially fixed (19–21).

Thermoplastic modifiers, analogous to their elastomeric counterparts, initially form a homogeneous mixture with the uncured epoxy resin before undergoing an in situ phase-separation process during network formation. The compatibility of the modifier with the epoxy resin is an important variant because the modifier must be soluble in the initial stages of the reaction and then phase-separate as curing takes place. Compatibility can be controlled by varying the structure, molecular weight, or end-group functionality of the modifier.

In the functionalized PSF-modified epoxy resin networks (Scheme I) discussed in the previous section, a reproducible, well-defined two-phase morphology was achieved depending on composition. At low weight percents (10–15%) under typical curing conditions for epoxy resins, the PSF-modified systems possessed discrete spheres (ca. 0.1–0.6 μm) of PSF evenly dispersed

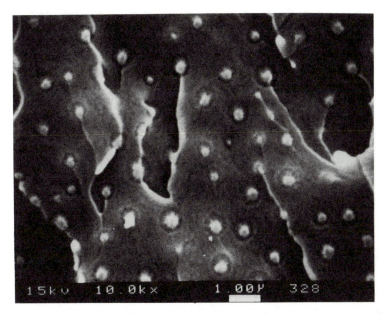

*Figure 1. SEM micrograph of an epoxy resin network containing 15 wt % of a 16,000-g/mol amino-functionalized polysulfone.*

in an epoxy resin matrix as shown in Figure 1. However, at higher thermoplastic content (30–40%), phase inversion resulted, and the morphology consisted of discrete spheres (ca. 4–6 μm) of epoxy resin embedded in a PSF matrix as evident in Figure 2 (9–11).

Recently, Inoue and co-workers (22, 23) modeled the phase-separation behavior of modified epoxy resins by utilizing curing agents of different reactivity and by varying cure temperature. By adopting these techniques, control of the morphology was achieved by accelerating the approach to the gel point and thus arresting morphological development before complete phase separation. Similar control over the morphology has been achieved by McGrath et al. (24–26) with amino-functional PSF-modified epoxy resins using microwave radiation to accelerate the network cure.

## Toughening Mechanisms

A great deal of controversy exists on the nature of the toughening mechanisms in modified epoxy resin networks (27–31). Much of the dispute is centered around whether the modifier or the matrix absorbs most of the fracture energy. The issue is extremely important because once the mechanisms responsible for increased toughness are clearly identified, then the design parameters can be modified to optimize properties and performance.

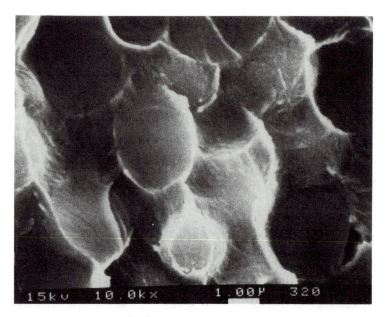

*Figure 2. SEM micrograph of an epoxy resin network containing 30 wt % of a
16,000-g/mol amino-functionalized polysulfone.*

Most of the investigations on the toughening mechanisms in modified
epoxy resin thermosets have been performed on rubber-toughened systems.
Kunz-Douglass et al. (27) proposed a crack-bridging model, suggesting that
the toughness increase is related mainly to the volume fraction and the tear
energy of the phase-separated rubber particles. Alternatively, Kinloch et al.
(29) proposed that the mechanism of rubber toughening involves void
formation in either the particles or at the particle–matrix interface that
initiates plastic flow in the matrix in the vicinity of the crack tip. Kinloch et al.
suggested that the main source of energy dissipation involved plastic shear
yielding of the epoxy resin matrix, thereby increasing the size of the plastic
zone preceding the crack tip and inducing the blunting of the propagating
crack (27).

Kim and Brown (31) utilized SEM, DMA and optical microscopy to
investigate the morphology and toughening mechanisms in a commercially
available thermoplastic-modified epoxy resin system. At low weight percent
modifier incorporations (10%), Kim and Brown proposed that the toughening
mechanism involved localized yielding in the epoxy matrix initiated when the
discrete modifier particles deform. At high weight percent incorporations
(30%), above the phase-inversion composition, the toughening mechanism
was attributed to the ductile yielding of the oligomeric modifier phase (i.e.,
the continuous phase).

## Summary

The need for high-performance materials in structural applications has accelerated the development of advanced epoxy resin material systems. The combined requirements for high modulus and thermal stability along with adequate toughness and durability have led to the development of thermoplastic-toughened epoxy resin networks. Indeed, several commercial products based on this technology exist. With the growing demands of the aerospace industry, thermoplastic toughening technology is being extended to other thermosetting systems. Ongoing investigations into the toughening mechanisms, control of phase separation, and the formulation of functionalized modifiers are necessary for future material development.

## Acknowledgment

The authors thank the National Science Foundation Science and Technology Center: High Performance Polymeric Adhesives and Composites for funding under contract DMR–88–09714.

## References

1. Potter, W. G. *Epoxy Resins*; Springer-Verlag: New York, 1970.
2. *Epoxy Resin Chemistry and Technology*; 2nd Ed.; May, C. A., Ed.; Marcel Dekker: New York, 1988.
3. Bucknall, C. B. *Toughened Plastics*; Wiley: New York, 1977.
4. Kinloch, A. J.; Young, R. J. *Fracture Behavior of Polymers*; Applied Science Publishers: London, 1983.
5. Riew, C. K.; Rowe, E. H.; Siebert, A. R. In *Toughness and Brittleness of Plastics*; Deanin, R. D.; Crugnola, A. M., Eds.; Advances in Chemistry 154; American Chemical Society: Washington, DC, 1976; pp 326–343.
6. Raghava, R. S. *Soc. Adv. Mater. Proc. Eng. 28th Nat. Symp.*, 1983; pp 367–375.
7. Bucknall, C. B.; Partridge, I. K. *Polymer* **1983**, *24*, 639.
8. Raghava, R. S. *J. Polym. Sci. Polym. Phys. Ed.* **1987**, *25*, 1017.
9. Hedrick, J. L.; Yilgor, I.; Wilkes, G. L.; McGrath, J. E. *Polym. Bull.* **1985**, *13*, 201.
10. Hedrick, J. L.; Yilgor, I.; Hedrick, J. C.; Wilkes, G. L.; McGrath, J. E. *Soc. Adv. Mater. Proc. Eng. 30th Nat. Symp.*, 1985; pp 947–958.
11. Hedrick, J. L.; Yilgor, I.; Jurek, M.; Hedrick, J. C.; Wilkes, G. L.; McGrath, J. E. *Polymer* **1991**, *32(11)*, 2020.
12. Chu, S. G.; Jabloner, H.; Swetlin, B. J. European Patent 0193082, 1986; U.S. Patent 4,822,832, 1989.
13. Cecere, J. A.; Hedrick, J. L.; McGrath, J. E. *Soc. Adv. Proc. Eng. 31st Nat. Symp.*, 1986; pp 580–588.
14. Cecere, J. A.; McGrath, J. E. *Polym. Prepr.* **1986**, *27*, 299.
15. Bucknall, C. B.; Gilbert, A. H. *Polymer* **1989**, *30*, 213.
16. Recker, H. G.; Allspach, T.; Altstadt, V.; Folda, T.; Heckmann, W.; Itteman, P.; Tesch, H.; Weber, T. *SAMPE Quart.* **1989**, *21(1)*, 46.

17. Recker, H. G.; Altstadt, V.; Eberle, W.; Folda, T.; Gerth, D.; Heckmann, W.; Itteman, P.; Tesch, H.; Weber, T. *Soc. Adv. Mater. Proc. Eng. 21st Internat. Tech. Conf.*, 1989; pp 283–293.

18. Almen, G. R.; Maskell, R. K.; Malhotra, V.; Sefton, M. S.; McGrail, P. T.; Wilkinson, S. P. *Soc. Adv. Mater. Proc. Eng. 33rd Internat. Symp.*, 1988; pp 979–989.

19. Williams, R. J. J.; Borrajo, J.; Adabbo, H. E.; Rojas, A. J. In *Rubber Modified Thermoset Resins*; Riew, C. K.; Gillham, J. K., Eds.; Advances in Chemistry 208; American Chemical Society: Washington, DC, 1984; pp 195–213.

20. Verchere, D.; Sautereau, H.; Pascault, J. P.; Moschiar, S. M.; Riccardi, C. C.; Williams, R. J. J. *J. Appl. Polym. Sci.* **1990**, *41*, 467, 701; **1991**, *42*, 717.

21. Manzione, L. T.; Gillham, J. K.; McPherson, C. A. *J. Appl. Polym. Sci.* **1981**, *26*, 889, 907.

22. Yamanaka, K.; Inoue, T. *J. Mater. Sci.* **1990**, *25*, 241.

23. Yamanaka, K.; Takagi, Y.; Inoue, T. *Polymer* **1989**, *30*, 662, 1839.

24. Hedrick, J. C.; Lewis, D. A.; Lyle, G. D.; Ward, T. C.; McGrath, J. E. *Proceedings of the American Society of Composites, 4th Technical Conference*; Technomic Publication: Lancaster, PA, 1989; pp 167–176.

25. Hedrick, J. C.; Lewis, D. A.; Ward, T. C.; McGrath, J. E. *Polym. Prep.* **1988**, *29(1)*, 363.

26. Hedrick, J. C.; Lewis, D. A.; Ward, T. C.; McGrath, J. E. *Proc. Mater. Res. Soc.* **1992**, *189*, 421.

27. Kunz-Douglass, S.; Beaumont, P. W. R.; Ashby, M. F. *J. Mater. Sci.* **1980**, *15*, 1109; **1981**, *16*, 3141.

28. Bascom, W. D.; Cottington, R. L.; Jones, R. L.; Peyser, P. *J. J. Appl. Polym. Sci.* **1975**, *19*, 2545.

29. Kinloch, A. J.; Shaw, S. J.; Tod, D. A.; Hunston, D. L. *Polymer* **1983**, *24*, 1341, 1355.

30. Yee, A. F.; Pearson, R. A. *J. Mater. Sci.* **1986**, *21*, 2462, 2475.

31. Kim, S. C.; Brown, H. R. *J. Mater. Sci.* **1987**, *22*, 2589.

RECEIVED for review March 6, 1991. ACCEPTED revised manuscript November 13, 1991.

# Epoxy–Rubber Interactions

F. J. McGarry and R. B. Rosner

Department of Materials Science and Engineering, Massachusetts Institute of Technology, Cambridge, MA 02139

*Films containing amine-terminated butadiene–acrylonitrile (ATBN) rubber and diglycidal ether of bisphenol A (DGEBA) epoxy, cross-linked with amine curing agent, exhibit tensile extensibility over the composition range of 50–600 parts by weight rubber to 100 parts by weight epoxy. This tensile extensibility suggests the presence of ductile behavior in the second-phase particles of ATBN rubber-toughened DGEBA epoxy systems, even if the particles contain substantial amounts of epoxy. Such cured films also are capable of absorbing large additional amounts of liquid epoxy that contains the cure agent. When the epoxy is cured in situ, the film tensile behavior is consistent with the overall proportions of rubber and epoxy present. The solubility behavior also suggests that the glassy epoxy matrix immediately surrounding a precipated particle contains rubber in solid solution and thereby can plastically yield under shear–stress action. As observations confirm, such flow would be heat recoverable.*

Cross-linked epoxy resins offer excellent properties at moderate cost, and because of this they are widely used in coatings, adhesives, and fiber-reinforced composites. The same cross-linking that produces strength, stiffness, and chemical resistance often results in brittleness. However, several decades ago it was found that the addition of certain low-molecular-weight elastomers could confer additional cracking resistance and fracture toughness (1). These elastomers are carboxy- or amine-terminated copolymers of butadiene and acrylonitrile (CTBN and ATBN, respectively) that are liquid at room temperature and soluble in most liquid epoxies. The reactive end groups form chemical bonds with the epoxy, and as the system cures, the rubber separates out to form small particles in the epoxy matrix. The

0065–2393/93/0233–0305$06.00/0

two-phase morphology results in improved fracture toughness without reducing the other attractive properties of the glassy matrix. (If the rubber remains in solid solution, reductions in modulus, strength, hardness, and softening temperature take place.) Another route to the same end product involves prereacting the rubber and epoxy with certain catalysts such that better control over the (preformed) particle size can be realized (2). In both instances, it is necessary to achieve the discrete second phase. Furthermore, it is believed that chemical bonding between the rubber particles and the surrounding epoxy matrix is essential to the toughening action.

Progress has been made in understanding the toughening mechanism and how it relates to the morphology of rubber-modified epoxies. The stress-whitening observed in failed samples suggested the occurrence of plastic flow, possibly in the form of crazing (3), but further studies showed that the areas around the rubber particles undergo shear-yielding and cavitation at the fracture surface (4, 5). Other work demonstrated that crazing is unlikely because of the high strand density present in the epoxy (6). A theory that explains the shearing and cavitation as the result of a two-step mechanism has been presented (7, 8). Initially, the dilatational stresses that develop near the crack tip cause debonding at the particle–matrix interface and microcavitation around each rubber particle. The failure of the particles induces further shear-yielding in the adjacent matrix. The theory emphasizes that the particles also keep the shear forces localized, thereby delaying catastrophic failure, but it is not clear why the adjacent matrix epoxy seems especially prone to such large-scale plastic flow. Further, the flow is recoverable if the material is heated above its glass-transition temperature.

Most research in the area has concentrated on the relationships between the morphology and the physical properties of rubber-toughened thermosets. Comparatively little has been done to investigate the properties of the rubber particles themselves, although the rubber particles are the critical part of the toughening mechanism. For example, estimates of the composition of the particles range from a high rubber content of 80% (9) to less than 30% by weight (10). A curious observation is also found in studies of the rubber–matrix interface: nuclear magnetic resonance relaxation data indicate that the interface is very sharp (11), yet under moderately different curing conditions the rubber may never precipitate at all (12).

To better understand the nature of the rubbery occlusions, a number of elastomer films were cast from mixtures of ATBN and catalyzed epoxy resin. The homogenous single-phase films were used as model materials to determine the relationship between the composition and the mechanical properties of the rubber particles. The solubility of the rubber in the epoxy was also investigated by observing the swelling behavior of the films in liquid epoxy. These results may be useful in better understanding the interfacial region in such rubber-modified thermosets.

## Experimental Details

The epoxy resin used in all the experiments was a diglycidal ether of bisphenol A (DGEBA; Epon 828; Shell Chemical Company) that had an average molecular weight of 380 g/mol. The rubber modifier was an amine-terminated butadiene acrylonitrile, copolymer (ATBN) that contained 17% bound acrylonitrile, and had a number average molecular weight of 3400 g/mol. ATBN is sold under the trade name of Hycar (BFGoodrich Chemical Company). The curing agent was tris(dimethyl aminomethyl) phenol tri(2-ethyl hexoate), an amine salt catalyst (Ancamine K61B; Pacific Anchor Chemical Company). Film formulations are given in Table I. When possible, the amount of curing agent was adjusted to ensure that the ratio of amine to epoxy groups was maintained at unity.

All three components were dissolved in toluene to make a 30% solids solution, which was cast on Teflon-coated plates. The films stood for 12 h under ambient conditions to allow the solvent to evaporate, were dried under room temperature vacuum for 1 h, and cured in a vacuum oven (0.5 torr) for 1 h at 60 °C and 2 h at 120 °C. The thicknesses of the final films ranged from 12 to 20 μm, but the thickness of any individual did not vary by more than 0.5 μm. Studies of the infrared spectra of these formulations taken at various points in the cure cycle showed that the samples were fully reacted at the end of the cure. In all cases the epoxy absorption at 863 cm$^{-1}$, measured with a Fourier transform infrared spectrometer (Cygnus model 100), completely disappeared. Furthermore, no changes in the spectra were observed after curing for longer times or at higher temperatures. (Most of the films were made simultaneously to minimize any effects that could arise from inconsistencies in sample preparation.)

At room temperature, the films were die cut into dog-bone-shaped specimens to measure their tensile strengths. Room temperature stress–strain curves to failure were recorded on a tensile tester (Instron model 1122) at a crosshead speed of 5 mm/min. Absorption experiments were performed by immersing samples in a liquid and weighing them at various times until they reached saturation. Several swelling media were used: (1) a 50/50 mixture of acetone and methyl ethyl ketone (MEK); (2) 100% toluene; and (3) solutions of the 50/50 acetone–MEK mixture that contained varying amounts of

**Table I. Film Formulations**

| Component | Formulae Based on 100 pbw Epoxy | | | | | | | | |
|---|---|---|---|---|---|---|---|---|---|
| Epoxy | 100 | 100 | 100 | 100 | 100 | 100 | 100 | 100 | 100 |
| ATBN (×16) | 0 | 50 | 100 | 200 | 300 | 400 | 475 | 550 | 600 |
| Curing agent | 10 | 9.0 | 7.9 | 5.8 | 3.7 | 1.6 | 0 | 0 | 0 |
| Amine–Epoxy ratio (A/E) | 1.0 | 1.0 | 1.0 | 1.0 | 1.0 | 1.0 | 1.0 | 1.16 | 1.26 |

catalyzed epoxy resin in dissolution (Table II). This last series of solutions, (3), was used to impregnate the films with increasing amounts of epoxy and curing agent to observe the degree of absorption that could be expected from the pure catalyzed resin and what effect this could have on the physical properties of the films. Dog-bone specimens that had been infused with liquid resin and catalyst were cured a second time following the same procedure described earlier and also were tested on the tensile tester.

## Results

The stress–strain curves for the initial cast films containing between 50 and 600 parts by weight (pbw) of the ATBN modifier are shown in Figure 1. The behavior ranges from a leathery plastic to a soft rubber, depending on

**Table II. Composition of Swelling Media (By Weight Percent)**

| Solvent 1 | Solvent 2 | Solute |
|-----------|-----------|--------|
| Toluene (100%) | — | — |
| Acetone (50%) | MEK (50%) | — |
| Acetone (45%) | MEK (45%) | Catalyzed resin[a] (10%) |
| Acetone (35%) | MEK (35%) | Catalyzed resin[a] (30%) |
| Acetone (20%) | MEK (20%) | Catalyzed resin[a] (60%) |

[a] Catalyzed resin = 91% DGEBA liquid epoxy (Epon 828), Ancamine K61 B.

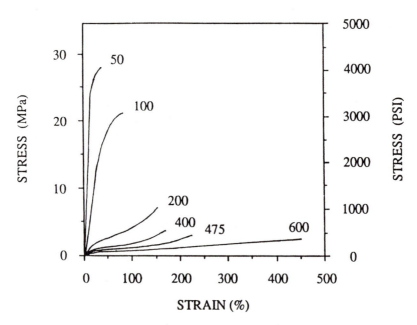

Figure 1. Stress–strain curves for films of various ATBN content.

composition, but in all cases a substantial amount of ductility is present. Plots of individual parameters, such as modulus, ultimate strain, and ultimate tensile strength (UTS), as functions of the ATBN content convey the relationship between composition and properties and are shown in Figures 2, 3, and 4, respectively.

Three regions of behavior are observed. Transitions occur at ATBN concentrations of 150 and 475 pbw. In the first region, where the ATBN concentrations are less than 150 pbw, the film acts like a flexibilized plastic: both modulus and strength are relatively high, but the samples are tough and

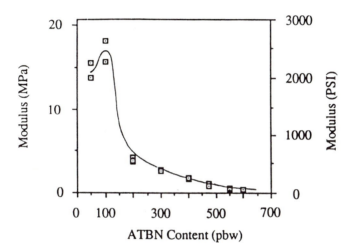

Figure 2. Modulus at 75–100% strain as a function of ATBN rubber content.

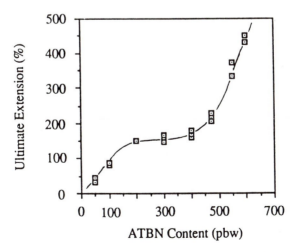

Figure 3. Ultimate extension as a function of the ATBN rubber content.

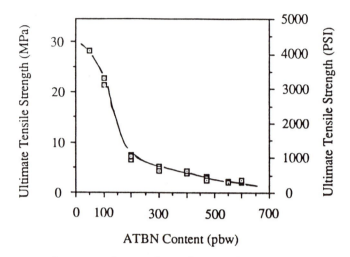

*Figure 4. Ultimate tensile strength as a function of ATBN rubber content.*

leathery. Other studies have reported similar properties when the rubber modifier content ranges from 20 to 100 pbw (*13*). Films with intermediate ATBN content, between 150 and 475 pbw, behave like a rubber. The stress–strain curves have the familiar shape of an elastomer, and the modulus and UTS decrease linearly with increasing ATBN content whereas the ultimate extension increases slightly. The third region starts at ATBN concentrations > 475 pbw, when the ratio of amine-to-epoxy units in the formulation exceeds unity. Under these conditions the modulus and UTS level off whereas the ultimate extension undergoes a large increase. One explanation for this behavior is the nonstoichiometric nature of the formulations used: As the amine-to-epoxy ratio departs from unity, more of the functional end groups of the ATBN chains remain unreacted after all the epoxy is consumed. The unreacted chains are not chemically incorporated into the cross-linked network of the film and thus do not contribute to the film strength. Further, it is probable that the unreacted chains act as plasticizers and promote intermolecular slippage, which results in the higher strains.

Returning now to the two-phase systems of rubber-toughened epoxy, if the elastomer particles in the glassy matrix have compositions similar to the corresponding films, then it is possible to estimate the elastomer properties by the data on the films. Although the exact composition of the particles is not known, the rubber within them has been estimated to range between 30 and 80% by weight (as mentioned previously). This estimate corresponds to ATBN contents of approximately 45 and 400 pbw in the homogenous films. Thus, the particle properties probably correspond to regions 1 and 2 in the film behavior: a transition from a flexibilized glass to a typical rubber. In all cases, considerable ductility seems to exist; the particles would be highly deformable.

Now consider the rubber particle–epoxy interface. How probable is a sharp compositional discontinuity here? Some idea of the thermodynamic interaction may be inferred from the swelling behavior of the films in various solvents. The ability of a liquid to act as a solvent for the rubber in the particle is shown by the amount of that liquid absorbed by the cross-linked film. The better the solvent, the greater the degree of absorption expected. Cured homogenous films containing 300 pbw of ATBN in two organic solvents swelled rapidly and significantly. Figure 5 shows that room temperature immersion in a 50/50 mixture of acetone and MEK caused a doubling in weight; immersion in toluene produced a 200% increase. This weight gain demonstrates that both mixtures are excellent solvents for the DGEBA–ATBN rubber. Immersion of the samples in hot liquid epoxy (uncatalyzed DGEBA at 105 °C) produced different but no less dramatic results. Unlike the organic solvents where the films reached saturation within minutes, the samples in the epoxy gained weight more slowly without approaching saturation (*see* Figure 6). The films continued to absorb for up to 12 h before finally disintegrating: At elevated temperatures, the resin is a good solvent. The temperature dependence is shown in Figure 6. If the swelling of the film by the hot epoxy is considered as a kinetic process, then the slope of the plot of rate vs. inverse temperature can be used with the Arrhenius equation to determine the activation energy. The data give a value of 20 kJ/mol, which is lower than the value for a chemical reaction (*14*), but reasonable for a thermodynamic interaction.

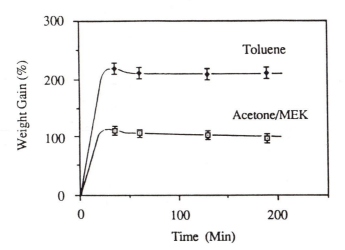

Figure 5. *Weight gain of films containing 300-pbw ATBN immersed in organic solvents at 25 °C.*

Previous studies of the phase separation of solutions of liquid epoxy and CTBN indicated that the acrylonitrile content of the rubber is an important factor (*10*). Absorption experiments performed with cured films that contain various amounts of acrylonitrile confirm these findings. Figure 7 shows the liquid epoxy uptake at 105 °C of three films that each contain 300-pbw ATBN with 0-, 10- and 17-wt % acrylonitrile. Despite the differences in thickness,

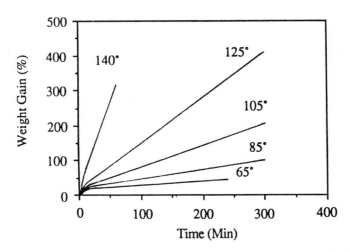

*Figure 6. Weight gain of films containing 300-pbw ATBN immersed in DGEBA (Epon 828) at various temperatures as a function of time.*

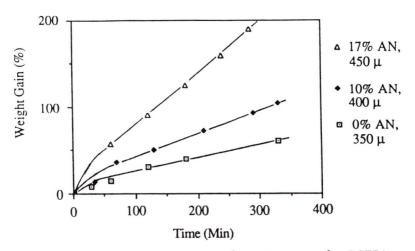

*Figure 7. Weight gain of films containing 300-pbw ATBN immersed in DGEBA (Epon 828) at 105 °C as a function of time. Acrylonitrile content of the ATBN and film thickness are given in the legend.*

the data show that the acrylonitrile content is the dominant factor in the swelling behavior and, therefore, it is also influential on the solubility of the rubber in liquid epoxy. From these results it is apparent that, during the curing of a rubber-toughened epoxy, a considerable amount of local mixing and interdiffusion between the rubber and the epoxy matrix is probable before gelation. This means that the boundary that develops between the two phases in the rubber-modified formulation will be broad on a molecular scale and the surrounding matrix will contain a finite amount of the rubber in solid solution (*15*). Such a situation could explain the large local shear deformations in the epoxy when a crack passes through the region: The rubber-plasticized epoxy yields and flows, and absorbs substantial work that is manifest as increased fracture toughness. The situation also explains why the cavity in the epoxy, which is usually much larger than the rubber particle resident in the epoxy, retracts and collapses when the material is heated above its (locally reduced) glass-transition temperature.

In Figures 1–4, the properties of epoxy–rubber formulations cured to a single-phase structure were presented and discussed. Figures 6 and 7 show that similar, cured-rubber films can absorb substantial amounts of liquid epoxy. In Figure 8, the results of another set of absorption experiments with a rubber film are presented. The film contained 300-pbw ATBN, 100-pbw DGEBA (Epon 828), and 3.7-pbw amine salt catalyst (Ancamine) curing agent. The film was dried and cured in the manner already described, and then it was immersed at room temperature in various solutions of acetone–MEK–liquid DGEBA resin catalyzed by the amine salt (Ancamine), as listed in Table II. The solution uptake is shown by the upper curve in

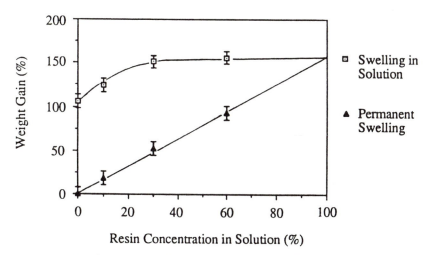

*Figure 8. Weight gain of a film containing 300-pbw ATBN rubber in solution and after drying.*

Figure 8. Then the samples were removed from the solvent solution and the acetone–Mek solvent was desorbed and evaporated. The residual permanent weight gain was due to the epoxy–curing agent and it was linearly proportional to the concentration of epoxy–curing agent in the immersion solution. (Extrapolating the two curves shows an expected uptake of 160% weight gain in 100% liquid epoxy at 25 °C.) The new composition of each film specimen can be determined from the amount of resin and curing agent absorbed. For example, if a film originally composed of 300-pbw ATBN and 100-pbw DGEBA showed a permanent weight increase of 52% due to the addition of epoxy and curing agent, the new composition would be 300-pbw ATBN and 291-pbw DGEBA. If the formulation is again normalized by the epoxy content, the new composition becomes 103-pbw ATBN and 100-pbw epoxy; the ATBN content has been effectively lowered by nearly two-thirds.

After a second vacuum heat treatment to cure the newly absorbed epoxy that had diffused into the film, the system appears to form a structure very similar to those produced for Figures 1–4. The samples were tested in the same way as the original tensile specimens. Figure 9 compares the moduli of the original films from Figure 2 with the moduli of films modified with the imbibed catalyzed resin. Both samples pass through the flexibilized plastic–elastomer transition at about the same point, but the original films appear to be somewhat stiffer. Despite this difference, the results suggest that fully cured ATBN-based elastomers are capable of absorbing large quantities of epoxy and curing agent, which can then be subsequently incorporated into the network of the film by a second cure. It is tempting to

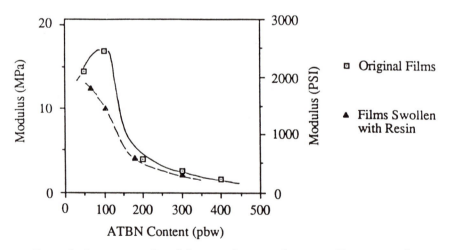

*Figure 9. Comparison of modulus as a function of ATBN rubber content for original films and films swollen by resin and catalyst.*

speculate that an interpenetrating network is produced in such a manner, but that remains to be proven.

## Conclusions

The solubility of reactive liquid rubbers in liquid DGEBA is widely recognized. This solubility can be exploited to produce two-phase cured solid mixtures in which the rubber-rich particles increase the fracture toughness of the glassy epoxy matrix by causing plastic shear deformation around the particles. Even if the particles contain substantial amounts of epoxy, they can still exhibit rubbery properties: low tensile modulus and considerable extensibility before fracture. The solubility of liquid epoxy in cured films of rubber–epoxy also is substantial. Liquid epoxy can be cured in situ to produce films with properties consistent with the overall proportions of epoxy and rubber present. Because of such solubility, it is postulated that the epoxy immediately surrounding a particle in the two-phase toughened mixture contains rubber in solid solution or reacted with the epoxy, enhancing its ability to plastically deform. Solubility considerations strongly suggest the existence of an interphase rather than a sharp boundary in the rubber particle–glassy epoxy systems.

## References

1. McGarry, F. J. *Proc. R. Soc. London A* **1970**, *319*, 59.
2. *Rubber-Toughened Plastics*; Riew, C. K., Ed.; Advances in Chemistry 222; American Chemical Society: Washington, DC, 1989.
3. Rowe, E. H.; Riew, C. K. *Plast. Eng.* **1975**, *31*, 45.
4. Bascom, W. D.; Ting, R. Y.; Moulton, R. J.; Riew, C. K.; Siebert, A. R. *J. Mater. Sci.* **1981**, *16*, 2657.
5. Sultan, J.; McGarry, F. J. *Polym. Eng. Sci.* **1973**, *13(1)*, 29.
6. Glad, M. D. Ph.D. Thesis, Cornell University, 1986.
7. Kinloch, A. J.; Shaw, S. J.; Tod, D. A.; Hunston, D. L. *Polymer* **1983**, *24*, 1341.
8. Kinloch, A. J.; Shaw, S. J.; Hunston, D. L. *Polymer* **1983**, *24*, 1355.
9. Kalfoglou, N. K.; Williams, H. L. *J. Appl. Polym. Sci.* **1973**, *17*, 1377.
10. Wang, T. T.; Zupko, H. M. *J. Appl. Polym. Sci.* **1981**, *26*, 2391.
11. Sayre, J. A.; Assink, R. A.; Lagasse, R. R. *Polymer* **1981**, *22*, 87.
12. Manzione, L. T.; Gillham, J. K.; McPherson, C. A. *J. Appl. Polym. Sci.* **1981**, *26*, 889.
13. Riew, C. K. *Rubber Chem. Technol.* **1981**, *54*, 374.
14. Barrows, G. M. *Physical Chemistry*; 4th ed.; McGraw-Hill: New York, 1979; p 683.
15. Bascom, W. D.; Cottinger, R. L.; Jones, R. L.; Peyser, P. *J. Appl. Polym. Sci.* **1975**, *19*, 2545.

Received for review March 6, 1991. Accepted revised manuscript June 17, 1992.

# The Synergistic Effect of Cross-Link Density and Rubber Additions on the Fracture Toughness of Polymers

Walter L. Bradley[1], W. Schultz[2], Carlos Corleto[1], and S. Komatsu[3]

[1]Polymer Science and Engineering Program, Texas A & M University, College Station, TX 77843-3123
[2]3M Corporation, 3M Center Building, 201–W–28, Saint Paul, MN 55144-1000
[3]Faculty of Engineering, Kinki University, Kure, 72917 Japan

*The effect of cross-link density on the efficacy of rubber toughening in thermosetting resins has been studied. Increasing the cross-link density changes the fracture toughness at a given temperature by both shifting the glass transition temperature (and thus the proximity of the test temperature to the glass transition temperature) and limiting the total crack-tip strain that can be realized at any given temperature. The temperature at which peak toughness is observed is shifted 50–120 °C lower in rubber-toughened systems compared to untoughened systems, with the peak of the rubber-toughened systems being four times the value observed in nontoughened systems. High cross-link density systems were found to be unresponsive to rubber toughening at any temperature.*

$I$NCREASING THE CROSS-LINK DENSITY may directly affect fracture toughness at a given temperature (e.g., ambient temperature) by limiting the degrees of freedom for shear deformation. Alternatively, an increase in cross-link density has a potentially indirect effect on the fracture toughness at a given temperature by changing the glass-transition temperature ($T_g$) and, thus, the proximity of the service or testing temperature to $T_g$.

0065–2393/93/0233–0317$06.00/0

The conventional wisdom is that only thermosets with relatively low cross-link density can be effectively toughened by the addition of rubber particles. Sue (1) suggested that only epoxies with a low cross-link density and a high monomer weight between cross-links are effectively toughened by rubber additions. Yee and Pearson (2–4) indicated that diglycidyl ether of bisphenol A (DGEBA) epoxy resin cured with 4,4'-diaminodiphenyl sulfone to a high cross-link density cannot be effectively rubber-toughened. Kinloch (5) recently noted that rubber toughening in highly cross-linked thermosets is limited because crazing and shear banding are not possible with the limited toughening observed in such systems, which results from crack pinning or crack bridging by rubber particles (if suitably large particles are included in the system microstructure).

When the cross-link density is high (giving a $T_g = 200$ °C or greater), fracture-toughness testing at ambient temperature is at least 180 °C below $T_g$. For such a circumstance, it is generally observed in DGEBA epoxies and other similar systems that the absolute increase in toughness that results from the addition of 10% carboxyl-terminated butadiene–acrylonitrile rubber is on the order of 0.3–0.4 kJ/m$^2$. The relative increase is typically no more than 300% (1, 6). Alternatively, when the cross-link density is relatively low (giving a $T_g$ of 100 °C), ambient-temperature fracture-toughness measurements are made within 80 °C of $T_g$, with an observed toughness enhancement through rubber additions that is on the order of 2.0 kJ/m$^2$, which corresponds to a relative increase of at least 1000% (7, 8).

Because almost all fracture-toughness measurements in the literature have been made at ambient temperature, it is unclear whether the low toughness and the relatively ineffective toughness enhancement that result from rubber additions in highly cross-linked resins are entirely because of a reduction in the intrinsic toughness of the matrix (as is usually assumed) or whether part of this reduction is because of the large shift in $T_g$ that accompanies such an increase in cross-link density.

Glad (9) and Kinloch (10) recently observed shear yielding and strain softening in highly cross-linked epoxies loaded in compression. The critical crack-tip opening displacement for highly cross-linked, brittle epoxies is 0.5 μm when the fracture energy $G_{Ic} = 70$ J/m$^2$. Such a crack-tip opening displacement suggests significant crack-tip strain, even in such brittle epoxies. Hibbs and Bradley (11), using direct measurements in the scanning electron microscope (see Figure 1), observed longitudinal strains greater than 15% around the tip of a growing crack in Hercules 3502 epoxy, which is usually thought to be "brittle" because it has a tensile elongation of less than 2% at ambient temperature. Thus, it is clear that significant shear straining is observed at the crack tip, even in low-toughness epoxies.

Fracture in thermosetting resins is most likely rate-limited by chain scission. Chain scission, in turn, depends on the axial stress that may develop

## 3501-6 Neat Resin

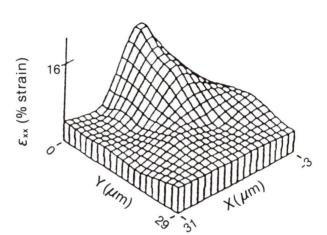

*Figure 1. Strain distribution around the tip of a growing crack in a higher cross-linked epoxy (11).*

in individual molecules. The crack-tip shear-straining magnitude necessary to develop sufficient stress in individual molecules for chain scission to proceed depends on both the cross-link density and the resin viscosity. When the resin viscosity is low, shear deformation sufficient to exhaust the capacity of the molecular structure to shear-deform without chain scission must occur before a significant buildup of stresses in individual molecules can take place. The shear strain required to give molecular stress sufficient to allow chain scission decreases as the molecular weight between cross-links decreases. However, at the much higher resin viscosity that occurs at temperatures of 125 °C or more below $T_g$, local viscosity considerations may predominate and the exhaustion of the various degrees of freedom for shear strain to occur on a molecular level is rendered unnecessary. Thus, the degree of crack-tip shear strain necessary to provide suitable molecular stressing to allow chain scission should depend on both the cross-link density and the temperature relative to $T_g$.

The toughening produced by rubber particle addition to thermosetting resins is the consequence of (1) the stress concentration adjacent to the rubber particles, which allows local yield ahead of the crack tip prior to crack advance, and (2) the increased shear stresses that result from the relaxation of the hydrostatic component of stress at the crack tip when the particles cavitate or debond (4, 5). The much larger shear stress that results from an increase in stress $\sigma_1$ and a decrease in stress $\sigma_3$ (recalling $\sigma_{max} = (\sigma_1 - $

$\sigma_3)/2)$ will cause either a much larger shear-deformation zone to develop ahead of the crack tip at a given temperature or the same degree of shear deformation to occur at a much lower temperature. Greater crack-tip deformation causes crack-tip blunting and greater load redistribution ahead of the crack tip when stress levels sufficient to give nonlinear constitutive behavior are reached over a significant volume of material ahead of the crack tip.

The purpose of this research program is to study the toughness of several epoxy systems as a function of test temperature relative to $T_g$. The $T_g$s of the systems were varied by changing the cross-link density and by incorporating a fluorene group into the network as a $T_g$-enhancing agent. The effect of 10% rubber additions to each system was studied over a range of temperatures relative to $T_g$ to allow separation of the indirect effect of cross-link density on rubber toughening (through shifting $T_g$) from the direct effect of cross-link density, which reduces the intrinsic shear-deformation capacity of the resin.

## Experimental Procedures and Results

The following resin systems were chosen for this study:

(A) Diglycidyl ether of bisphenol A (DGEBA) cured stoichiometrically with the diprimary amine (9,9-bis(3-methyl-4-amino-phenyl)fluorene) (OTBAF).

(B) Same resin as A, except with 10-wt % core-shell rubber added.

(C) DGEBA cured stoichiometrically with 0.60 equivalents of NH supplied by a disecondary amine chain extending agent, (4-methyl-9,9-bis[4-aminophenyl]fluorene) (BMAF) and 0.4 equivalents of NH supplied by the diprimary amine cross-linking agent OTBAF.

(D) Same resin as C, but with 10-wt % core-shell rubber added.

(E) Thirty-five weight percent DGEBA and 65-wt % fluorene epoxy (9,9-bis-4-[2,3-epoxy propoxy]phenyl fluorene), cured stoichiometrically with 0.6 NH equivalents of BMAF and 0.4 equivalents of OTBAF.

(F) Same resin as E, but with 10-wt % core-shell rubber added.

For simplicity, these systems will hereafter be referred to by letter. It should be remembered that A, C, and E have no rubber additions. System A is a highly cross-linked network (a stoichiometric cure of a diprimary amine with a glycidyl ether epoxy), whereas C and E are lightly cross-linked networks because of the addition of a BMAF, a disecondary amine chain extension agent. The molecular weight between cross-links for A is approximately 400; C and E are approximately 2100. Highly cross-linked A and lightly cross-linked E have similar $T_g$ values (220 and 225 °C). The $T_g$ of A is achieved through high levels of cross-linking, whereas the $T_g$ of E is achieved through the incorporation of the $T_g$-enhancing fluorene group. System C, the lightly cross-linked DGEBA resin, has a $T_g$ of 189 °C. The addition of the prereacted rubbery phase to A, C, and E had very little effect on the $T_g$ of these resins (B, $T_g$ = 218 °C; D, $T_g$ = 185 °C; F, $T_g$ = 220 °C). All glass transitions were determined from the peak values of tan δ, which was determined using a dynamic mechanical spectrometer (Rheometrics).

Yield strengths were measured using compression tests of specimens cut from the same 0.4-in.-thick rectangular panels as the compact tension fracture-toughness test specimens. Specimens with a 2:1 aspect ratio and a rectangular cross section were used. Tests were conducted over a wide range of temperatures. The resins were found to strain-soften (load drop observed even though tests were in compression rather than tension), and the maximum load was used to calculate the compressive yield strength of each specimen. All materials at all test temperatures were strained 16%, at which point the tests were terminated. Typical results for all six thermosetting materials are illustrated by the test results for material A presented in Figure 2. Material A is quite ductile for a so-called brittle epoxy resin with 100% stoichiometric cross-linking. The measured yield strengths over a range of temperatures for each of the six materials in the study are presented in Figure 3. The expected decrease in yield strength with increasing temperature is observed. Careful examination of the data points indicates that there are really two bands rather than one. The rubber-toughened specimens (filled symbols) consistently have yield strengths that are less than the resins without rubber additions (open symbols). The 35% difference at ambient temperature corresponds to a temperature shift of 50 °C.

The fracture-toughness measurements initially were made using a combination of linear elastic fracture mechanics (LEFM) and elastic–plastic fracture mechanics (i.e., a J-integral approach), depending on the material and the temperature. These results were presented at the "Toughened Plastics Symposium" sponsored by the American Chemical Society in Washington, DC in August, 1990 and published in the conference proceedings (*12*). Subsequent work indicates that some of the J-integral results presented at the ACS meeting were incorrect. The difficulty comes in determining what portion of the stress-whitened zone on the fracture surface ahead of the crack is crack growth and what portion is the result of rubber particle cavitation or debonding ahead of the crack tip. For this reason, the smaller the crack growth is, the greater is the uncertainty in the crack extension measurement. Determination of a valid plane-strain fracture toughness ($J_{Ic}$) requires some

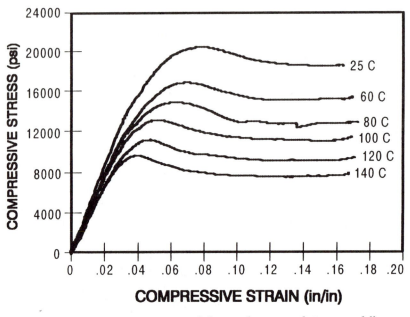

*Figure 2. Compressive stress–strain behavior for material A at six different temperatures.*

points in the region of small crack extension, which is exactly where the experimental problem with determining crack extension is most severe.

The fracture-toughness values were subsequently redetermined using a second batch of each of the six material systems, again using compact tension specimens, but with width $W = 5$ cm rather than the $W = 2.5$ cm specimens used initially. The initial and retesting both were done in accordance with the new, American Society for Testing and Materials (ASTM) standard for fracture-toughness determination in plastics using a linear elastic approach, ASTM D 5045–91a (13). The new ASTM standard allows for sharp notches to be introduced by machining with subsequent razor sharpening, that was used in the retesting. Fatigue precracking was used in the initial tests according to the ASTM E–399 standard for metals. Cayard and Bradley (14) have shown that toughness values measured using properly fatigue-precracked compact tension specimens are similar to results for compact tension specimens with razor-sharpened machined-in notches. The compact tension specimens were standard size in the initial tests and the retests, except for thickness. For $W = 2.5$ cm, the thickness was only 0.88 cm and for $W = 5$ cm, the specimens were 1.25 cm thick. The razor notching was made with a specially modified razor blade lightly drawn (rather than pressed) across the machined-in notch. The razor blade was modified by cutting many

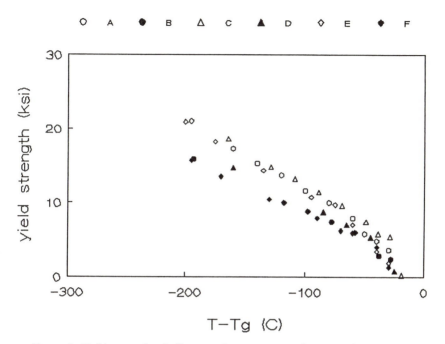

*Figure 3. Yield strength of all material systems as a function of temperature relative to $T_g$.*

small sections from the tip using a diamond saw. Thus, the blade functions like a very sharp cutting tool.

The 5% secant offset values of stress intensity, $K_q$, as defined in ASTM D 5045–91a were measured. Subsequently, minimum specimen requirements were calculated for each material at each test temperature using the usual relationship $2.5 \times (K_q/S_y)^2$, where $S_y$ is yield strength. When either the size or the load ratio requirement $(P_{max}/P_q) > 1.1$ was not clearly met, multiple-specimen J-integral tests were run using the procedure specified in the protocol of the European Fracture Group (March, 1991).

The initial compliance of the compact tension specimens, as determined from the measured load-displacement curves, was utilized to back-calculate the effective modulus ($E'$) for the various thermosetting resins at each test temperature. The $E'$ values were similar to but consistently 15–20% lower than the modulus values determined in compression tests. The lower values occurred because either the specimens were loaded in tension or the effective loading rate was somewhat slower. The measured fracture-toughness ($K_{Ic}$) values were used in combination with $E'$ to calculate the fracture energy $G_{Ic} = K_{Ic}^2/E'$. Generally, the J-integral was necessary for epoxy specimens with $W = 5.0$ cm that were tested within 50 °C of $T_g$.

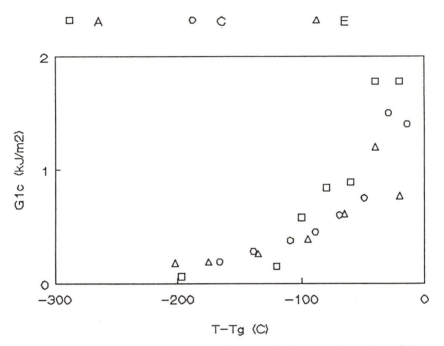

*Figure 4. Toughness vs. temperature relative to* $T_g$ *for materials A, C, and E*
*(no rubber).*

Fracture-toughness results measured by $G_{Ic}$ (or $J_{Ic}$) for thermosetting materials A, C, and E are presented in Figure 4. It is clear that an initial increase in toughness begins at 125 °C below $T_g$ in all three systems, and a more dramatic increase in toughness occurs at 60 °C below $T_g$. All three curves fall within a relatively narrow band. If the same data were plotted as a function of temperature $T$ rather than temperature relative to the glass transition ($T - T_g$), all the circles for system C would be shifted to the left by 30–35 °C, relative to A and E. This shift indicates that material C has a higher fracture toughness at all temperatures. It is clear in these untoughened systems (no rubber additions) that changes in $T_g$ are indirectly and partially responsible for the observed changes in toughness at a given temperature.

Fracture-toughness results measured by $G_{Ic}$ (or $J_{Ic}$) for thermosetting materials B, D, and F, each of which has 10-wt % rubber additions, are presented in Figure 5. Materials D and F, which are only lightly cross-linked, follow a similar pattern in which significant toughness is observed at temperatures as low as 200 °C below $T_g$. Material system B, which has a high cross-linking density, behaves in a fashion very similar to the systems with no rubber additions: significant toughness is observed only at tempera-

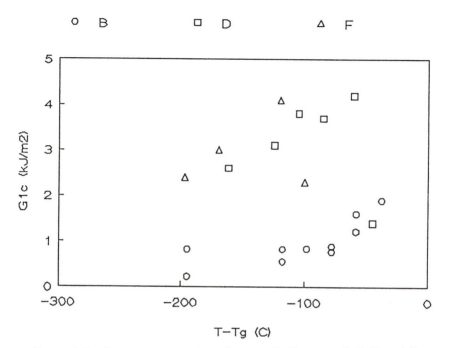

*Figure 5. Toughness vs. temperature relative to* $T_g$ *for materials B, D, and F, each with 10% rubber additions.*

tures within 60 °C of $T_g$. The contention that highly cross-linked thermosetting materials are not as responsive to rubber toughening as low cross-link density thermosets is clearly supported by a comparison of Figures 4 and 5. Toughenable thermosets with a high $T_g$ are more easily obtained using $T_g$-enhancing groups in the backbone than by increasing the cross-link density.

A more detailed comparison of the various systems with and without rubber particle additions has been made in Figures 6–8. These figures exemplify, in another way, the hypothesis that the moderately cross-linked thermosetting materials C and E are extremely responsive to rubber toughening in the temperature range of 50–200 °C below $T_g$. By contrast, the highly cross-linked thermosetting material A shows much less responsiveness to rubber additions.

Only at room temperature ($-195$ °C) do the results suggest significant rubber toughening in the highly cross-linked system A (*see* Figure 6). However, the room-temperature results are somewhat ambiguous. Initial test results on fatigue-precracked compact tension (CT) specimens ($W = 2.5$ cm) indicated $G_{Ic} = 0.22$ kJ/m$^2$ for system B, but subsequent testing of razor-sharpened CT specimens with $W = 5.0$ cm indicated a toughness of 0.82

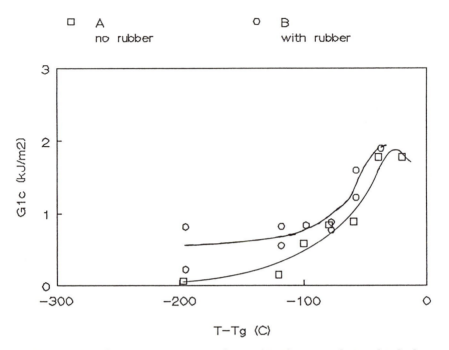

*Figure 6. Toughness vs. temperature relative to* T_g *for materials A and B, both with a high cross-link density and B with 10% rubber addition.*

kJ/m$^2$. The enhanced toughness that results from the addition of rubber to system A is 0.06–0.22 kJ/m$^2$, which is not very impressive, or 0.06–0.82 kJ/m$^2$, which is proportionately the same increase as that observed for low cross-link density systems.

One might suspect that the razor sharpening did not produce sufficiently sharp starter cracks in the retesting of system B, giving $G_{Ic}$ = 0.82 kJ/m$^2$ rather than the 0.22 kJ/m$^2$ observed originally using smaller fatigue-precracked specimens. However, the initial crack growth of the razor-notched specimens did not result in catastrophic failure, and subsequent growth from these "natural" cracks also gave $G_{Ic}$ values of approximately 0.8 kJ/m$^2$. Results at higher temperatures for the smaller ($W$ = 2.5 cm) fatigue-precracked specimens and the larger ($W$ = 5.0 cm) razor-notched specimens used during retesting were generally more consistent than the room-temperature results seen in Figure 5.

The rubber-toughened systems with light cross-linking (D and F) have a peak toughness at temperatures of 50–120 °C below $T_g$ (Figure 5). A peak toughness value for the untoughened thermosetting resins occurs within 30–40 °C of $T_g$, based on the results in Figure 4. These peaks represent an optimal compromise between decreased shear-yield strength as the tempera-

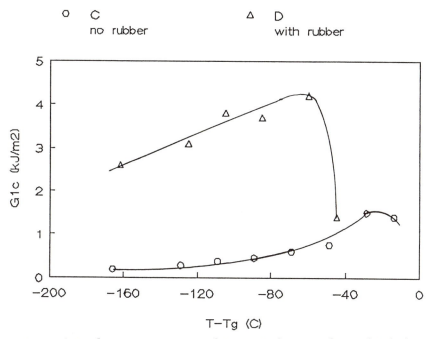

Figure 7. Toughness vs. temperature relative to $T_g$ for materials C and D, both with a light cross-link density and D with 10% rubber addition.

ture rises and increased critical crack-tip opening displacement (and crack-tip strain-to-failure), as seen schematically in Figure 9. Recall from earlier discussion that the crack-tip strain required to develop sufficient axial stress in the molecules to allow chain scission is a function of temperature, both because of the availability of thermal energy and because of the temperature dependence of the viscosity of the glassy polymer. The effect of rubber toughening in the moderately cross-linked systems is to shift this peak by 40–50 °C in the direction of a lower temperature, as illustrated in Figure 9 and supported by a comparison of Figures 4 and 5. The magnitude of this shift is predictable from knowledge of the material's shear-yield strength at peak toughness in the absence of rubber additions and a determination of the shear-yield strength as a function of temperature, in combination with appropriate stress analysis for the system with rubber particle cavitation. The magnitudes of the temperature shift seen for the peaks in Figures 7 and 8 are similar to the values noted for yield strength in Figure 3.

A second effect suggested by Figures 7 and 8 is that the maximum toughness without rubber is at least the magnitude of the maximum toughness with rubber additions. A comparison of Figure 6 with Figures 7 and 8 indicates the very different behavior of high cross-link density thermosets compared to low cross-link density thermosets.

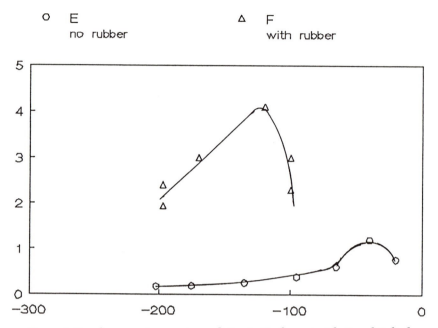

*Figure 8. Toughness vs. temperature relative to* $T_g$ *for materials E and F, both with light cross-link density and F with 10% rubber addition.*

## Discussion and Conclusions

The results clearly support the contention that the degree of cross-linking is an important variable in the susceptibility of the thermoset to rubber toughening. A window of opportunity is also suggested for rubber toughening of low to moderately cross-linked thermosets. At temperatures within 60 °C of $T_g$, rubber toughening is probably unnecessary because the material is quite capable of crack-tip shear deformation without rubber additions. From 60 to 150 °C (possibly 200 °C) below $T_g$, the toughness enhancement monotonically increases in the moderately cross-linked thermosetting materials used in this study from a 5-fold to a 13-fold increase, with a maximum increase of 3.5 kJ/m$^2$. By contrast, toughness enhancement over the same temperature range in the heavily cross-linked material A increases from 1-fold to 3-fold (or 13-fold if $G_{Ic} = 0.82$ kJ/m$^2$ rather than 0.22 kJ/m$^2$), with a maximum actual increase of only 0.4 kJ/m$^2$ (or 0.75 kJ/m$^2$ if $G_{Ic} = 0.82$ kJ/m$^2$). At temperatures of more than 200 °C below $T_g$, one would expect to find eventually a temperature range where rubber toughening is again ineffective because of the lack of incipient toughness in the net resin system and the consequent inability of the resin to deform at the crack tip, even adjacent to cavitated rubber particles. Recent results (17) indicate that particle cavitation occurs in systems D and F before crack growth begins, whereas crack growth

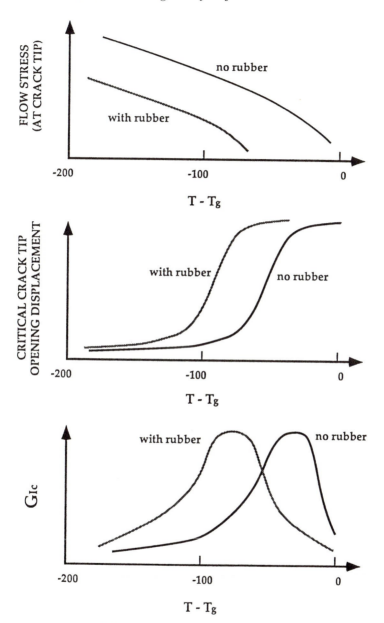

*Figure 9. Conceptual basis for peak in* $G_{Ic}$ *vs.* $T - T_g$. *Top: Flow stress (principal normal stress at yield at crack tip) as a function of temperature with constraint (no rubber) and with constraint relaxed (with cavitated rubber particles); middle: critical crack tip opening displacement at crack extension; bottom: critical energy release rate* ($G_{Ic}$) *as a function of* $T - T_g$ *where* $G_{Ic}$ = *flow stress (at crack tip)* × *crack-tip opening displacement. Note: The bottom graph is constructed as a point-by-point product of the top two graphs.*

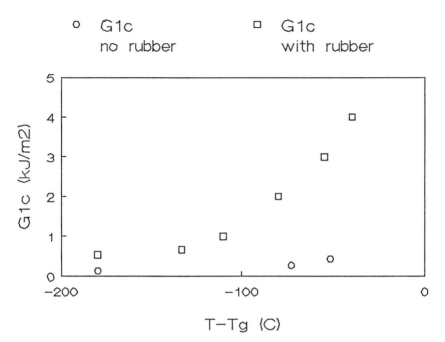

Figure 10. $G_{Ic}$ vs. $T - T_g$ for epoxy with and without rubber, calculated from results of Hunston et al. (7).

in system B occurs at a lower value of $K_1$ than that required to give particle cavitation. Thus, the toughening in the highly cross-linked system B is because of the stress concentrating effect of the rubber particles alone, whereas the toughening in the lower cross-link density systems D and F combines the stress concentrating effect with the relaxation of $\sigma_3$ because of particle cavitation. System B might show more rubber toughening if particle cavitation or debonding could occur before crack growth begins.

Comparison of the results of several other investigators to the results obtained in this study is interesting. Results published by Hunston et al. (7), Kinloch et al. (15), and Murakami et al. (16) are replotted in Figures 10–12 to facilitate comparison with the present results. It is particularly significant that these curves are similar to each other as well as to our results, even though each research group obtained $T - T_g$ variation in very different ways. Hunston et al. (7) used variations in loading rate, in part, to vary the effective temperature. Kinloch et al. (15) varied $T - T_g$ by varying the test temperature (T), whereas Murakami et al. (16) varied $T - T_g$ by varying the glass-transition temperature ($T_g$) and performed all of their fracture toughness testing at ambient temperature.

It is also interesting to plot the ratio of fracture toughnesses ($G_{Ic}$) for resins with and without rubber as a function of $T - T_g$, as seen in Figure 13.

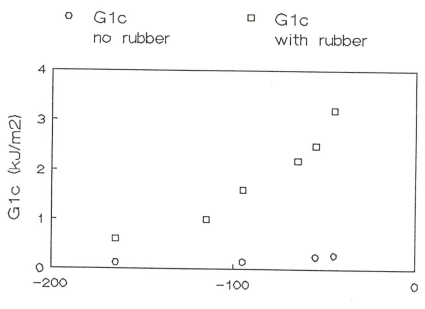

Figure 11. $G_{Ic}$ vs. $T - T_g$ for epoxy with and without rubber, from results of Kinloch et al. (15).

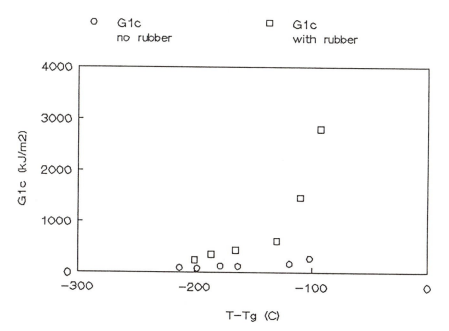

Figure 12. $G_{Ic}$ vs. $T - T_g$ for epoxy with and without rubber, calculated from results of Murakami et al. (16.)

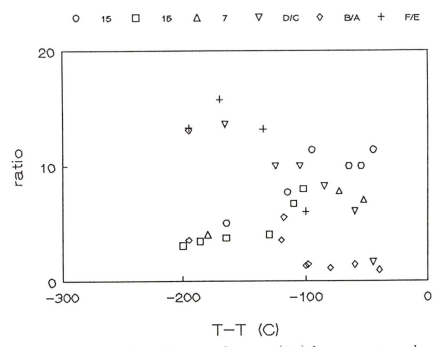

*Figure 13. Ratio of critical energy release rate $(G_{Ic})$ for epoxy resins and without rubber additions as a function of $T - T_g$, calculated from the present results and results taken from references 7, 15, and 16.*

The values for systems A and B are consistently low over the whole range of temperatures (except at $T - T_g = -195$ °C, where ratios using $G_{Ic}$ values of 0.22 and 0.82 are both included). The remaining results taken from this and other studies are for lower cross-link density epoxies. The ratios vary with temperature, as would be expected based on the conceptual model presented in Figure 9, and peak values occur between $-50$ and $-200$ °C. The only results that fall clearly outside this trend are those for systems E and F, which achieve a high $T_g$ in combination with lower cross-linking by the addition of a $T_g$-enhancing fluorene group. This approach is seen in Figure 8 as well as Figure 13 to result in a system that is toughenable at temperatures as low as 200 °C below $T_g$, where more conventional, high cross-link density systems have limited susceptibility to rubber toughening (7, 15, 16).

Two areas deserve additional study. First, the apparent lack of effective rubber toughening in systems with higher cross-link density requires better understanding. It may be that crack growth at $K_1$ values lower than required for particle cavitation or debonding is the key, which suggests adjustments in the chemistry of the rubber or the interface. Second, the much smaller process zone (on the order of the crack-tip opening displacement) that is developed in such systems might actually be smaller than the distance between conventional size rubber particles, which renders the rubber relatively ineffective to alter the state of stress at the crack tip. The addition of 5 vol % rubber with a radius of 0.1 μm results in an average interparticle spacing of about 0.5 μm, which is similar to the crack-tip opening displacement in the more brittle epoxy systems. Thus, the rubber particle size (and spacing) and critical cavitation−debonding stresses may require adjustment for high cross-link density epoxies if the same proportionate rubber toughening (approximately 10 × ) is to be effected as that found in lower cross-link density resins at 50 to 125 °C below $T_g$. Finally, the behavior postulated in Figure 9 for the variation of the ratio of $G_{Ic}$(rubber) to $G_{Ic}$(no rubber) with $T - T_g$ requires further study to establish its generality.

## Acknowledgments

The assistance of Inga Lax, Wesley Cantwell, and Lu Fan in various aspects of the experimental program is gratefully acknowledged. A portion of this work was performed in the Polymers Laboratory of Dr. H. H. Kausch at Ecolé Polytechnique Federale de Lausanne (EPFL), Switzerland. A more comprehensive presentation of work done at EPFL will be presented separately (17).

## References

1. Sue, H.-J. *Polym. Sci. Eng.* in press.
2. Yee, A. F.; Pearson, R. A. *J. Mater. Sci.* **1986**, *21*, 2462.
3. Pearson, R. A.; Yee, A. F. *J. Mater. Sci.* **1986**, *21*, 2475.
4. Pearson, R. A.; Yee, A. F. *J. Mater. Sci.* **1989**, *24*, 2571.
5. Kinloch, A. J. In *Rubber-Toughened Plastics*; Riew, C. K., Ed.; Advances in Chemistry 222; American Chemical Society: Washington, DC, 1989; p 67.
6. Zengli, F.; Yishi, S.; Wei, Y. *Proceedings of "Benibana" International Symposium on How to Improve the Toughness of Polymers*; Yamagata University: Yamagata, Japan, October 1990; p 84.
7. Hunston, D. L.; Kinloch, A. J.; Shaw, S. F.; Wang, S. S. *Adhesive Joints*; Mittal, K. L., Ed.; Plenum Publishing Co.: New York, 1984; p 789.
8. Bascom, W. D.; Cottington, R. L.; Jones, R. L.; Peyser, P. *J. Appl. Polym. Sci.* **1975**, *19*, 2545.
9. Glad, M. D., Ph.D. Dissertation, Cornell University, 1986.
10. Kinloch, A. J. *Polym. Commun.* **1987**, *28*, 322.
11. Hibbs, M.; Bradley, W. L. *Proceedings of the Annual Meeting of the Society of Experimental Mechanics*; Society of Experimental Mechanics: Bethel, CT, 1987.
12. Bradley, W. L.; Lax, I. *Proceedings of the ACS Division of Polymeric Materials: Science and Engineering*; American Chemical Society: Washington, DC, 1990; p 104.
13. *Standard Test Method for Plane-Strain Fracture Toughness and Strain Energy Release Rate of Plastic Materials*; American Society for Testing and Materials: Philadelphia, PA, 1992; D 5045−91a, p. 752.

14. Cayard, M.; Bradley, W. L. *Proceedings of the Seventh International Conference on Fracture*; Pergamon: New York, March 1989; p 2713.

15  Kinloch, A. J.; Maxwell, D.; Young, R. J. *J. Mat. Sci.* **1985**, *20*, 4169.

16. Murakami, A.; Ioku, T.; Saunders, D.; Aoki, H.; Yoshiki, T.; Murakami, S.; Watanabe, O.; Saito, M.; Inoue, H. *Proceedings of "Benibana" International Symposium on How to Improve Toughness of Polymers*; Yamagata University: Yamagata, Japan, October 1990; p 65.

17. Bradley, W. L.; Fan, L.; Bradley, S.; Cantwell, W.; Kausch, H. H. in preparation.

RECEIVED for review March 6, 1991. ACCEPTED revised manuscript November 13, 1991.

# Rubber-Modified Epoxies

## Analysis of the Phase-Separation Process

Didier Verchère[1], Henry Sautereau[1], Jean-Pierre Pascault[1], S. M. Moschiar[2], C. C. Riccardi[2], and R. J. J. Williams[2]

[1]Laboratoire des Matériaux Macromoléculaires, Unité Recherche Associée–Centre National de la Recherche Scientifique no. 507, Institut National des Sciences Appliquées de Lyon, Bât. 403, 20 Avenue Albert Einstein, 69621 Villeurbanne Cédex, France
[2]Institute of Materials Science and Technology (INTEMA), University of Mar del Plata and National Research Council (CONICET), J. B. Justo 4302, 7600 Mar del Plata, Argentina

*The phase-separation process of a diepoxide based on bisphenol A diglycidyl ether cured with a cycloaliphatic diamine in the presence of an epoxy-terminated butadiene–acrylonitrile random copolymer (ETBN) was experimentally studied and theoretically simulated. The increase in the average molecular weight of the epoxy–amine polymer is shown to be the main thermodynamic factor leading to phase separation. The competition between nucleation–growth and spinodal decomposition is analyzed. Low values of interfacial tensions and polymerization rates favor the first mechanism in most cases. A secondary phase separation must occur inside the dispersed particles, leading to a segregated epoxy–amine phase. The theoretical simulation explains the increase of the average diameter of dispersed-phase particles with increasing polymerization temperature. The volume fraction of dispersed phase is the dominating factor affecting fracture toughness.*

To IMPROVE CRACK RESISTANCE AND IMPACT STRENGTH, low levels of carboxyl- or epoxy-terminated butadiene–acrylonitrile random copolymers (CTBN or ETBN) are often incorporated into normally brittle epoxy resins. This enhancement in toughness results from the separation during cure of a randomly dispersed rubbery phase (1–7).

A qualitative description of the phase-separation process was first reported by Visconti and Marchessault (8). A CTBN was dissolved in a mixture of a cycloaliphatic epoxy resin with an anhydride hardener, to yield a homogeneous solution. At a certain point of the polymerization reaction small spherical domains of an elastomeric phase were segregated. Small-angle light scattering was used to analyze the variation in size and shape of the rubbery domains. Phase separation took place well before gelation, and a phase inversion occurred beyond a CTBN concentration of 20% by weight.

Manzione et al. (9, 10) showed that a variety of different morphologies could be obtained from a single rubber-modified epoxy formulation by varying the cure temperature. This property is a consequence of the competing effects of nucleation and growth rates, on the one hand, and the polymerization rate, on the other.

For a particular system, Montarnal et al. (11) confirmed that the phase-separation process was arrested well before gelation or vitrification. They also found that the average size of dispersed domains increased with cure temperature, but the volume fraction of dispersed phase did not change significantly.

Williams and co-workers (12, 13) proposed a theoretical model to predict the fraction, composition, and particle size distribution of dispersed domains. The model is based on a thermodynamic description through a Flory–Huggins equation and constitutive equations for the polymerization and phase-separation rates. The latter include expressions for nucleation, growth, and coalescence rates. The simulation gives a correct prediction of the possibility of phase inversion, the arrest of phase separation at the gel point (although evolution of the morphology in dispersed-phase domains may continue after gelation), the presence of epoxy copolymer in the dispersed phase, the increase in the volume fraction and average size of dispersed domain with the initial rubber concentration, the presence of a maximum in the average size of particles as a function of cure temperature, and the possibility of generating bimodal distributions of particle sizes. Similar arguments explaining the phase separation process in terms of a nucleation–growth mechanism were employed by Roginskaya and co-workers (14).

Kinloch (15) reviewed the relationships between the microstructure and fracture behavior of rubber-toughened thermosetting polymers. On the basis of his own results (16, 17) and those of Yee and Pearson (18), Kinloch clearly established that shear-yielding in the matrix is the main source of energy dissipation and increased toughness. Plastic deformation is caused by interactions between the stress fields ahead of the crack and around rubbery particles.

In the described context, a 3-year cooperative program between our laboratories was carried out to analyze the factors that control the morphology development in rubber-modified epoxies. A particular system consisting of an epoxy based on bisphenol A diglycidyl ether (DGEBA) and cured with a cycloaliphatic diamine was selected. A careful experimental study of the polymerization kinetics was carried out. Gelation and vitrification were de-scribed in a time versus temperature transformation (TTT) diagram (*19*). Formulations containing an ETBN (arising from a CTBN pre-reacted with an epoxy excess), were studied. Both experimental results and a theoretical simulation of the phase-separation process were published as a series of four papers (*20–23*). In this chapter, a review of the most significant findings is presented and compared with the present knowledge on these systems. Also, some suggestions for future experimental work are made.

## Materials and Cure Cycles

Chart I shows the structural formulas of the different monomers and poly-mers. The DGEBA-based epoxy is DER 332 (Dow), with $\bar{n} = 0.03$ ($\bar{n}$ is shown in Chart I). For comparison, the following DGEBA-based epoxies were also used: Bakelite 164 ($\bar{n} = 0.15$), DER 337 (Dow, $\bar{n} = 0.49$), and Araldite GY280 (Ciba-Geigy, $\bar{n} = 0.74$).

The diamine is 4,4′-diamine-3,3′-dimethyldicyclohexylmethane (3DCM; Laromin C260, BASF). Results obtained with 1.8 *p*-methanediamine (MNDA, Aldrich) are used for comparison.

The CTBN rubber is Hycar $1300 \times 8$ (Goodrich), with a number-average molecular weight close to 3600, an 18% acrylonitrile content (18% AN), and a COOH functionality equal to 1.8. Unless otherwise stated, this is the rubber referred to in the following text. Other rubbers, CTBN $\times$ 13 and CTBN $\times$ 9, that differ in acrylonitrile content (26% AN) and COOH func-tionality (equal to 2.3), respectively, are also mentioned.

ETBN adducts with DGEBA ($\bar{n} = 0.03$) were prepared by reacting carboxyl groups with epoxides, using a carboxyl-to-epoxy ratio equal to 0.065, at 85 °C, in the presence of a 0.18% by weight triphenylphosphine. Because of the large excess of DGEBA, most of the ETBN is a solution of a triblock copolymer (DGEBA–CTBN–DGEBA) in the epoxy monomer.

Formulations were precured at a constant temperature for a time dura-tion necessary to arrest the phase separation by gelation. Four different precure temperatures were selected: 29, 50, 75, and 100 °C. Corresponding precure times were 6 days, 360–400 min, 90–120 min, and 50 min, respec-tively. Samples were postcured for 14 h at 190 °C to obtain maximum conversion of the epoxy–amine matrix without degradation reactions (*19*).

## Miscibility in Unreactive Formulations

Figure 1 (*24*) shows cloud-point curves (CPC) for mixtures of CTBN $\times$ 8 with epoxy monomers of different molecular masses (or different $\bar{n}$ values). An upper-critical-solution-temperature behavior (UCST) is observed for

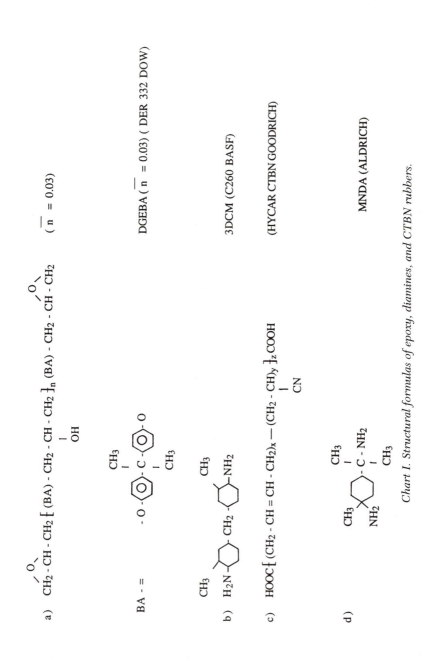

a)

$$CH_2 - CH - CH_2 \{ (BA) - CH_2 - CH - CH_2 \}_n (BA) - CH_2 - CH - CH_2$$
$$\overset{O}{\diagup}\diagdown \qquad\qquad\qquad\qquad\qquad \underset{|}{OH} \qquad\qquad\qquad\qquad \overset{O}{\diagup}\diagdown$$

( $\overline{n}$ = 0.03)

DGEBA ( $\overline{n}$ = 0.03) ( DER 332 DOW)

BA - =   $- O - \bigcirc - \overset{CH_3}{\underset{CH_3}{C}} - \bigcirc - O$

b)

$$H_2N - \bigcirc\text{-}CH_2 - \bigcirc - NH_2$$
CH₃ ... CH₃

3DCM (C260 BASF)

c)   $HOOC \{ (CH_2 - CH = CH - CH_2)_x - (CH_2 - CH)_y \}_z COOH$
$$\underset{CN}{|}$$

(HYCAR CTBN GOODRICH)

d)

$$\bigcirc \overset{CH_3}{\underset{NH_2}{C}} \quad \overset{CH_3}{\underset{CH_3}{C}} - NH_2$$

MNDA (ALDRICH)

*Chart 1. Structural formulas of epoxy, diamines, and CTBN rubbers.*

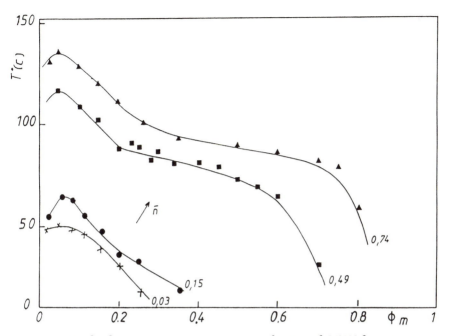

*Figure 1. Cloud-point temperature versus mass fraction of CTBN for mixtures of CTBN × 8 with epoxy monomers of different n values. (Reproduced with permission from reference 24. Copyright 1989 Butterworth.)*

every case (25–29).

The most significant feature is the high sensitivity of the location of miscibility gaps as a function of the molar mass of the epoxy monomer. A small increase in $\overline{M_n}$ from 349 g/mol ($\overline{n} = 0.03$) to 383 g/mol ($\overline{n} = 0.15$) leads to an increase in the precipitation threshold temperature (maximum of CPC) of some 14 °C. Because the solubility parameter has the same value for every epoxy monomer (24), it is indeed the increase in the molar mass that produces a significant shift in the miscibility gap.

The Gibbs free energy per unit volume, $\Delta G$, associated with the mixing of components E (epoxy) and R (CTBN rubber) with molar volumes $V_E$ and $V_R$, respectively, is given by (30),

$$\Delta G = R_g T [(\phi_E/V_E)\ln \phi_E + (\phi_R/V_R)\ln \phi_R] + \Lambda \phi_E \phi_R \qquad (1)$$

where $R_g$ is the gas constant, $T$ is temperature, $\phi_i$ ($i$ = E or R) represents the volume fraction of component $i$, and $\Lambda$ is the interaction parameter in units of energy per unit volume.

The first term in equation 1 is the combinatorial part of the free energy of mixing that arises from the Flory–Huggins model; the second term represents the residual free energy of mixing. Equation 1 also may be regarded as a definition of the interaction parameter, $\Lambda$. Frequently, the interaction between both components is expressed by using a $\chi$ parameter. When the composition dependence may be neglected, $\Lambda$ and $\chi$ are related by

$$\chi = \Lambda V_r / R_g T \tag{2}$$

where $V_r$ is a reference volume, which is normally defined as the molar volume of the smallest component (the epoxy monomer in this case).

Because $V_E \ll V_R$, small changes in $V_E$ lead to high relative variations in the free energy of mixing and, consequently, in the location of the miscibility gap.

The effect of CTBN polydispersity on the miscibility with epoxy monomers was reported previously (31). A precipitation threshold at low rubber volume fractions ($\phi_{Rth} = 0.07-0.08$) was predicted in agreement with experimental observations. The critical point is located at $\phi_{Rc}$ close to 0.20. A phase inversion is produced for $\phi_R > \phi_{Rc}$.

## *Miscibility in Reactive Formulations*

The conversion at the cloud point for DGEBA–3DCM stoichiometric formulations containing different ETBN concentrations was determined for cure temperatures 50 and 75 °C. Results are shown in Figure 2. Although significant scattering is seen in experimental results, the conversion at the cloud point does not depend significantly on temperature (in the range 50–75 °C) but does depend markedly on rubber concentration.

Particle size distributions for DGEBA–3DCM samples precured at 50 °C were also measured during polymerization. Figure 3 shows the evolution of the average particle diameter ($\bar{D}$) after polymerization times such that the glass-transition temperature ($T_g$) of the sample is higher than room temperature (after 3 h at 50 °C; $T_g = 26$ °C). This last condition is necessary for the metallization of the surface. Also shown in Figure 3 are the times at the beginning ($t_{cp}$) and end ($t_{cp+\Delta cp}$) of phase separation, vitrification time ($t_{vit}$) and gelation time ($t_{gel}$). The corresponding conversions ($p$) are shown on another scale. Clearly, the final morphology is attained well before gelation or vitrification. The point corresponding to $\bar{D} = 0.1$ μm indicates the detection limit of the cloud-point device. The experimental value of $t_{cp+\Delta cp}$ shows that the final morphology is produced in a very narrow conversion range. The postcure step has no effect on the morphology.

A simple thermodynamic analysis that describes the miscibility of the reactive mixture may be carried out. The initial system is a solution of two components: the epoxy–amine copolymer (taken as a pure component) and the rubber (taken as the CTBN block of the ETBN triblock copolymer). Thus, a pseudobinary system is defined in such a way that component E

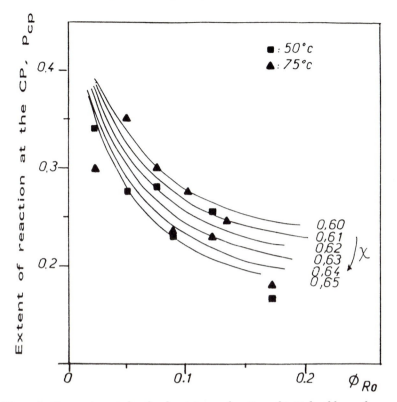

*Figure 2. Conversion at the cloud point as a function of initial rubber volume fraction for cure temperatures 50 and 75 °C. Binodal curves for different values are also plotted. (Reproduced with permission from reference 22. Copyright 1991 John Wiley & Sons, Inc.)*

(epoxy–amine copolymer) and component R (rubber) keep a constant overall volume fraction independently of the chemical bonds between the epoxy blocks of ETBN and the diamine. This last reaction has a very low probability because these epoxies make a very low fraction of the total amount, that is, 2.4% for a formulation containing a 15% mass fraction of rubber. Statistical calculations show that the average mass attached to ETBN becomes important at conversions higher than the range at which most of the phase separation has taken place.

We take $V_r = V_{E0}$, the initial molar volume of the epoxy–amine copolymer, defined as

$$V_{E0} = (M_{A4} + 2M_{B2})/3\rho_E = 276.6 \text{ cm}^3/\text{mol}$$

where $M_{A4}$ and $M_{B2}$ are the molecular weights of diamine and diepoxide,

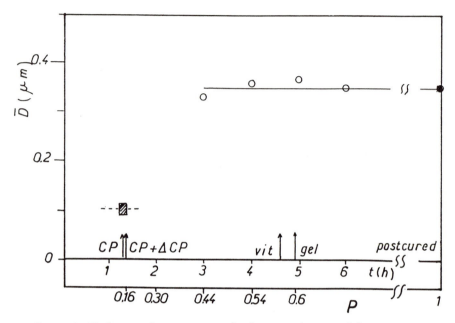

*Figure 3. Evolution of average particle diameter determined by scanning electron microscopy as a function of epoxy conversion (and time). (Reproduced with permission from reference 21. Copyright 1991 John Wiley & Sons, Inc.)*

respectively, and $\rho_E = 1.127$ g/cm, is the density of the epoxy–amine copolymer.

Substituting in equations 1 and 2 yields

$$\Delta G = (R_g T/V_{E0})[(\phi_E/z_E)\ln \phi_E + (\phi_R/z_R)\ln \phi_R + \chi\phi_E\phi_R] \quad (3)$$

where $z_E$ and $z_R$ are the ratios of molar volumes of both components with respect to the reference:

$$z_E = V_E/V_{E0} \quad (4)$$

$$z_R = V_R/V_{E0} \quad (5)$$

$z_R$ remains constant during polymerization ($z_R = 13.73$) and $z_E$ increases with conversion. For simplicity and for explanation of trends, number-average values are taken to define the size increase. Therefore,

$$z_E = V_E/V_{E0} = \frac{\overline{M_n}}{\overline{M_{n0}}} \quad (6)$$

The number-average molecular weight of the epoxy–amine copolymer may be calculated as

$$\overline{M_n} = (\text{total mass})/(\text{total number of moles})$$

$$= (A_4 M_{A4} + B_2 M_{B2})/(A_4 + B_2 - 4pA_4) \tag{7}$$

where the total number of moles is calculated as the initial number less the number of reacted diamine equivalents and $p$ is the extent of reaction (each epoxy–amine bond decreases the number of molecules in one unit if there are no intramolecular cycles).

Because the epoxy–amine formulation is stoichiometric, $B_2 = 2A_4$. Then,

$$Z_E = \overline{M_n} / \overline{M_{n0}} = (1 - 4p/3)^{-1} \tag{8}$$

Because small changes in $z_E$ lead to high relative variations in the free energy of mixing, the increase of $z_E$ during polymerization is the main factor leading to phase separation (a possible change of $\chi$ with conversion has a less significant effect on the free energy variation). For example, for a cloud-point conversion $p_{cp} = 0.2$, $z_{E,cp} = 1.364$. Thus, a 36% increase in $z_E$ is sufficient to promote phase separation.

Binodal and spinodal curves in $p$ versus $\phi_R$ coordinates may be calculated by starting from equation 3, selecting a particular value for the interaction parameter, and using standard procedures, for example, equality of chemical potentials (*12, 13*). Figure 2 shows several binodal curves in the range of $\chi$ values going from 0.60 to 0.65. Because of the scattering in experimental results, an average value of $\chi = 0.63$ was taken for both cure temperatures.

Figure 4 shows a conversion versus rubber concentration phase diagram that indicates the location of binodal and spinodal curves for $\chi = 0.63$. The solution remains homogeneous at conversions below the binodal curve; it becomes metastable between both curves and unstable above the spinodal curve. The diagram is valid for cure temperature, $T$, between 50 and 75 °C for our particular system. Increasing temperature leads to an increase in miscibility and a shift of the diagram to higher reaction levels.

The critical rubber volume fraction is $\phi_{Rc} = 0.24$; expressed as a mass fraction, $(\%R)_c = 0.21$. For higher initial rubber concentrations, a phase inversion is predicted. An experimental verification of this value arises from the analysis of the glass-transition temperature of the epoxy network, $T_{g\infty}$, after a complete cure cycle. Figure 5 shows these values as a function of the initial rubber mass fraction in the formulation. Experimental points are represented by two straight lines. The intersection of these lines defines a critical mass fraction equal to $(\%R)_c = 0.24$. Therefore, the thermodynamic analysis based on the Flory–Huggins equation seems adequate to describe the stability region of rubber–epoxy mixtures.

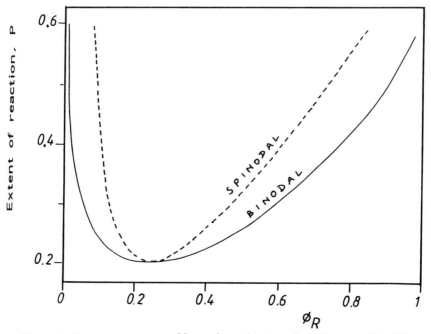

*Figure 4. Conversion versus rubber volume fraction phase diagram. Binodal and spinodal curves for a cure temperature in the range 50–75 °C are shown.*

Figure 5 shows that $T_{g\infty}$ decreases with an increase in the initial rubber concentration. This decrease must be attributed to the increase in the amount of rubber dissolved in the matrix at the end of the polymerization. However, precure temperature has practically no effect on $T_{g\infty}$ for formulations containing the same initial rubber content. This observation agrees with results reported for other systems (*11, 32*).

## *Nucleation–Growth vs. Spinodal Decomposition*

Figure 6 shows scanning electron micrographs (SEMs) of samples containing a mass fraction of rubber %R = 15, precured at different temperatures. The morphology consists of a set of spherical domains dispersed in a continuous matrix. This kind of morphology has been observed in most of the rubber-modified epoxies analyzed in the literature. Therefore, it is natural that many authors have used a nucleation–growth mechanism to describe the phase-separation process (*9–14, 22*). Moreover, the possibility of coalescence may be excluded, at least in those systems in which dispersed domains are not very close to one another. Indeed, this possibility has been theoretically predicted (*12*).

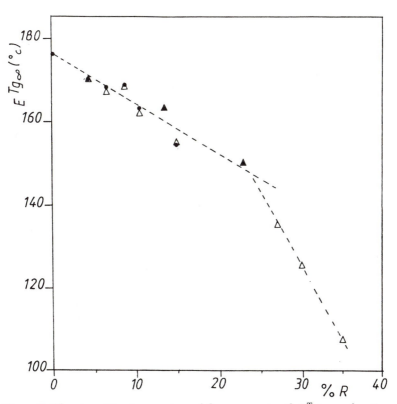

*Figure 5. Glass transition temperature of the epoxy network $E^{Tg\infty}$ as a function of the initial rubber mass fraction, %R, for formulations precured at 50 ( ● ) and 75 ( △ ) °C and postcured at 190 °C for 14 h. (Reproduced with permission from reference 20. Copyright 1990 John Wiley & Sons, Inc.)*

Conclusions from a final inspection of an electron micrograph are certainly insufficient and may be grossly misleading. Only a detailed analysis of thermodynamics in phase diagrams and the kinetic process can show whether, in a given case, we are indeed dealing with spinodal decomposition.

Figure 7 shows three possible trajectories in the metastable region of the phase diagram, starting from a composition typical of commercial formulations. The trajectory is determined by the value of the ratio:

$$K = (\text{phase-separation rate})/(\text{polymerization rate})$$

The phase-separation rate includes constitutive equations for nucleation and growth (12, 13, 22), and the polymerization rate is given by the particular kinetic expression of the epoxy–amine system in the presence of rubber (19, 20).

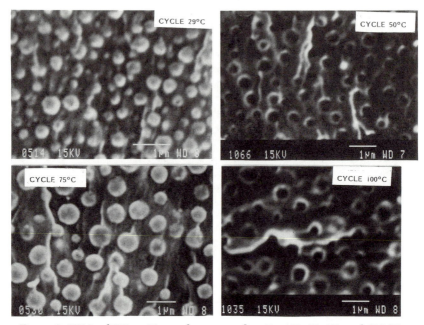

*Figure 6. SEMs of %R = 15 samples precured at* T = 29, 50, 75, *and 100 °C.*
*(Reproduced with permission from reference 21. Copyright 1991 John Wiley &*
*Sons, Inc.)*

If $K \to \infty$, equilibrium is instantaneously reached and the system evolves
along the binodal curve (trajectory a). The segregated phase composition is
given by the other branch of the binodal curve using a horizontal tie line
(trajectory a'). Alternatively, if $K \to 0$, no phase separation will be produced
until the spinodal curve is reached (trajectory c), at which point separation
takes place by a continuous and spontaneous process. Because the solution is
initially uniform in composition, separation occurs by a diffusional flux against
the concentration gradient (uphill diffusion with a negative diffusion coefficient). This condition leads to morphologies that display some degree of
connectivity (usually a co-continuous structure). At later stages of spinodal
decomposition the domain structure becomes large, that is, a coarsening of
the texture takes place.

Trajectory b represents the general case in which phase separation takes
place by nucleation and growth at a rate insufficient to achieve equilibrium
conditions. Compositions of the dispersed phase are indicated by trajectory
b', which lies outside the metastable region (12, 13, 22). This trajectory
represents the composition of generated nuclei (nucleation) as well as the
composition of the material incorporated to preexisting nuclei (growth) at the
particular conversion $p$. Dependent on $K$ value, a particular system
may start phase separation by nucleation and growth and end by spinodal

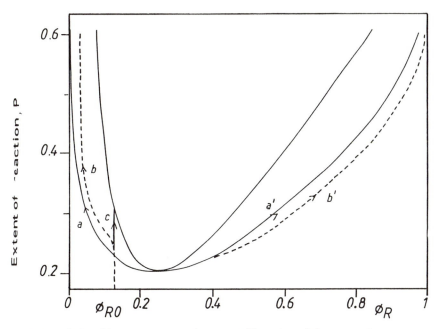

*Figure 7. Possible trajectories in the metastable region of the conversion versus composition phase diagram; a' and b' represent the compositions of the dispersed phase corresponding to trajectories a and b, respectively.*

demixing. However, experimental morphologies reveal that $K$ values must be sufficiently high so that nucleation–growth is the dominant mode of phase separation.

Recently, Yamanaka et al. (*33, 34*) suggested that phase separation proceeds via spinodal decomposition and that the spherical domain structure results from the fixation of the morphology at a late stage of spinodal demixing. This speculation is based on the statement (*33*): "A nucleation–growth mechanism is not conceivable because, first, it is recognized to be a very slow process and it may be skipped (i.e., nothing happens at the metastable region), and second, nucleation occurs accidentally and the subsequent growth results in an irregular domain structure (it never results in the regular structure that gives a light-scattering peak)."

Although the second argument is not self-sustaining (why should the nucleation–growth mechanism give an irregular structure?), the first argument deserves a further analysis.

The necessary condition to have a nucleation–growth mechanism is that the second phase can nucleate from the homogeneous solution. The nucleation rate, NR, is proportional to (*12,13, 22*):

$$NR \approx \exp(-\Delta G_c/kT) \tag{9}$$

where $k$ is the Boltzmann constant and

$$\Delta G_c = 16 \, \pi \sigma_0^3/3|\Delta G_N|^2 \tag{10}$$

$\sigma_0$ is the interfacial tension between both phases, and $\Delta G_N$ is the free energy change per unit volume.

A very high dependence of the nucleation rate on the interfacial tension is expected. Figure 8 shows trajectories for several values of the interfacial tension $\sigma_0$ in the metastable region that are predicted by using a nucleation–growth model (22). For $\sigma_0 \leq 0.01$ mN/m (dyne/cm), all the trajectories are equivalent, that is, the value of $\sigma_0$ has no effect on the phase separation process. This conclusion was previously reported by Williams and co-workers (12, 13) using values in the range where no influence is actually

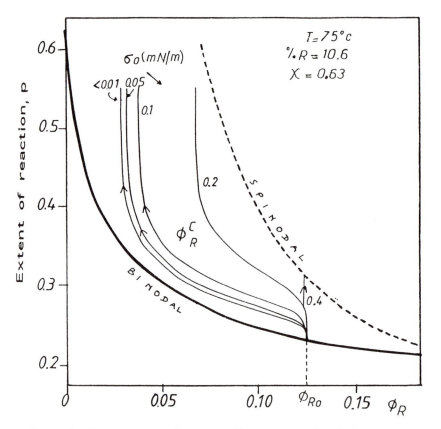

*Figure 8. Trajectories in the metastable region predicted by using a nucleation–growth model for several values of the interfacial tension $\sigma_0$. (Reproduced with permission from reference 22. Copyright 1991 John Wiley & Sons, Inc.)*

observed (although the stated values of $\sigma_0$ were higher, as was correctly shown by Sohn et al. (35)). However, if $\sigma_0$ is higher than 0.1 mN/m, spinodal decomposition is rapidly attained. For $\sigma_0 = 0.4$ mN/m practically no phase separation occurs by the nucleation–growth mechanism, and demixing proceeds by spinodal decomposition.

Consider the range of possible $\sigma_0$ values in rubber-modified epoxies. Sohn et al. (35) recently reported interfacial tension values between $\alpha,\omega$-methylcarboxylatobutadiene–acrylonitrile copolymers and an epoxy pre-polymer as a function of temperature and copolymer composition. For a copolymer containing the same acrylonitrile content used here, they reported a value of $\sigma_0 = 0.58$ mN/m at 55 °C and a temperature dependence $d\sigma_0/dT = -(0.01-0.02)$ mN/m °C. However, a sharp decrease in interfacial tension is observed (36, 37) with the use of block copolymers like the ETBNs normally used. This decrease arises mainly from the energetically preferred orientation of the blocks at the interface into their respective compatible phases. Also, interfacial tensions of demixed polymer solutions derived from polymer–polymer–solvent systems were reported (38, 39) in the range of $10^{-4}-10^{-1}$ mN/m.

Therefore, the very low values of the interfacial tensions in these systems enable phase separation by a nucleation–growth mechanism, provided the polymerization rate is not too fast. Indeed, quenching an unreactive CTBN–epoxy solution to the unstable region by a sudden temperature decrease leads to spinodal demixing in spite of the very low values of interfacial tension (29). This demixing gives rise to an interconnected-domain structure.

Observation of the conversion versus composition phase diagram plotted in Figure 7 indicates that the condition favoring spinodal demixing is $K \rightarrow 0$ or begin with a composition located close to the critical point (in this case the unstable region can be avoided only by a very fast separation). Precisely under these conditions, morphologies resembling those obtained by spinodal demixing are observed in rubber-modified epoxies (28, 33). In both cases the curing agent was a polyaminoamide (Versamid, Henkel), which has a high viscosity and a short gel time (both properties lead to a low $K$ value). Rubber concentrations were %R = 21 (28) and %R = 23 (33), that is, values very close to the critical point. This last condition seems to be the very reason for the observed morphologies, because when the rubber concentration was lowered to %R = 11.8, the typical dispersion of spherical domains was observed (33).

In conclusion, the very low values of interfacial tension in rubber-modified epoxies lead to a nucleation–growth mechanism for phase separation. This mechanism is unusual for polymer blends (mixtures of unreactive, high molecular weight thermoplastics) showing interfacial tensions lying in the range 0.5–11 mN/m (40).

To confirm this assertion, we introduced an initially more miscible rubber (24), CTBN × 13 (26AN) instead of CTBN × 8 (17AN), into the

DGEBA–3DCM mixture. The precured temperature was 75 °C. The average size of dispersed domains for the system based on CTBN × 8 is 0.43 μm, whereas it is only 0.18 μm for the system based on CTBN × 13. Similar experiments performed by Hsich (28) with another diamine, Jeffamine D230, display similar reactivity to 3DCM. This result confirms that the mechanism of phase separation is the result of nucleation and growth. The higher initial miscibility of CTBN × 13 (24) results in the occurrence of phase separation at a higher epoxy conversion than with CTBN × 8 under similar cure conditions. Experimental values are $p_{cp}$ = 0.30 (for CTBN × 13) and $p_{cp}$ = 0.17 (for CTBN × 8). Thus the expectation that the average size of dispersed domains is larger with CTBN × 8 than with CTBN × 13 is confirmed.

In the spinodal decomposition mechanism, as observed by Hsich with a more viscous and more reactive system (28), the opposite result will occur because the driving force is inversely proportional to the square root of the difference between cure temperature, $T_i$, and spinodal temperature $T_s$. The value of $(T_i - T_s)^{-0.5}$ for the system based on CTBN × 8 is greater than that based on CTBN × 13. Consequently the particle size of the dispersed domains should be smaller with CTBN × 8, a fact contrary to the experimental observation.

A definitive argument supporting the nucleation–growth mechanism was recently provided by Rozenberg (41): Diffusion coefficients measured in situ during phase separation were always positive, a result that is consistent with a nucleation–growth mechanism but not with spinodal demixing.

## Evolution of the Average Size of Dispersed Domains

Figure 9 shows a theoretical simulation (22) of the average diameter of dispersed-phase particles as a function of conversion for samples containing different initial rubber concentrations and polymerized at 50 °C. Points represent experimental values determined from SEMs. Phase separation beyond $p$ = 0.35 is severely retarded because of the high viscosity increase (the diffusion coefficient is inversely proportional to viscosity through the Stokes–Einstein equation). Therefore, most of the morphology is generated in a narrow conversion range well before gelation (or vitrification).

Yu and Von Meerwall (42) measured the self-diffusion of molecules of two epoxy systems during curing, well beyond the gel point. Consistent with the trends presently under discussion, the molecular motions begin to decline around $p$ = 0.3; then the diffusion decrease is continuous across the gel point up to $p$ ≈ 0.7.

For %R = 10.6 and 15, the observed decrease in the average size with conversion before attaining the constant value is the consequence of the

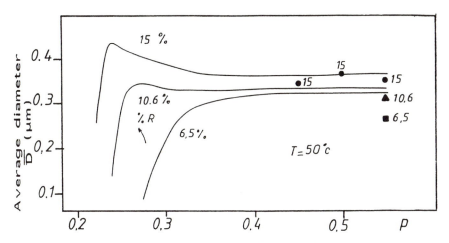

*Figure 9. Average diameter of dispersed phase particles as a function of conversion for polymerization carried out at 50 °C starting from different initial rubber mass fractions. Points represent experimental values, and full curves are model simulations. (Reproduced with permission from reference 22. Copyright 1991 John Wiley & Sons, Inc.)*

predominance of nucleation over growth in this conversion range. This factor produces a decrease in the average diameter of the whole population and leads to a bimodal distribution of sizes, that is, the smaller particles are those born at higher conversions.

## Particle Size Distribution

Figure 10 shows a comparison between experimental and predicted particle size distributions for a sample containing %R = 10.6 and polymerized at 50 °C. The simulation, based on a nucleation–growth mechanism to describe phase separation (*12, 13, 22*), gives a reasonable prediction of the observed range of particle sizes, but does not fit the actual shape of the distribution. The model also predicts that decreasing particle size corresponds to increasing conversion at which they were born.

## Composition of Both Phases

Values of the glass-transition temperature of the epoxy network, $E^{Tg}$, are shown in Figure 5 as a function of the initial rubber mass fraction, %R. The decrease in $E^{Tg}$ with %R is explained by the fact that some of the rubber remains dissolved in the matrix at the end of the cure.

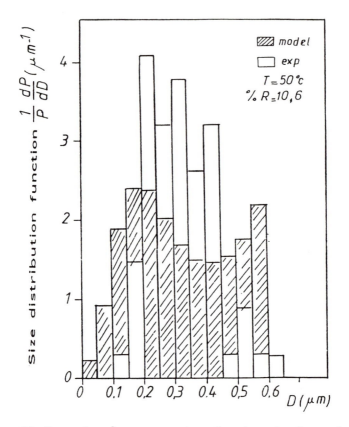

*Figure 10. Comparison between experimental and predicted particle-size distributions for a sample containing %R = 10.6, polymerized at 50 °C (the larger fraction is depicted behind the smaller one). (Reproduced with permission from reference 22. Copyright 1991 John Wiley & Sons, Inc.)*

The mass fraction of rubber dissolved in the matrix, $w_R^c$, may be calculated by assuming the validity of the Fox equation (*43*):

$$\frac{1}{(E^{Tg})} = \frac{(1 - w_R^c)}{(E^{Tg, pure})} + \frac{w_R^c}{(R^{Tg, pure})} \tag{11}$$

where $E^{Tg, pure} = 180$ °C (*19*) and $R^{Tg, pure} = -60$ °C (*20*).

The volume fraction of rubber dissolved in the matrix is calculated from

$$\phi_R^c = (w_R^c/\rho_R)/[w_R^c/\rho_R + (1 - w_R^c)/\rho_E] \tag{12}$$

where $\rho_R = 0.948$ g/cm$^{-3}$ and $\rho_E = 1.127$ g/cm$^{-3}$.

The composition of the dispersed phase may be obtained by stating a mass balance of rubber in the overall system. By calling $\phi_{R0}$, the initial volume fraction of rubber added to the formulation, it must be verified that

$$\phi_{R0} = V_D \overline{\phi_R^D} + (1 - V_D)\phi_R^c \tag{13}$$

where $V_D$ is the volume fraction of dispersed phase (measured by SEM) and $\overline{\phi_R^D}$ is the average volume fraction of rubber in dispersed domains. Then, the average volume fraction of epoxy–amine polymer in dispersed domains is given by

$$\overline{\phi_E^D} = 1 - \overline{\phi_R^D} \tag{14}$$

Figure 11 shows experimental values of $\overline{\phi_E^D}$ and $\phi_R^c$ compared with model simulations (22). Taking into account the uncertainty in experimental

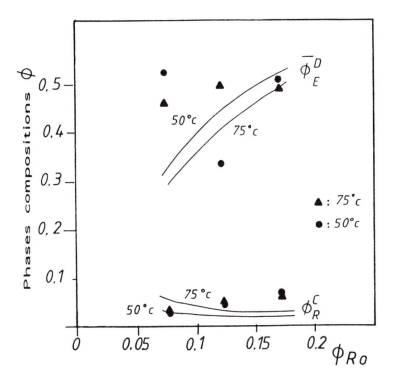

*Figure 11. Average volume fraction of epoxy–amine polymer in dispersed domains, $\overline{\phi_E^D}$, and rubber volume fraction remaining in the matrix, $\phi_R^c$, as a function of the initial rubber concentration, $\phi_{R0}$. Points represent experimental values, and full curves are model simulations. (Reproduced with permission from reference 22. Copyright 1991 John Wiley & Sons, Inc.)*

values, the fitting may be considered reasonable. Both the model and experimental results show that a high fraction (from 30 to 50%) of the volume of dispersed domains is composed of epoxy–amine polymer. Several authors have reported the presence of an unstained region inside the dispersed domains observed in transmission electron micrographs (44–46). These reports provide direct evidence of the presence of epoxy polymer in the dispersed phase. Using $T_g$ shifts and mass balances, Hwang et al. (47) recently showed, for two different formulations of rubber-toughened epoxies, that about one-half of the composition of the rubber phase is epoxy.

## Nature of Dispersed Domains

From observation of the trajectories a' and b' depicted in Figure 7, it may be inferred that the largest particles (formed at low conversions) are richer in epoxy copolymer than the smallest particles (formed at high conversions). The smallest particles have a very high rubber content.

The average volume fraction of epoxy–amine polymer in dispersed domains, $\overline{\phi_E^D}$, lies in the unstable region at $p_{gel}$ as seen in Figure 7. This condition means that a secondary phase separation has to take place inside dispersed particles (at least in those particles with compositions located inside the unstable region at gelation). If equilibrium is attained, compositions of both phases are read at both branches of the binodal curve, that is, almost pure rubber and pure epoxy–amine polymer are segregated. This segregation explains why the glass-transition temperature ascribed to the rubber phase, $R^{Tg}$, is close to that of pure rubber. In fact, experimental values of $R^{Tg}$ are lower than values of $R^{Tg, pure}$ (9, 21, 46), a fact attributed to differences in the coefficients of thermal expansion between the glassy epoxy matrix and the rubber-rich phase. The larger coefficient of thermal expansion of the rubber results in constraint of the rubber domains upon cooling below the glass-transition temperature of the matrix. Supporting this statement, Romanchick et al. (46) showed that $R^{Tg}$ of the rubber contained in dispersed domains, initially present in their uncured system, decreased when the epoxy matrix was cross-linked.

With regard to the epoxy–amine polymer contained in dispersed domains, some differences may occur in the relaxation temperature with respect to the relaxation temperature in the matrix. To determine this, the Rheometrics dynamic analyzer was used at an oscillation frequency of 0.016 Hz. Because the $\alpha$ relaxation associated with the glass transition of the rubbery phase is close to the $\beta$ relaxation of the epoxy network, a deconvolution technique was used to separate both peaks.

Figure 12a shows a comparison of the relaxations appearing in samples with and without rubber. The $\alpha$ relaxation of the epoxy copolymer, $E^{T\alpha}$, shifts to lower temperature by increasing the initial rubber concentration.

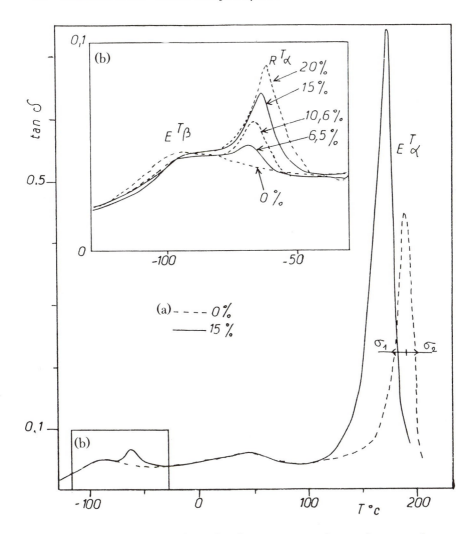

*Figure 12. a, Dynamic mechanical analysis at 0.016 Hz for samples precured at 50 °C and containing %R = 0 and 15. b, Magnification of the low-temperature region showing $E^{T\beta}$ and $R^{T\alpha}$ for samples precured at 50 °C and containing %R8 = 0, 6.5, 10.6, 15, and 20. (Reproduced with permission from reference 21. Copyright 1991 John Wiley & Sons, Inc.)*

The relaxation is asymmetrical, that is, $\sigma_1$ and $\sigma_2$ that measure the half-widths of the peak at half of its total height are not equal.

Figure 13 shows the correlation between the asymmetry of the $E^{T\alpha}$ relaxation, measured by $(\sigma_1 - \sigma_2)/\sigma_2$, and the overall volume fraction of epoxy polymer in dispersed domains, $V_D \overline{\phi_E^D}$. It may be inferred that the epoxy present in the dispersed phase relaxes in the lower temperature region

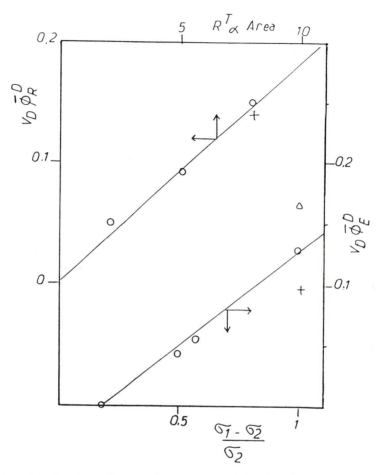

*Figure 13. Correlations between the asymmetry of the $E^{T\alpha}$ relaxation, measured by $(\sigma_1 - \sigma_2)/\sigma_2$, and the overall volume fraction of epoxy polymer in dispersed domains, $V_D\overline{\phi_E^D}$, and correlations between the area under the $R^{T\alpha}$ relaxation after deconvolution and the overall volume fraction of rubber in dispersed domains, $V_D\overline{\phi_R^D}$. Samples were precured at 29 ( △ ), 50 (○), and 100 ( + ) °C. (Reproduced with permission from reference 21. Copyright 1991 John Wiley & Sons, Inc.)*

of the $E_\alpha^T$ relaxation. This relaxation may be due to the covalently bonded CTBN blocks or to a lower degree of cross-linking because of a nonstoichiometric segregation of epoxy and amine functionalities from the matrix.

The secondary phase separation taking place inside dispersed domains may possibly continue after gelation of the matrix. This feature would explain the decay in light transmission after gelation observed by Wang and Zupko (27) and attributed to localized changes in composition.

Figure 12b shows that the $R^{T\alpha}$ relaxation increases with the initial rubber concentration whereas Figure 13 shows a linear correlation between the area under the relaxation peak and the volume fraction of rubber in dispersed domains, $V_D \overline{\phi_R^D}$.

A conclusion is that two phases occur inside dispersed domains. The rubbery phase relaxes at a lower temperature than pure rubber because of the presence of unrelaxed thermal stresses that result from differences in the coefficients of expansion during cooling from the cure temperature. The epoxy copolymer relaxes at a temperature very close to, but lower than, the matrix. This condition may be due to the bonds between epoxy and CTBN blocks or to a lower epoxy copolymer cross-link density.

## *Influence of Rubber Concentration*

Both experimental results and model simulations show that the volume fraction of dispersed phase increases with the initial rubber concentration in the formulation (in the usual range of $\phi_R < \phi_{Rc}$). This increase is due to the decrease in the extent of reaction at the beginning of phase separation (*see* Figures 2 and 4) and the consequent decrease of viscosity and increase in nucleation and growth rates.

Figure 14 shows the viscosity ($\eta$) rise as a function of time, at 50 °C, for formulations containing different rubber amounts. Clearly, the higher the rubber concentration, the lower the viscosity at the cloud point (indicated by an arrow). Of interest, also, is the fact that after phase separation the rate of viscosity increase slows down, yielding to a crossover of curves. Thus, increasing the rubber amount increases the viscosity of the initial homogeneous solution but decreases the viscosity of the rubber–epoxy dispersion. The crossover of curves occurs at a time close to the experimental cloud point.

## *Influence of Cure Temperature*

Figure 15 shows the increase in the average particle size with cure temperature observed for three rubber-modified epoxies. This trend may be explained by both kinetic and thermodynamic arguments.

Assume that the location of the binodal curve does not change significantly in the temperature range of interest (fixed thermodynamic conditions). In the metastable region, the trajectory depends on the value of the factor $K$ = phase separation rate/polymerization rate. Because the activation energy for the reaction kinetics $E_{\alpha,\,polym} = 6921K$ (*19*) is higher than the activation energy for mass transfer by diffusion $E_{\alpha,\,diff} = 4037K$ (*20*), an increase in temperature leads to a decrease in $K$. As a consequence, the trajectory in the metastable region deviates from the binodal curve, that is, in

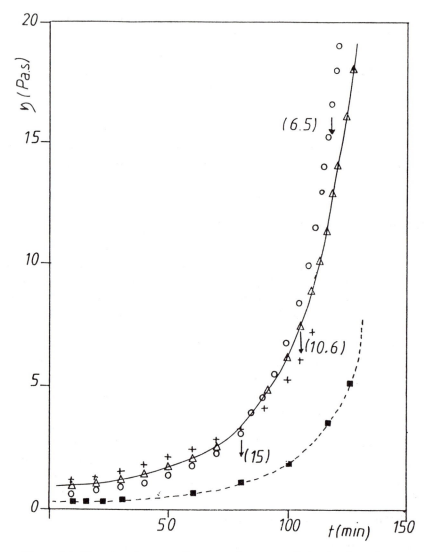

*Figure 14. Viscosity rise as a function of time at 50 °C for formulations containing different amounts of rubber (shear rate = 100/s). Arrows indicate the viscosity value at the cloud point. %R = 0 ( ■ ), 6.5 (○), 10.6 ( △ ), and 15 ( + ). (Reproduced with permission from reference 20. Copyright 1990 John Wiley & Sons, Inc.)*

Figure 7 trajectory b corresponds to a higher temperature than trajectory a. This fact has two implications: a decrease in the concentration of particles and an increase in their average size because of the higher driving force for growth (composition difference between the trajectory and the binodal

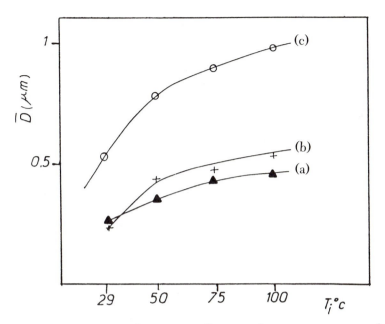

*Figure 15. Average particle size as a function of precure temperature for different systems: a, DGEBA(n̄ = 0.03)–3DCM–15%R8; b, DGEBA (n̄ = 0.03)–3DCM–15%R9; and c, DGEBA(n̄ = 0.15)–MNDA–15%R8. (Reproduced with permission from reference 21. Copyright 1991 John Wiley & Sons, Inc.)*

curve). The counterbalance between these two effects makes the volume fraction of dispersed phase vary insignificantly with temperature (*11, 12*).

If the temperature range is broad enough, the thermodynamic argument must be added to the kinetic argument. In this case, it must be taken into account that miscibility increases with temperature because the interaction parameter follows an equation of the type $\chi = A + B/T$ with $B > 0$. Therefore, the system will enter the metastable region at a higher conversion. Because viscosity is much more sensitive to the conversion value than the polymerization rate, a decrease in the value of the factor $K$ will result and again produce fewer but bigger particles.

If temperature is so high that miscibility is complete up to gelation, no phase separation will be observed. This result indicates that a maximum dependence of the average particle size with temperature is expected, as has been theoretically predicted (*12*) and experimentally confirmed (*6, 48*).

## Fracture Behavior

For our system, increasing the rubber mass fraction from 0 to 20%, led to a decrease in the Young modulus from 2.4 to 1.3 GPa and in the yield strength from 128 to 69 MPa. These trends arise from the increase in both the volume fraction of dispersed phase and the amount of rubber remaining in the matrix at the end of phase separation. The plastification of the neat epoxy–amine thermoset is clearly reflected by the decrease in $E^{Tg}$ shown in Figure 5.

Figure 16 shows a linear correlation between the fracture energy, $G_{Ic}$, and the volume fraction of dispersed phase, in agreement with experimental results reported by other authors (18, 45). In our case, both the increase in $V_D$ and the simultaneous decrease in $E^{Tg}$ with %R favor the plastic shear yielding in the matrix and the consequent increase in fracture energy.

However, as pointed out by Kinloch (15), the relationship between toughness and volume fraction of rubbery particles depends on the value of $t_f/a_T$ ($t_f$ is the time to failure and $a_T$ is the time–temperature shift factor). This condition means that the parametric sensitivity of $G_{Ic}$ on $V_D$ may be changed by varying the test temperature (49).

## Suggestions for Future Work

To obtain experimental conclusive evidence of the mechanism of phase separation, it is necessary to follow the development of morphology as a function of conversion with electron microscopy, that is, selecting a system of

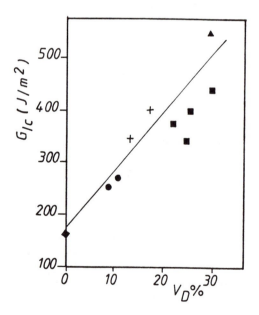

*Figure 16. Fracture energy at room temperature, $G_{Ic}$, as a function of the volume fraction of dispersed phase, $V_D$, for different initial mass fractions of rubber: %R = 0 ( ♦ ), 6.5 ( ● ), 10.6 ( + ), 15 ( ■ ), and 20 ( ▲ ). (Reproduced with permission from reference 23. Copyright 1991 John Wiley & Sons, Inc.)*

high $T_{g0}$ and quenching the morphology by a temperature jump to room temperature (provided the system does not go through the unstable region during the quenching). In our system, the first experimental value shown in Figure 9 corresponds to a conversion giving a $T_g$ slightly higher than room temperature.

Another interesting possibility is to use a system where the polymerization rate may be varied in a broad range by using a small catalyst concentration. This approach would enable different trajectories to be followed in the metastable region at constant temperature. Eventually, spinodal decomposition should result for very fast polymerizations.

Two-stage polymerizations may be used to vary and control morphologies derived from the same formulation. Phase separation may be produced during the formation of a prepolymer obtained by using an excess of one of the monomers (first stage), and then the stable dispersion may be stoichiometrically balanced and polymerized in a second stage to obtain the final material which exhibits the morphology generated in the first stage.

## Acknowledgments

This work was performed within the framework of a cooperation program between the National Research Councils of France (CNRS) and Argentina (CONICET).

## References

1. McGarry, F. J.; Willner, A. M., research report, School of Engineering, Massachusetts Institute of Technology: Cambridge, MA, 1966.
2. McGarry, F. J.; Willner, A. M., research report R 68–8, School of Engineering, Massachusetts Institute of Technology: Cambridge, MA, 1966.
3. McGarry, F. J.; Willner, A. M., research report R 69–36, School of Engineering, Massachusetts Institute of Technology: Cambridge, MA, 1969.
4. Drake, R. S.; McCarthy, W. J. *Rubber World* **October, 1968**, 159.
5. Rowe, E. H.; Siebert, A. R.; Drake, R. S. *Mod. Plast.* **1970**, 47, 100.
6. Burhans, A. S.; Soldatos, A. C. Presented at the 25th Annual Technical Conference, RP/Compos. Div. SPI, 1970; paper 3-C.
7. Burhans, A. S.; Soldatos, A. C. In *Multicomponent Polymer Systems*; Platzer, N. A. J., Ed.; Advances in Chemistry 99; American Chemical Society: Washington, DC, 1971; pp 531–546.
8. Visconti, S.; Marchessault, R. H. *Macromolecules* **1974**, 7, 913.
9. Manzione, L. T.; Gillham, J. K.; McPherson, C. A. *J. Appl. Polym. Sci.* **1981**, 26, 889.
10. Manzione, L. T.; Gillham, J. K.; McPherson, C. A. *J. Appl. Polym. Sci.* **1981**, 26, 907.
11. Montarnal, S.; Pascault, J. P.; Sautereau, H. In *Rubber-Toughened Plastics*; Riew, C. K., Ed.; Advances in Chemistry 222; American Chemical Society: Washington, DC, 1989; pp 193–223.

12. Williams, R. J. J.; Borrajo, J.; Adabbo, H. E.; Rojas, A. J. In *Rubber-Modified Thermoset Resins*; Riew, C. K.; Gillham, J. K., Eds.; Advances in Chemistry 208; American Chemical Society: Washington, DC, 1984; pp 195–213.
13. Vazquez, A.; Rojas, A. J.; Adabbo, H. E.; Borrajo, J.; Williams, R. J. J. *Polymer* **1987**, *28*, 1156.
14. Roginskaya, G. F.; Volkov, V. P.; Bogdanova, L. M.; Chalykh, A. E.; Rozenberg, B. A. *Polym. Sci. USSR* **1983**, *25*, 2305.
15. Kinloch, A. J. In *Rubber-Toughened Plastics*; Riew, C. K., Ed.; Advances in Chemistry 222; American Chemical Society: Washington, DC, 1989; pp 67–91.
16. Kinloch, A. J.; Shaw, S. J.; Tod, D. A.; Hunston, L. D. *Polymer* **1983**, *24*, 1341.
17. Kinloch, A. J.; Shaw, S. J.; Hunston, D. L. *Polymer* **1983**, *24*, 1355.
18. Yee, A. F.; Pearson, R. A. *J. Mater. Sci.* **1986**, *21*, 2462.
19. Verchère, D.; Sautereau, H.; Pascault, J. P.; Riccardi, C. C.; Moschiar, S. M.; Williams, R. J. J. *Macromolecules* **1990**, *23*, 725.
20. Verchère, D.; Sautereau, H.; Pascault, J. P.; Moschiar, S. M.; Riccardi, C. C.; Williams, R. J. J. *J. Appl. Polym. Sci.* **1990**, *41*, 467.
21. Verchère, D.; Pascault, J. P.; Sautereau, H.; Moschiar, S. M.; Riccardi, C. C.; Williams, R. J. J. *J. Appl. Polym. Sci.* **1991**, *42*, 701.
22. Moschiar, S. M.; Riccardi, C. C.; Williams, R. J. J.; Verchère, D.; Sautereau, H.; Pascault, J. P. *J. Appl. Polym. Sci.* **1991**, *42*, 717.
23. Verchère, D.; Pascault, J. P.; Sautereau, H.; Moschiar, S. M.; Riccardi, C. C.; Williams, R. J. J. *J. Appl. Polym. Sci.* **1991**, *43*, 293.
24. Verchère, D.; Pascault, J. P.; Sautereau, H.; Moschiar, S. M.; Riccardi, C. C.; Williams, R. J. J. *Polymer* **1989**, *30*, 107.
25. Roginskaya, G. F.; Volkov, V. P.; Chalykh, A. E.; Avdeev, N. N.; Rozenberg, B. A. *Doklad. Akad. Nauk. SSSR* **1979**, *21*, 2111.
26. Roginskaya, G. F.; Volkov, V. P.; Chalykh, A. E.; Mavdeev, V. V.; Rozenberg, B. A. *Dokl. Akad. Nauk. SSSR* **1980**, *252*, 402.
27. Wang, T. T.; Zupko, H. M. *J. Appl. Polym. Sci.* **1981**, *26*, 2391.
28. Hsieh, H. S. Y. *Polym. Eng. Sci.* **1990**, *30*, 493.
29. Lee, H. S.; Kyu, T. *Macromolecules* **1990**, *23*, 459.
30. Roe, H. J.; Zin, W. C. H. *Macromolecules* **1980**, *13*, 1221.
31. Borrajo, J.; Riccardi, C. C.; Moschiar, S. M.; Williams, R. J. J. In *Rubber-Toughened Plastics*; Riew, C. K., Ed.; Advances in Chemistry 222; American Chemical Society: Washington, DC, 1989; pp 319–328.
32. Chan, L. C.; Gillham, J. K.; Kinloch, A. J.; Shaw, S. J. In *Rubber-Modified Thermoset Resins*; Riew, C. K.; Gillham, J. K., Eds.; Advances in Chemistry 208; American Chemical Society: Washington, DC, 1984; pp 235–260.
33. Yamanaka, K.; Takagi, Y.; Inoue, T. *Polymer* **1989**, *60*, 1839.
34. Yamanaka, K.; Inoue, T. *J. Mater. Sci.* **1990**, *25*, 241.
35. Sohn, J. E.; Emerson, J. A.; Thompson, P. A.; Koberstein, J. T. *J. Appl. Polym. Sci.* **1989**, *37*, 2627.
36. Noolandi, J.; Hong, K. M. *Macromolecules* **1984**, *17*, 1731.
37. Anastasiadis, S. H.; Gancarz, I.; Koberstein, J. T. *Macromolecules* **1989**, *22*, 1449.
38. Langhammer, G.; Nester, L. *Makromol. Chem.* **1965**, *88*, 179.
39. Riess, G. In *Initiation à la Chimie et à la Physicochimie Macromoléculaires. Mélanges des Polymères*; Groupe Français des Polymères (GFP), Ed.; Strasbourg: 1986; Vol. 6, p 73.
40. Gaines, G. L., Jr. *Polym. Eng. Sci.* **1972**, *12*, 1.
41. Rozenberg, B. A. *Makromol. Chem., Macromol. Symp.* **1991**, *41*, 165.
42. Yu, W.; Von Meerwall, E. D. *Macromolecules* **1990**, *23*, 882.
43. Fox, T. G. *Bull. Am. Phys. Soc.* **1956**, *1*, 123.

44. Kunz, S. C.; Sayre, J. A.; Assink, R. A. *Polymer* **1982**, *23*, 1897.
45. Bucknall, C. B.; Yoshii, T. *Br. Polym. J.* **1978**, *10*, 53.
46. Romanchick, W. A.; Sohn, J. E.; Geibel, J. F. In *Epoxy Resin Chemistry II*; Bauer, R. S., Ed.; ACS Symposium Series 221; American Chemical Society: Washington, DC, 1983; p 85.
47. Hwang, J. F.; Manson, J. A.; Hertzberg, R. W.; Miller, G. A.; Sperling, L. H. *Polym. Eng. Sci.* **1989**, *29*, 1466.
48. Butta, E.; Levita, G.; Marchetti, A.; Lazzeri, A. *Polym. Eng. Sci.* **1986**, *26*, 63.
49. Kinloch, A. J.; Hunston, D. L. *J. Mater. Sci. Lett.* **1987**, *6*, 137.

RECEIVED for review March 6, 1991. ACCEPTED revised manuscript August 29, 1991.

# Thermal Shock Resistance of Toughened Epoxy Resins

Masatoshi Kubouchi[1] and Hidemitsu Hojo[2]

[1]Department of Chemical Engineering, Tokyo Institute of Technology, O-okayama, Meguro-ku, Tokyo 152, Japan
[2]Department of Industrial Engineering and Management, Nihon University, College of Industrial Technology, 1-2-1 Izumi-cho, Narashino-shi, Chiba-ken 275, Japan

*The thermal shock resistance of epoxy resin specimens toughened with carboxy–terminated poly(butadiene–acrylonitrile) (CTBN) and polyglycol were tested using a new notched disk-type specimen. The new thermal shock testing method consists of quenching a notched disk-type specimen and applying a theoretical analysis to the test results to determine crack propagation conditions. For both toughened epoxy resins, this test method evaluated improvements in thermal shock resistance. The thermal shock resistance of epoxy resin toughened with CTBN exhibited a maximum at a 35 parts per hundred resin content of CTBN. The epoxy resin toughened with polyglycol exhibited improved thermal shock resistance with increasing glycol content.*

THE PROBLEM OF CRACK RESISTANCE is increasing in importance with the demand for high reliability resin products. The uses for epoxy resin range from large electrical products to miniature electronic parts. The broad spectrum of use reflects the excellent insulation resistance of epoxy resins.

Practical methods for testing the thermal-stress crack resistance of epoxy resin are heat cycle tests, such as the Olyphant washer test (*1*), that investigate crack initiation. These relatively rapid tests use a specimen containing a metal insert: a washer or bolt embedded near the center of an encapsulating resin disk. In this type of specimen, the stress condition is

0065–2393/93/0233–0365$06.00/0

complicated and difficult to analyze, which makes comparison with failure mechanisms in commercial applications difficult.

A new test method is proposed that evaluates the crack resistance of epoxy resin by using a thermal shock method with a notched disk-type specimen (2), followed by application of an analytical method based on linear fracture mechanics (3). Because this testing method consists of simple fracture modes, a quantitative analysis can be applied. This chapter examines the applicability of the new testing and evaluation method to toughened epoxy resin modified with carboxy-terminated poly(butadiene–acrylonitrile) (CTBN) or plasticizer.

## Experimental Method

The new thermal shock tests are conducted by quickly cooling the preheated notched disk-type specimen. Consideration of simple specimen geometry and stress analysis follows the same procedures as Manson (4).

**Materials and Specimen**.    Two bisphenol-type epoxies (epoxy equiv wt = 184–194 and 370–435) are used as the matrix resin, and acid anhydride is added as a hardener. The disk-shaped specimen is made by casting. The shape and size (60-mm diameter; 10 mm thick) of the disk-type specimen are shown in Figure 1. A sharp notch is introduced at the base of the slot (0.3 mm wide) using a razor blade. Toughened epoxy specimens modified with carboxy-terminated poly(butadiene–acrylonitrile) (CTBN; mol wt = 3400) or polyglycol plasticizer were also tested. A larger specimen size (120 mm diameter; 10 mm thick) was used for testing polyglycol-toughened epoxy. The physical and mechanical properties at room temperature of the respective epoxy resins are shown in Table I.

**Test Method**.    As shown in Figure 1, the specimen is fixed both top and bottom with balsa insulators, which cause the specimen to suffer lateral thermal cooling shock. The disk specimen cools from the surface to the center along the radius. Then expansion difference between the outer and inner parts of the specimen creates mode-I fracture mechanics at the tip of the specimen notch.

The 5 min cooling period corresponds to the time necessary to maximize the stress intensity factor at the notch tip. For the lowest temperature, a dry ice–pentane cooling system (approximately 200 K) is used. Temperature difference is created by changing the initial temperature from 280 to 380 K. The crack initiation due to thermal shock is detected by visual observation after cooling.

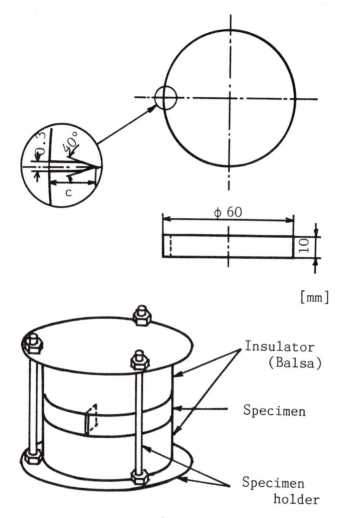

*Figure 1. The test specimen.*

## Theoretical Analysis

In this analysis it is assumed that the temperature distribution is not influenced by the notch and that the thermal stress distribution is a quasi-static state. Based on the elastic stress analysis and linear fracture mechanics, the stress intensity factor $K_I$ at the notch tip is obtained as follows:

$$K_I = 1.12 \frac{2\sqrt{c}}{\sqrt{\pi}} \int_0^R \frac{\sigma_\theta}{\sqrt{(c^2 - \xi^2)}} \, d\xi \qquad (1)$$

**Table I. Physical and Mechanical Properties of Epoxy Resin
at Room Temperature**

| Property | CTBN-Toughened, Modifier Content (phr)[a] | | | | Plasticizer-Modified, Modifier Content (phr)[b] | | |
|---|---|---|---|---|---|---|---|
| | 0 | 20 | 35 | 50 | 0 | 10 | 20 |
| Tensile strength (MPa) | 82.5 | 68.0 | 55.6 | 39.3 | 81.2 | 79.2 | 65.9 |
| Young's modulus (GPa) | 3.56 | 2.78 | 2.07 | 1.85 | 3.29 | 2.78 | 2.71 |
| Poisson's ratio (m/m) | 0.36 | 0.40 | 0.37 | 0.40 | 0.39 | 0.34 | 0.35 |
| Fracture toughness (MPa$\sqrt{m}$) | 1.24 | 2.06 | 2.18 | 1.51 | 1.87 | 1.93 | 2.16 |
| Rockwell hardness (H scale) | 75 | 57 | 27 | 19 | 81 | 64 | 58 |
| Thermal conductivity (J/m s K) | 0.24 | 0.23 | 0.23 | 0.23 | 0.20 | 0.20 | 0.20 |
| Coefficient of linear thermal expansion (1/K) | 5.24 | 7.69 | 8.93 | 8.72 | 4.91 | 7.08 | 6.80 |

[a] EP + Me–THPA; EP equivalent 184–194.
[b] EP + PA; EP equivalent 370–435.

$$\sigma_\theta = \frac{\alpha E \Delta T}{(1 - \nu)} \left\{ \frac{1}{R^2} \int_0^R T^* r \, dr + \frac{1}{r^2} \int_0^r T^* r \, dr - T^* \right\} \tag{2}$$

where $\nu$ is Poisson's ratio, $\alpha$ is the coefficient of thermal expansion, $E$ is the elastic modulus, $R$ is the radius of the disk specimen, $\Delta T$ is the temperature difference ($= T_i - T_f$, where $T$ is temperature and subscripts i and f indicate initial and final), $c$ is the notch length, $\xi$ is the coordinate onto the notch, $r$ is the distance from the center of the disk specimen, and $T^*$ is the nondimensional temperature ($= (T - T_i)/\Delta T$). Equation 1 is normalized by $\alpha E\sqrt{R}/(1 - \nu)$ and the nondimensional stress intensity factors ($K_I^*$) are defined as

$$K_{I_{exp}}^* = \frac{(1 - \nu) K_I}{\alpha E\sqrt{R}} \frac{1}{\Delta T} \tag{3}$$

$$K_{I_{cal}}^* = \frac{2.24}{\sqrt{\pi}} \sqrt{\frac{c}{R}} \int_0^r \frac{\sigma_{\theta_{cal}}^*}{\sqrt{(c^2 - \xi^2)}} \, d\xi \tag{4}$$

$$\sigma_{\theta_{cal}}^* = \frac{1}{R^2} \int_0^R T^* r \, dr + \frac{1}{r^2} \int_0^r T^* r \, dr - T^* \tag{5}$$

Equation 3 consists of material mechanical properties and temperature difference, whereas equation 4 can be calculated with thermal properties (which are almost constant for neat epoxy resins) and test conditions, such as cooling time, specimen size, cooling bath, and cooling method.

In equation 4 three nondimensional variables are involved: cooling time $\tau$ $(= at/R^2$; Fourier number), severity of thermal shock $\beta$ $(= hR/k$; Biot number), and normalized crack length $c/R$, where, $a$, $t$, $h$, and $k$ are thermal diffusion coefficient, cooling time, heat transfer coefficient, and thermal conductivity of the material. In this study, $\beta$ becomes 30 min (3) and disk diameter is 60 mm. The $K_{I_{cal}}^*$ are shown in Figure 2 for various cooling time and take maximum value when $\tau = 0.032$ ($t = 5$ min) at $c/R = 0.2$.

Considering the critical condition, critical temperature difference $\Delta T_c$ and fracture toughness $K_{Ic}$ were substituted in equation 3. As a result of the substitution, the criterion of crack propagation becomes

$$K_{I_{exp}}^* \leq K_{I_{cal}}^* \tag{6}$$

Test results were evaluated by comparing the nondimensional stress intensity factor $K_{I_{exp}}^*$ with the $K_{I_{cal}}^*$ theoretically calculated.

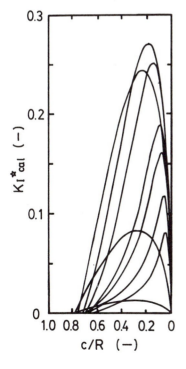

*Figure 2. Calculated nondimensional stress intensity factors as a function of notch length for several cooling times.*

## Results and Discussion

Typical thermal shock test results of brittle-type resin specimens are shown in Figure 3 (3). The open and solid circles in the figure represent data points and the solid line represents the calculated nondimensional stress intensity factor value, which gives the envelope shown in Figure 2. Good agreement with the critical condition that determines crack propagation is found.

**Ductile-Type Epoxy Resin.** An example of the relation between temperature difference and the normalized notch length for the ductile-type epoxy specimen test results is shown in Figure 4. In this figure, the open circles denote notch-cracking data and the solid circles denote no cracking. Figure 5 is a photograph of one of the cracked specimen data points in Figure 4. The crack initiated from the notch and propagated along the radius direction, which supports the analysis of a mode-I problem.

The solid triangle in Figure 4 represents the case when a crack did not propagate and the notch deformed plastically. A photograph of this deformed specimen is shown in Figure 6. A similar case was observed with another

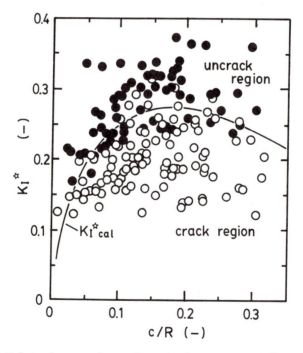

Figure 3. Relation between the nondimensional stress intensity factor $K_I^*$ and normalized notch length c/R of brittle-type epoxy resin. $\bigcirc$, cracked and $\bullet$, uncracked.

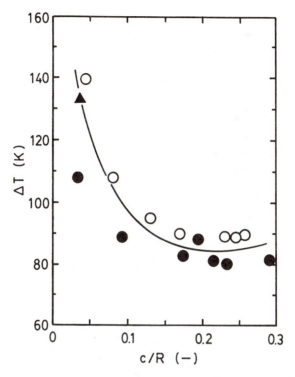

*Figure 4. Relation between temperature difference* $\Delta T$ *and normalized notch length* c/R *of bisphenol-type unmodified epoxy resin (epoxy equiv wt = 184–194).* ○, *cracked,* ●, *uncracked, and* ▲, *uncracked but deform plastically.*

*Figure 5. Photograph of a cracked specimen.*

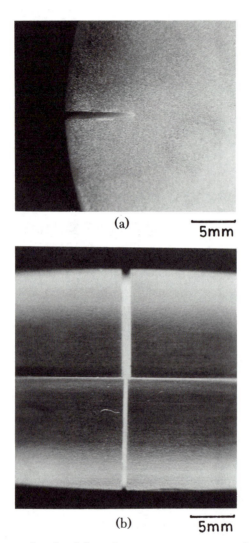

*Figure 6. Photographs of a deformed specimen, appearance of the deformed specimen (a), lateral side view (upper, after test; lower, before test) (b).*

resin system when the initial temperature was set rather high. The critical temperature difference (the lowest temperature at which cracks initiate from the notch) decreases with increasing $c/R$ and takes a minimum value at $c/R = 0.2$.

Figure 7 is a comparison of nondimensional stress intensity factors $K^*_{I_{cal}}$ and $K^*_{I_{exp}}$ calculated from the data points shown in Figure 5. The solid line

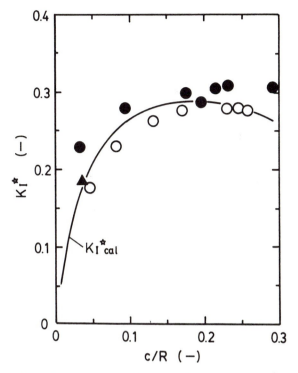

*Figure 7. Relation between the nondimensional stress intensity factor* $K_I^*$ *and normalized notch length* c/R *of unmodified epoxy resin (epoxy equiv wt = 184–194).* ○, *cracked,* ●, *uncracked, and* ▲, *uncracked but deform plastically.*

agrees well with the critical condition whether or not the crack propagates from the notch.

**CTBN-Toughened Epoxy.** To examine the effect of CTBN modification on the thermal shock resistance of epoxy, the test was made with CTBN-modified epoxy specimens. CTBN contents were changed from 0 to 50 phr in this study.

A plot of $K_{I_{exp}}^*$ for epoxy resin modified with 35 phr CTBN is shown in Figure 8. At higher temperature, the very small craze formed at the notch tip is recognizable. Neat epoxy resins or other highly cross-linked thermosets do not craze, but CTBN-modified epoxies formed crazes (5, 6) that disappeared or decreased in size when heated beyond the glass-transition temperature. Figure 9 is an example of these cases. Figure 9a indicates the small craze formation after thermal shock testing and Figure 9b indicates the disappear-

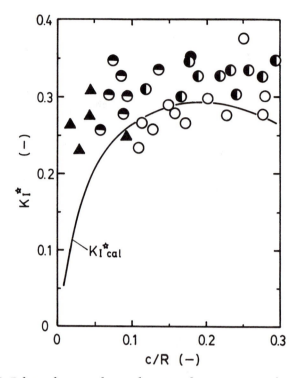

*Figure 8. Relation between the nondimensional stress intensity factor* $K_I^*$ *and normalized notch length* c/R *of 35 phr CTBN-modified epoxy resin.* ○, *cracked,* ◐, *form craze that disappears after heat treatment,* ◑, *form craze that decreases in size after heat treatment, and* ▲, *uncracked but deform plastically.*

ance of the craze in the same specimen after heat treatment. The critical cracking value is consistent with the calculated value of $K_{I_{cal}}^*$.

The results of thermal shock tests performed by changing the CTBN contents of the specimen are plotted in Figure 10. The optimum CTBN content for thermal shock resistance of CTBN-modified epoxy resin is at 35 phr.

This thermal shock resistance represented by minimum temperature difference to effect crack propagation and the fracture toughness for each CTBN-modified epoxy are shown in Figure 11. The resin modified with 35 phr CTBN also indicates a maximum $K_{Ic}$ value at or near this optimum content, a rubber–resin two-phase microstructure was formed (7). Above or below the 35 phr CTBN content, a single phase structure was observed. Therefore, the thermal shock resistance was strongly influenced by the variation of microstructure-dependent fracture toughness.

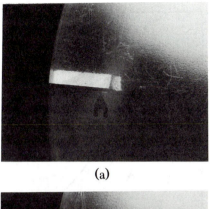

(a)

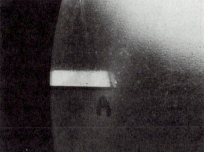

(b)

*Figure 9. Photographs of craze formation after test (a) and disappearance of the craze in the same specimen after heat treatment (b).*

**Plasticizer-Modified Epoxy.**  To improve the brittleness of the resin, a polyglycol plasticizer was added to epoxy. Modified resins containing 0 to 20 phr glycol indicated single phase. The test results for plasticizer-modified epoxy specimens are shown in Figure 12. The broken line indicates the critical temperature difference of the resin without plasticizer and the solid line shows the results of specimens with 10 phr of plasticizer. The improvement of thermal shock resistance was confirmed only to 10 K at $c/R = 0.2$. The data of plastic flow at the notch (indicated by the solid triangles) increased remarkably in comparison with CTBN-modified epoxy data.

The $K_I^*$ versus $c/R$ relation of 10 phr plasticized specimen is shown in Figure 13. This relation indicates that for the resin with plasticizer, experimental $K_{I_{exp}}^*$ is in good agreement with theoretical $K_{I_{cal}}^*$. Therefore, the same evaluation method is found to be applicable to the thermal shock resistance of low-plasticizer-content modified epoxy resin.

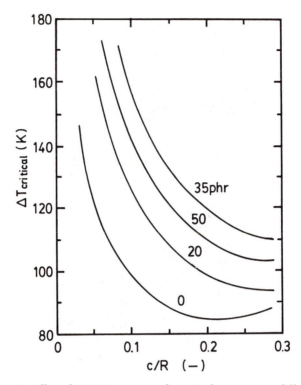

Figure 10. Effect of CTBN content on the critical temperature difference.

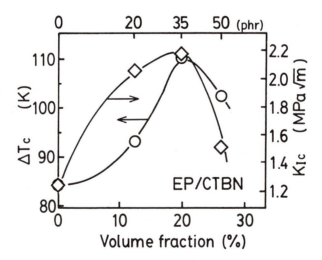

Figure 11. Effect of CTBN content on the thermal shock resistance and fracture toughness at room temperature.

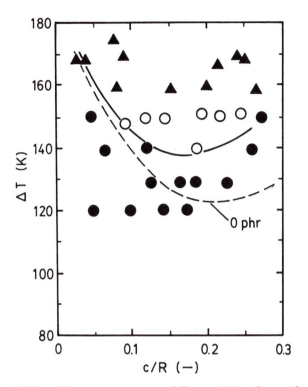

*Figure 12. Relation between temperature difference ΔT and normalized notch length c/R of 10 phr plasticizer-modified epoxy resin. ○, cracked, ●, uncracked, and ▲, uncracked but deform plastically. The solid line indicates the critical temperature difference of 10 phr modified resin and the broken line indicates unmodified resin (epoxy equiv wt = 370–435).*

With addition of over 20 phr of plasticizer, the critical temperature difference was not obtained. In this case, the 120-mm-diameter disk specimen was used to increase the severity of thermal shock, and the critical temperature difference was recognized. The test results with the large size specimen are plotted in Figure 14, where the broken line indicates the critical temperature difference of the resin with 10 phr plasticizer and the data points and the solid line show the results of specimens with 20 phr plasticizer. It is concluded that use of the larger size specimen is necessary for the test of fully plasticized resin.

## Summary

The results of this study are summarized as follows.

1. The thermal shock test using the notched disk-type specimen

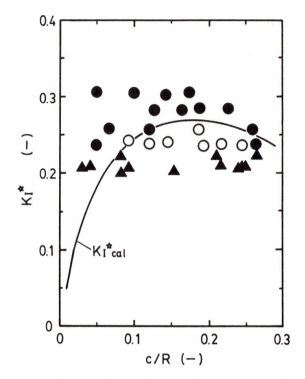

*Figure 13. Relation between the nondimensional stress intensity factor* $K_I^*$ *and normalized notch length* c/R *of 10 phr plasticizer-modified epoxy resin.* ○, *cracked,* ●, *uncracked, and* ▲ *, uncracked but deform plastically.*

is applicable to the epoxy resin toughened with CTBN or plasticizer modifier.

2. For the epoxy resin (epoxy equiv wt = 184–194) modified with CTBN, an optimum concentration for thermal shock resistance exists, and for the cracking condition, the experimental stress intensity factor $K_{I_{exp}}^*$ agrees well with the theoretical stress intensity factor $K_{I_{cal}}^*$.

3. For the epoxy resin (epoxy equiv wt = 370–435) modified with a low content of plasticizer, the same thermal shock resistance evaluation method was found to be valid, but with greater than 20 phr content, marked plastic flow occurred and the critical temperature difference was not obtained. In such a case, a larger (120-mm-diameter) specimen allowed the evaluation of the thermal shock resistance of the resin.

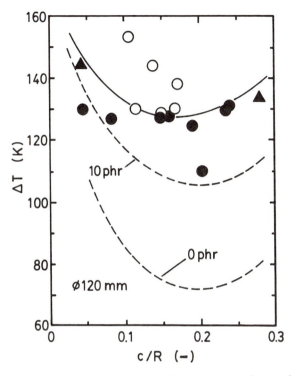

*Figure 14. Relation between temperature difference ΔT and normalized notch length c/R of 20 phr plasticizer-modified epoxy resin with large size specimen (120 mm). ○, cracked, ●, uncracked, and ▲, uncracked but deform plastically. The solid line indicates the critical temperature difference of 20 phr modified resin and the broken lines indicate 10 phr modified and unmodified resin (epoxy equiv wt = 370–435) with large size specimen.*

## References

1. Noshay, A. *Insulation Circuits* **1973**, 33–38.
2. Hojo, H.; Kubouchi, M.; Tamura, M.; Ichikawa, I. *J. Thermoset. Plastics Jpn.* **1988**, 9, 133–140.
3. Kubouchi, M.; Hojo, H. *J. Soc. Mater. Sci. Jpn.* **1990**, 39, 202–207.
4. Manson, S. S. *Thermal Stress and Low-Cycle Fatigue*; McGraw-Hill: New York, 1966; pp 275–312.
5. Sultan, J. N.; McGarry, F. J. *Phys. Eng. Sci.* **1973**, 13, 29.
6. Bucknall, C. B.; Yoshii, T. *Br. Poly. J.* **1978**, 10, 53–59.
7. Hojo, H. *J. Soc. Mater. Sci. Jpn.* **1982**, 18, 281–286.

RECEIVED for review March 6, 1991. ACCEPTED revised manuscript June 3, 1992.

# Toughening Epoxy Resin with Poly(methyl methacrylate)-Grafted Natural Rubber

**Amir H. Rezaifard[1], Kenn A. Hodd[1], and John M. Barton[2]**

[1]Department of Materials Technology, Brunel University, Uxbridge UB8 3PH, England
[2]Materials and Structures Department, Defence Research Agency (Aerospace Division), Farnborough GU14 6TD, England

*A novel rubber, poly(methyl methacrylate)-g-natural rubber (Heveaplus MG), has been studied as a toughening agent for bisphenol A diglycidyl ether (Shell 828 epoxy resin) cured with piperidine. Effective dispersions of the rubber, in concentrations of 2–10 parts per hundred parts resin, were achieved by adjusting the solubility parameter of the epoxy to approximate that of poly(methyl methacrylate) by adding bisphenol A. The fracture energy of the rubber-modified resin was determined by compact tension tests (in the temperature range −60 to +40 °C) and by Charpy impact tests. The poly(methyl methacrylate)-g-natural rubber was found to be an effective toughening agent for the epoxy resin at both low and high rates of strain. Possible fracture mechanisms are discussed.*

IN THE CURED STATE, EPOXY RESINS ARE BRITTLE MATERIALS that have fracture energies some 2 orders of magnitude lower than modern thermoplastics and other high-performance materials. To retain their position as materials of choice for many adhesives and composite matrix applications, epoxies must be formulated to improve fracture toughness. Epoxy resin toughness may be increased by plasticization, by adding a fortifier, or by blending with a rubber or a thermoplastic.

Rubber toughening by dispersing an elastomeric phase in the form of fine spherical particles in the epoxy matrix has been widely used and is

0065–2393/93/0233–0381$07.00/0

particularly effective. Carboxy-terminated butadiene–acrylonitrile random copolymers (CTBNs) (*1*) are the most frequently used rubbers, but a number of other types of rubber also have been evaluated.

None of the rubbers evaluated has included natural rubber (NR), although NR has a number of properties attractive from a toughening viewpoint. For example, NR has a low glass-transition temperature ($T_g$) and, in its high molecular weight form, has high tear and tensile strengths.

In this chapter, the use of a modified form of NR, (Heveaplus MG), to toughen the bisphenol A diglycidyl ether resin (*2*) is reported. Heveaplus MG is a commercial available graft copolymer of poly(methyl methacrylate) (PMM) and natural rubber. Table I compares properties of MG rubbers with other elastomeric toughening agents used with epoxies.

Earlier work at Brunel demonstrated that MG rubbers improve the fracture toughness of tetraglycidyldiaminodiphenylmethane (TGDDM; Ciba-Geigy MY720) cured with diaminodiphenylsulfone (DDS) (*3*).

## Experimental Details

**Materials.** Materials were used as supplied, except where reference is given to their modification.

**Table I. Properties of Some Elastomer Toughening Agents for Epoxy Resins**

| Elastomer | Source | Molecular Weight[a] | Properties or Comments |
|---|---|---|---|
| Hycar CTBN 300 × 8 | BFGoodrich | 3500[w] | Contains 17 wt % acrylonitrile |
| Hycar CTBN 300 × 13 | BFGoodrich | 3500[w] | Contains 27 wt % acrylonitrile |
| MG30 | MRPRA | 34,000–39,000[v] | Contains 35% ungrafted natural rubber, 12% unbound PMM |
| MG50 | MRPRA | 850,000–900,000[v] | Contains 24% ungrafted natural rubber, 25% unbound PMM |
| PNSR | – | 200,000–300,000[w] | Contains 30–40% unbound polystyrene |
| Hycar 1472 | BFGoodrich | 85,000[w] | Contains 27 wt % acrylonitrile, has pendant carboxyl groups |
| 1001 CG | BFGoodrich | 250,000[w] | No carboxyl groups |

[a] Weight average molecular weights are signified by a w superscript; viscosity average molecular weights are signified by a v superscript.

Bisphenol A diglycidyl ether (BADGE) resin (Ciba-Geigy MY750 or Shell Epikote 828) was modified with either carboxy-terminated butadiene–acrylonitrile copolymer (CTBN) rubber (BFGoodrich Hycar 1300 × 8) or poly(methyl methacrylate)-*g*-natural rubber (PMM-*g*-NR) Malaysian Rubber Producers Research Association (MRPRA) Heveaplus MG30 and MG50 (*see* Table I).

Prior to inclusion in the epoxy resin system, the MG rubbers were masticated on a cold two-roll mill for 5 min.

Before addition to the epoxy resin system, homopoly(methyl methacrylate) or *cis*-1,4-homopolyisoprene was removed from some samples of MG rubbers by extraction with appropriate solvents: acetone for PMM and *n*-hexane for NR.

Homogeneous distribution of MG rubber was obtained by diluting a "premix concentrate" of MG in epoxy with epoxy resin until the desired concentration was reached. The premix was prepared by dissolving 10 g of masticated MG in butan-2-one (methyl ethyl ketone, MEK) or methylene chloride (dichloromethane, DCM) adding 100 g of epoxy to the solution, and removing the solvent in vacuo.

Epoxy formulations containing CTBN were prepared by the admixture of epoxy and rubber in ratio appropriate to the desired concentration of rubber.

Some epoxy–rubber formulations were modified by the addition of bisphenol A (BPA), in which cases the resin and BPA were reacted in appropriate ratio at 130 °C for 30 min before the addition of rubber.

Piperidine (pip) was used as a curing agent for all epoxy formulations and the formulations were cast between glass plates at 120 °C for 16 h. The resulting void-free plaques (200 × 200 × 3, 6, or 9 mm) were cut as necessary and polished for use in mechanical and other tests.

**Characterization and Testing**.   The solubility parameters of the epoxy prepolymers were either determined experimentally utilizing the ternary technique of polymer–solvent–nonsolvent (*4*) or estimated by the summation of atomic and group contributions to the solubility parameter (*5*).

Thermomechanical spectra of the rubber–resin compositions were obtained using sample bars (30 × 6 × 2 mm) mounted in a dynamic mechanical thermal analyser (DMTA; Polymer Laboratories).

A transmission electron microscope (TEM; JEOL 100CX) was used to examine < 0.5-μm cross sections of the rubber-modified epoxy castings stained with osmium tetroxide. A microanalyzer (Quantimet) scanning via a macroscopic optical lens was employed for areal analysis (*6, 7*) of the volume fractions and particle size distributions of the rubbers.

Fracture surfaces from compact tension and from impact specimens were sputter-coated with a gold–palladium mixture and examined using scanning electron microscopy (SEM; Cambridge S250 or JEOL 850).

The Young's moduli of the cured 3-mm-thick resin plaques were measured by three-point flexural loading in accordance with British Standards (BS) 2782 method 335A using a screw-driven tensile tester (Instron) and a 1-mm/min crosshead speed.

Plane-strain-compression yield-stress measurements were made on rectangular plaques (50 × 14 × 1.5 mm) that were compressed between two parallel 3.2-mm-wide polished dies (8).

The fracture toughnesses of the resin formulations were determined using compact tension specimens 6 mm thick and a 1-mm/min crosshead speed. Test conditions satisfied ASTM 399 and BS 5447. a fan-assisted environmental chamber enabled measurements to be conducted in the temperature range $-60$ to $+40$ °C.

Resin samples (80 × 10 mm) were notched (radius 0.25 mm), a sharp crack was initiated using a razor blade, and they were impacted on an impact machine (Ceast Charpy). The mode-I critical strain energy release rates ($G_{iIc}$) were deduced by the method of Plati and Williams (9).

## Results and Discussion

### The Nature of Poly(methyl methacrylate)-g-Natural Rubber.
MG rubbers are prepared by the free radical grafting of methyl methacrylate to natural rubber in its latex state using a hydroperoxide–polyamine initiator (10). The resulting graft copolymer comprises a NR main chain with a molecular weight of about 200,000 and an average one or two grafted PMM side chains with molecular weights in excess of 100,000 (11).

MG rubbers are available as latex or slab stock, and in a range of levels of PMM associated with the NR. For this study, two grafted rubbers were used: MG30 and MG50, which have 30 and 50 wt % of associated PMM, respectively. MG rubbers have three constituents: PMM-g-NR, homoPMM, and ungrafted NR (12).

Extraction of the grafted rubber with acetone removed ungrafted PMM, and extraction with n-hexane removed ungrafted NR. Three new types of MG rubber were thus produced: one with no associated homopoly(methyl methacrylate), one with no associated natural rubber, and one with neither homoPMM nor NR. Table II gives the amounts of homopolymers extracted and the designations of the modified rubbers.

### The Formulations Studied.
The effects on the toughness of the BADGE–pip resin by the addition of MG rubbers with and without homoPMM, or homoNR, and without both homopolymers were studied. By further modification with BPA, the effects on fracture toughness and impact resistance also were studied. For comparison purposes the toughness of BADGE–pip–CTBN resin formulations were measured.

## The Morphologies of MG-Modified Epoxy Resin.

MG rubbers are phase-separating rubbers in which the dominant phase is determined by the environment in which the phase separation is produced. MG rubber assumes a stiff plastic form (i.e., PMM-dominated) when cast from a good solvent for PMM; conversely, a flexible rubber form (i.e., NR-dominated) is assumed when cast from a good solvent for NR. This behavior is illustrated by the data in Table III that show some common solvents, solvent solubility parameters in relation to PMM and NR parameters, and the nature of the cast MG rubber films produced.

The morphology of MG rubber dispersed in epoxy resin is controlled by the solubility parameter of the resin. Increasing the solubility parameter ensures a cleanly phase-separated rubber with a PMM outer phase. Clean phase separation improves the dispersion of the rubber and reduces the average size of the rubber particles.

BPA addition to the BADGE–pip formulation raised the solubility parameter of the resulting resin. Table IV shows that the solubility value is increased from 18.5 to 19.2 $J/m^3$ by the addition of 24 parts per hundred parts resin (phr) BPA. At this level of BPA addition the solubility parameter of the resulting epoxy resin is very close to that of PMM (compare Tables III and IV).

**Table II. Compositional Changes in MG30 and MG50 after Extraction of Homopolymer PMM or NR**

| Original Graft Rubber | HomoPMM Extracted[a] (%) | NR Extracted[a] (%) | Both Polymers Extracted[a] (%) |
|---|---|---|---|
| MG30 | 19 (MG30A) | 28 (MG30B) | 47 (MG30C) |
| MG50 | 28 (MG50A) | 23 (MG50B) | 51(MG50C) |

[a] The new name codes are given in parentheses.

**Table III. Solvents Used for Casting Films of MG Rubbers and Their Solubility Parameters**

| Solvent | Solubility Parameter ($MPa^{1/2}$) | Film Formed |
|---|---|---|
| Acetone | 20.4 | Hard and brittle |
| Methylene chloride | 19.8 | |
| Methyl ethyl ketone | 19.0 | Flexible and elastic |
| n-Hexane | 14.9 | Tacky and elastic |
| Poly(methyl methacrylate) | 18.2 | |
| Natural rubber | 16.5 | |
| Carboxy-terminated butadiene–acrylonitrile copolymer | 17.9 | |

Figures 1 and 2 are TEM micrographs of two cured MG50–BADGE–pip formulations. In Figure 2, 24 phr BPA were added. Because of the lack of BPA in the formulation in Figure 1 and the consequent low solubility parameter of the system, the rubber is dispersed throughout the resin in aggregated particular domains. The distinctively lighter areas that surround the rubber particles in Figure 1 are regions rich in PMM. To confirm this observation, homopoly(methyl methacrylate) was added to a MG–BADGE–pip formulation to enhance the bright penumbra of the rubber as seen in the TEM. Figure 3 shows the effect of adding 2.9 wt % of PMM to 2.1 wt % MG30–BADGE–pip. Note that the rubber particles remain agglomerated.

**Table IV. The Solubility Parameter of Bisphenol A-Modified BADGE,
Calculated Using Small's Method (5)**

| Amount of BPA (phr) | Solubility Parameter ($MPa^{1/2}$) | |
| --- | --- | --- |
| | Method 1[a] | Method 2[b] |
| 0 | 18.45 | 18.45 |
| 3 | 18.50 | 18.60 |
| 12 | 18.70 | 18.90 |
| 24 | 19.00 | 19.30 |
| 36 | 19.20 | 19.60 |

[a] Method 1: Rule of mixtures applied.
[b] Method 2: 2 moles of BADGE reacted with 1 mole of BPA to yield a chain-extended molecule.

*Figure 1. $OsO_4$ stained TEM micrograph of 5 phr MG50–BADGE–pip, showing
poor dispersion of rubber particles.*

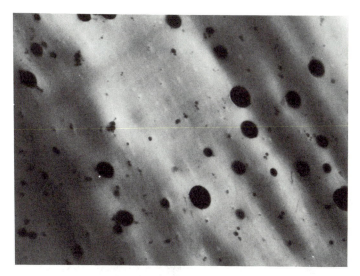

*Figure 2. TEM micrograph of 5 phr MG50–BADGE–pip–24 phr BPA.*

Figure 2, which demonstrates the effect of adding BPA, shows the rubber evenly dispersed and finely divided, with all particle diameters less than 1 μm. Little or no brightening of the rubber–epoxy interface is observed because the PMM is miscible with the epoxy–BPA phase.

The aforementioned features of the morphology of MG rubber–epoxy systems are also observable in the 2.9% PMM–2.1% MG30–BADGE–pip formulation when BPA is added, as may be seen in Figure 4.

The bright penumbra is not observed in TEM micrographs of MG50–epoxy formulations derived from homoPMM-extracted MG50 rubber (e.g., MG50A in Table II). Figure 5 shows a typical micrograph.

**The Fracture Toughness of MG-Modified Epoxy Resin.**   The addition of MG30 or MG50 to Epikote 828–pip formulations produced significant improvements in resin toughness, as measured by the mode-I critical strain energy release rate ($G_{Ic}$) and the mode-I critical stress intensity factor ($K_{Ic}$) obtained from compact tension measurements. These improvements are evidenced in Figure 6, in which the observed values of $G_{Ic}$ for MG–BADGE–pip formulations are plotted against MG rubber concentration. At a concentration of 10 phr of MG30 rubber there is a threefold increase in the critical strain energy release rate of the resin system. The energy release values are less than 1 kJ/m$^2$ for the unmodified resin and greater than 3 kJ/m$^2$ for the rubber-modified resin. With 5 phr of MG50 a lesser, but still significant, increase is observed.

*Figure 3. TEM micrograph of 2.9 phr PMM−2.1 phr MG30−BADGE−pip, showing the effect of excess PMM.*

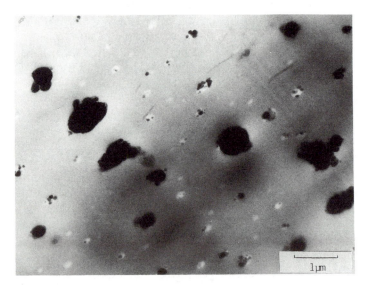

*Figure 4. TEM micrograph of 2.9 phr PMM−2.1 phr MG30−BADGE−pip−24 phr BPA.*

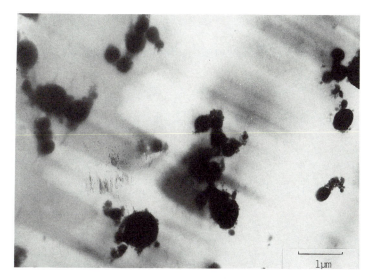

*Figure 5. TEM micrograph of 5 phr MG50A–BADGE–pip.*

The addition of BPA to the resin formulations transforms the toughening ability of the MG rubbers as judged from the measured $K_{Ic}$ and $G_{Ic}$ values. A plot of $G_{Ic}$ versus weight percent addition of BPA (Figure 7) shows the fracture toughness of the MG–BADGE–pip–24 phr BPA is increased sixfold to greater than 6 kJ/m² as the MG content is increased from 0 to 10 phr. This is a twofold improvement on the best value recorded for the MG-modified resin without BPA. It is significant that the values obtained for both MG30 and MG50 rubbers for the BPA-modified resins are similar.

These fracture toughness findings are in accord with the morphologies of the rubber–resin formulations described previously: As the dispersity of the rubber is improved by the addition of BPA, so is the fracture energy increased.

The fracture toughness effects of extracting the homopolymers (singly or together) from the MG rubbers are summarized in Figures 8 and 9. Extraction of homoPMM improved the fracture toughness, whereas removal of NR impaired toughness. This suggests that there are optimum concentrations of PMM for fracture toughness performance in the presence and absence of BPA. Without the addition of BPA the best values for fracture toughness are found for graft copolymers with PMM concentrations in the range 14–30% as the results in Figure 8 indicate by reference to MG30, MG30A, and MG50A. In the presence of 24 phr BPA, the formulation containing 5 phr MG rubber with 29% grafted PMM is found to give the highest fracture energy at 5385 J/m², as may be seen from Figure 9 and Table V. The optimum concentration or concentration range of PMM probably relates to

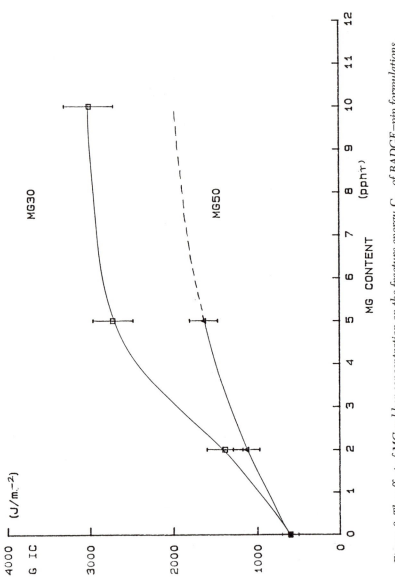

Figure 6. The effect of MG rubber concentration on the fracture energy $G_{Ic}$ of BADGE–pip formulations.

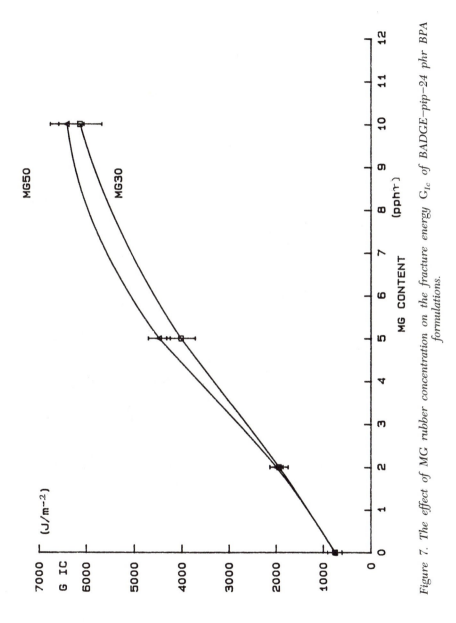

Figure 7. *The effect of MG rubber concentration on the fracture energy G$_{Ic}$ of BADGE–pip–24 phr BPA formulations.*

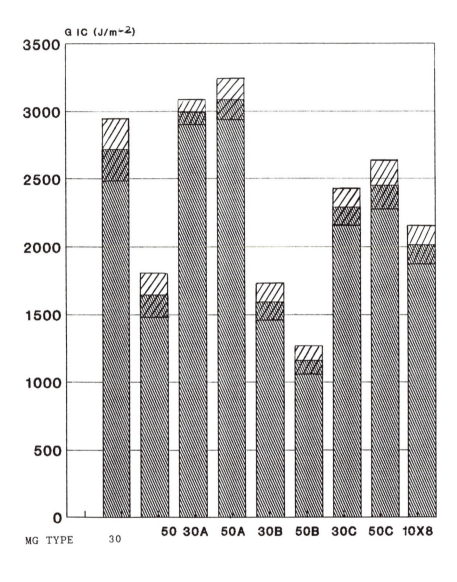

Standard Deviation

*Figure 8. The effect of 5 phr additions of extracted MG rubber on the fracture energy $G_{Ic}$ of BADGE–pip formulations.*

the level of phase separation of PMM observed at the epoxy–rubber inter-face. Excessive phase separation results in aggregation of the rubber particles into colonies, which curtails their efficient dispersion. This effect may be seen in the TEM micrographs shown in Figures 2 and 3.

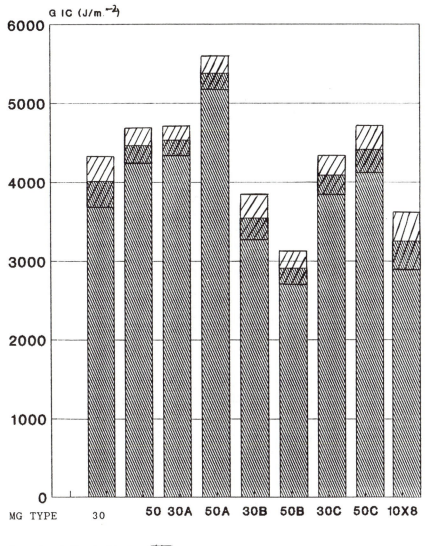

Figure 9. The effect of 5 phr additions of extracted MG rubber on the fracture energy $G_{Ic}$ of BADGE–pip–24 phr BPA formulations.

The effects of temperature on the fracture behavior of the epoxy resin modified with the novel toughening agent also have been examined. Using compact tension tests, $K_{Ic}$ was determined in the temperature range $-60$ to $+40$ °C. Figures 10 and 11 show the collected results of these measurements for the 5 phr MG30- and the 5 phr MG50-toughened BADGE–pip systems

**Table V. The Effects of the Addition of Different MG Rubbers and of Bisphenol A on the Fracture Toughness of BADGE–pip**

| Rubber Type[a] | Rubber Content (phr) | BPA Content (phr) | $K_{Ic}$ ($MN/m^{3/2}$) | $G_{Ic}$ ($J/m^2$) |
|---|---|---|---|---|
| None | 0.0 | 0.0 | 1.34 | 602 |
| MG30 | 2.0 | 0.0 | 2.0 | 1389 |
| MG50 | 2.0 | 0.0 | 1.81 | 1130 |
| MG30 | 5.0 | 0.0 | 2.77 | 2720 |
| MG50 | 5.0 | 0.0 | 2.20 | 1640 |
| MG30 | 10.0 | 0.0 | 2.88 | 3005 |
| CTBN 1300 × 8 | 10.0 | 0.0 | 2.28 | 2014 |
| None | 0.0 | 24.0 | 1.42 | 752 |
| MG30 | 2.0 | 24.0 | 2.25 | 1936 |
| MG50 | 2.0 | 24.0 | 2.27 | 1990 |
| MG30 | 5.0 | 24.0 | 3.13 | 4015 |
| MG50 | 5.0 | 24.0 | 3.35 | 4471 |
| MG30A | 5.0 | 24.0 | 3.34 | 4535 |
| MG50A | 5.0 | 24.0 | 3.58 | 5385 |
| MG30B | 5.0 | 24.0 | 2.95 | 3547 |
| MG50B | 5.0 | 24.0 | 2.73 | 2906 |
| MG30C | 5.0 | 24.0 | 3.10 | 4089 |
| MG50C | 5.0 | 24.0 | 3.25 | 4419 |
| MG30 | 10.0 | 24.0 | 3.75 | 6114 |
| MG50 | 10.0 | 24.0 | 3.85 | 6389 |
| CTBN 1300 × 8 | 10.0 | 24.0 | 2.81 | 3249 |

[a] Refer to Table II for name codes.

without and with BPA modification, respectively. For comparison, the observed $K_{Ic}$ values in the same temperature range for the unmodified and the 10 phr CTBN-toughened resin systems are included.

In the BADGE–pip system the MG30 modification exhibits the best overall temperature performance, whereas in the BPA-containing formulations both MG rubber types show improved and comparable results, which are above those for the same system toughened with CTBN.

The low-temperature toughening of the epoxy resin by the MG rubbers is also good, judging from the observed values of $K_{Ic}$. This toughening may be attributable to the low $T_g$ of NR ($-68$ °C), which is apparent from the DMTA thermograms of MG formulations where a secondary $T_g$ at $-55$ °C is detectable (see Figure 12).

**Impact Behavior of MG-Modified Epoxy Resin.** The Charpy impact energies of the novel systems were measured and the mode-I critical strain energy release rate under impact ($G_{Ic}$) was derived following the method of Plati and Williams (9).

Plati and Williams showed that

$$I = G_{iIc}(B \times D \times Z)$$

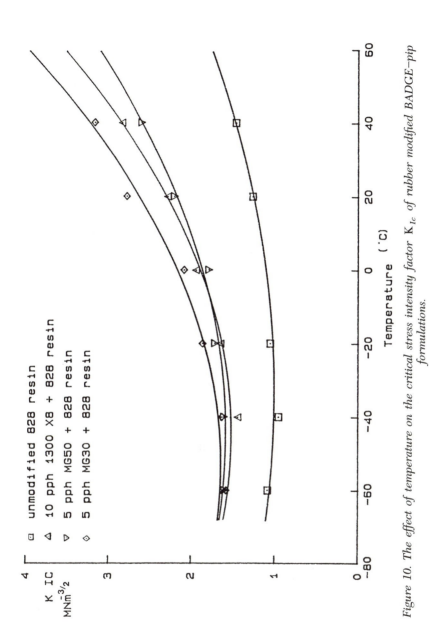

*Figure 10. The effect of temperature on the critical stress intensity factor $K_{Ic}$ of rubber modified BADGE–pip formulations.*

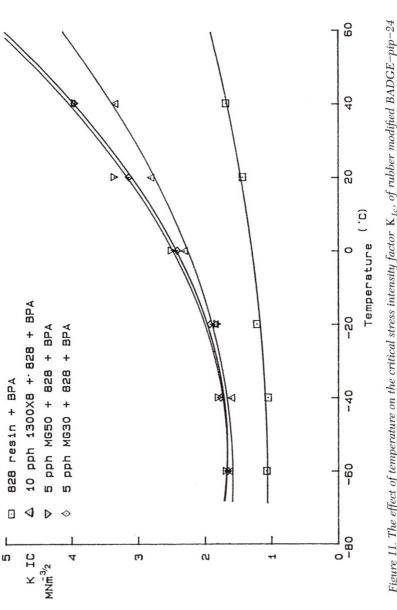

Figure 11. The effect of temperature on the critical stress intensity factor $K_{IC}$ of rubber modified BADGE–pip–24 phr BPA formulations.

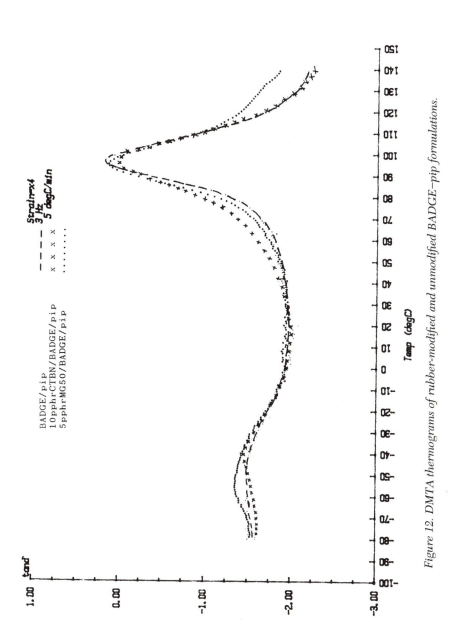

*Figure 12. DMTA thermograms of rubber-modified and unmodified BADGE–pip formulations.*

where $I$ is the dissipated impact energy (Joules), $B$ is the specimen width (meters), $D$ is the specimen thickness (meters), and $Z$ is a constant related to the specimen geometry [available from tables published by Plati and Williams (9)].

A plot of $I$ vs. $(B \times D \times Z)$ is a straight line with a slope of $G_{iIc}$. The values of the fracture energies derived for the MG–epoxy combinations are collected in Figure 13. An epoxy–CTBN formulation is included in Figure 13 for comparison. The values for $G_{iIc}$ for the novel rubber formulations obtained from the impact studies are up to 50% lower than those from the compact tension measurements. The $G_{Ic}$ and $G_{iIc}$ rates for 5 phr MG50–Epikote 828–pip–24 phr BPA were found to be 4471 and 2332 J/m$^2$, respectively. The decline in fracture toughness determined from impact measurements was less pronounced in the comparable CTBN formulation 10 phr CTBN–Epikote 828–pip–24 phr BPA, where $G_{iIc} = 2881$ J/m$^2$ as compared with $G_{Ic} = 3294$ J/m$^2$.

These results suggest that the MG–epoxy formulations are more sensitive to strain rate than the CTBN formulation. To confirm this impression the various formulations were subjected to plane-strain compression tests at a lower (0.5 mm/min) and a higher (100 mm/min) strain rate. The results showed the expected general increase in yield stress with increased strain rate, and, as may be seen in Figure 14, the trends observable in the yield-stress values mirror the impact results: The higher rubber loadings give lower yield stresses and hence higher impact resistance. Among the formulations discussed, the CTBN system had the lowest yield stress, but a formulation containing an increased quantity of MG rubber (10 phr MG50–Epikote 828–pip–24 phr BPA) was found to have both a lower yield stress at 0.5 and 100-mm/min strain rates and a higher impact fracture energy ($G_{iIc} = 3684$ J/m$^2$).

## Toughening Mechanisms in the Epoxy Resin–Novel Rubber Systems.

PMM-$g$-NR rubbers are observed to be effective toughening agents not only for BADGE–pip and BADGE–pip–BPA formulations, but also show promise with TGDDM–DDS formulation (3). It is of interest to identify the toughening mechanisms of the MG rubber types, and, where possible, to compare them with the toughening processes of other reinforcing agents such as CTBN.

The evidence suggests that the toughening of epoxy resins produced by the novel rubbers results from four mechanisms. The contribution of each mechanism to a given failure event is dependent on the stress detail involved (e.g., the strain rate). These four mechanisms are:

1. Local cavitation in the in situ rubber particles, caused by dilational triaxial stresses (13–15).

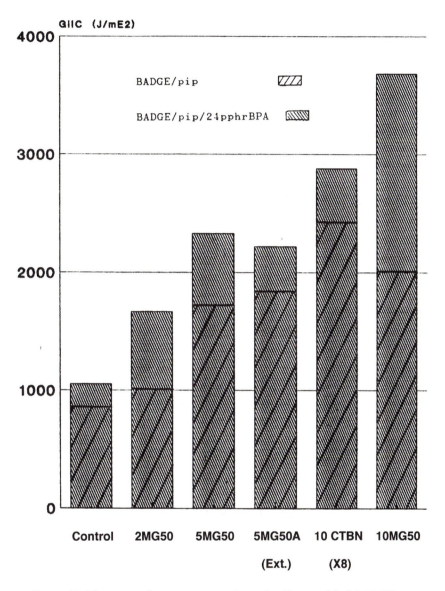

*Figure 13. The impact fracture energies $G_{ilc}$ of rubber-modified BADGE–pip and BADGE–pip–24 phr BPA formulations.*

2.  Plastic shear yielding in the resin matrix, enhanced by stress concentrations associated with the embedded soft particles (*13–17*).

3.  Stretching and tearing of embedded rubber particles (*18, 19*).

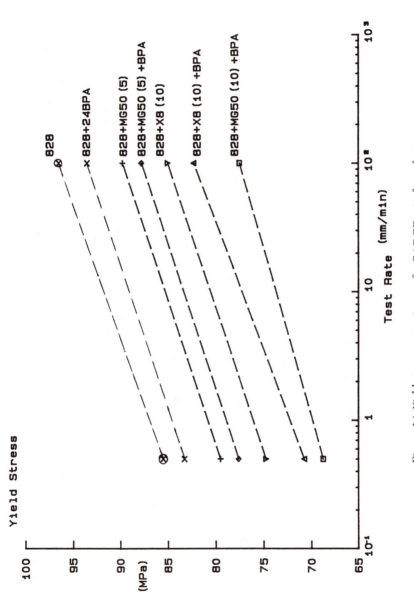

*Figure 14. Yield stress vs. strain rate for BADGE–pip formulations.*

   4. Induction of a multilevel fracture path, leading to enlargement
   of the fracture surface area (*20, 21*).

The high molecular weight of the natural rubber base of MG rubber imparts a greater tensile strength to the particles, which eliminates premature cavitation, but fracture surfaces exhibit widespread particle cavitation.

Shear yielding is an inherent deformation mechanism of epoxy resins. This form of deformation is stimulated by the presence of embedded soft spherical particles because points of maximum stress concentration develop around the equators of such particles when the resin is stressed (*17*). These points of stress concentration act as sites for the initiation of shear deformation in the resin. Further void formation is promoted and yielding constraints adjacent to a failed particle are reduced. The resulting plastic deformation helps to blunt a propagating crack tip and suppress fracture. A larger quantity of particles generates more sites for plastic deformation. Sultan and McGarry (*22*) showed that smaller particles (< 0.1 μm) promote the shear-yielding capabilities of an epoxy resin matrix to macroscopic levels, whereas the presence of both large (> 0.5 μm) and small particles resulted in the best overall toughness. Riew et al. (*1*) concur that a bimodal particle size distribution in the rubber-reinforcing agent is advantageous for toughening epoxies.

In the MG system, the addition of BPA to the epoxy resin affords greater control over the particle size of the dispersed rubber and produces an apparent bimodal distribution (see Figure 2). Such dispersions gave the best fracture toughness values observed for the MG–epoxy systems (see Table V), and failed tensile test samples exhibited pronounced diagonal shear bands. These results suggest that the properly dispersed MG rubber system promotes failure in a manner analogous to that described by Sultan and McGarry (*22*) and by Riew et al. (*1*).

Crack development arrested by the stretching and tearing of rubber particles left behind an advancing crack tip. Bascom and Hunston (*16*) and Kunz and co-workers (*19*) suggest that this form of energy dissipation contributes significantly to the fracture energy of rubber-toughened epoxies. Other workers (*13, 15*) dispute the magnitude of the contribution this mechanism makes, but agree that during the course of a fracture it absorbs some energy. The high tensile strength and elongational stiffening (with the possibility of crystallization) of NR impart a higher tear energy to the MG system than may be found in other epoxy-toughening rubbers. Consequently tearing may be a more significant contributor to the failure energy of MG-toughened epoxies.

The process of cavitation interferes with tearing for a failed particle and requires little or no tearing energy. Where cavitation is dominant in the failure process, the importance of tearing will diminish. With MG–epoxy formulations the stress whitening ahead of the crack tip, which is indicative of cavitation, is observed in compact tension tests (i.e., at low strain rates).

Stress whitening disappears from impacted MG systems and intact rubber particles are able to bridge the fracture surfaces behind the moving crack. Under high strain rate conditions, the cavitation and shear-yielding processes diminish in importance. In epoxies modified with the novel rubber, it is possible that tearing is prominent in energy absorption, which explains the effective toughening observed in the 10 phr MG50 formulation (*see* Figure 13).

Bascom and Hunston (*21*) and Levita et al. (*20*) proposed the concept of a multilevel fracture path that develops where rubber particles are dispersed throughout a resin matrix. The consequence of such a fracture path is that the fracture surface is very broken and "choppy." The effect of the fracture path may be identified by comparing Figures 15 and 16.

Figure 15 shows the stress-whitened fracture surface typically observed for an epoxy resin modified with CTBN. The cavitation of the rubber particles and, more particularly, the broken fracture surface are apparent. In comparison with the fracture surface shown in Figure 16, which is a SEM micrograph of the stress-whitened fracture surface of a 5 phr MG50–BADGE–pip–BPA casting, the fracture surface in Figure 15 is smooth. A composition containing a rubber with a broadly distributed bimodal or polymodal particle size distribution may dissipate more energy during fracture than a composite containing a monomodal rubber distribution.

The difference in particle size distributions of the two formulations with

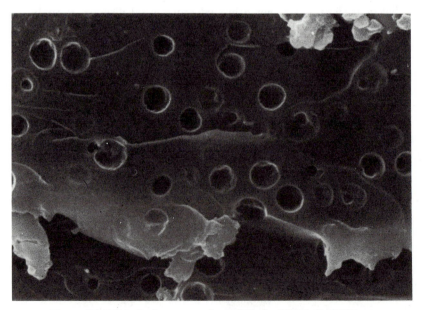

*Figure 15. SEM of a fracture surface of 10 phr CTBN–BADGE–pip.*

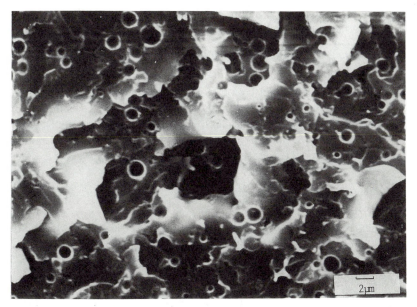

*Figure 16. SEM of a fracture surface of 5 phr MG50–BADGE–pip–24 phr BPA.*

bimodal and polymodal novel rubber and the CTBN more closely monomodal is noteworthy as is the cavitation apparent in the CTBN particles.

## Summary

Two Heveaplus MG PMM-*g*-NR formulations (MG30 and MG50) have been evaluated as toughening agents for Shell Epikote 828 cured with piperidine. On the basis of the fracture energies derived from compact tension $(G_{Ic})$ and Charpy impact $(G_{iIc})$ measurements, both MG30 and MG50 toughen the resin when added in the range 2–5 and 2–10 phr, respectively.

Control of the system morphology, which is imperative to attain the highest fracture toughnesses, is improved by adding bisphenol A to the epoxy resin. Optimum results are achieved by addition of 24 phr bisphenol A. The removal of homopoly(methyl methacrylate) from the grafted rubbers is also beneficial to the fracture toughness of the MG-modified epoxy resin.

The best toughening effects attained are comparable with those gained by adding 10 phr CTBN to the same resin system. The addition of the MG rubbers to the epoxy resin formulation increased fracture energy over the temperature range −60 to +40 °C.

Possible distinctions between the fracture mechanisms of CTBN–epoxy and MG–epoxy systems have been raised. In particular, the contributions of tearing and of multilevel fracture paths to the failure of the MG–epoxy system have been identified.

## Acknowledgments

We are grateful to the Ministry of Defence for financial support. We also acknowledge, with thanks, helpful discussions with Dr. A. Tinker of the Malaysian Rubber Producers Research Association and the supply of MG rubbers. We also appreciate the donation of other materials by BFGoodrich, Ciba-Geigy, and Shell (U.K.).

## References

1. Riew, C. K.; Rowe, E. H.; Siebert, A. R. In *Toughness and Brittleness of Plastics*; Deanin, R. D.; Crugnola, A. M., Eds.; Advances in Chemistry 154; American Chemical Society: Washington DC, 1976; p 326.
2. British Patent Application No. 8926181 (20.11.89) to MoD (U.K.).
3. Lee, W. H., Ph.D. Thesis, Brunel University, Uxbridge, England, 1986.
4. Bettleheim, F. A. In *Experimental Physical Chemistry*; W. B. Saunders Company: London, 1971; Chap. 46, p 425.
5. Small, P. A.; *J. Appl. Chem.* **1953**, 3, 71.
6. Hilliard, J. E. In *Quantitative Microscopy*; Dehoff, R. T.; Rhines, F. N., Eds.; McGraw-Hill: London, 1968; Chap. 3, p 45.
7. Underwood, E. E. *Quantitative Stereology*; Addison-Wesley: London, 1970; Chap. 1, p 1.
8. Williams, J. G.; Ford, H. *J. Mech. Eng. Sci.* **1964**, 6, 405.
9. Plati, E; Williams, J. G. *Polym. Eng. Sci.* **1975**, 15, 470.
10. Allen, P. E.; Merret, F. M. *J. Polym. Sci.* **1956**, 22, 193.
11. Merret, F. M. *The Separation and Characterisation of Graft Copolymers*; Malaysian Rubber Producers Research Association: Brickenhonbury, Herts, England, 1955; Publication No. 224.
12. Merret, F. M. *J. Polym. Sci.* **1957**, 24, 467.
13. Yee, A. F.; Pearson, R. A. *J. Mater. Sci.* **1986**, 21, 2462.
14. Yee, A. F.; Pearson, R. A. *J. Mater. Sci.* **1986**, 21, 2475.
15. Kinloch, A. J.; Shaw, S. J.; Tod, D. A. *Polymer* **1983**, 24, 1341.
16. Bascom, W. D.; Hunston, D. L. In *Adhesion 6*; Allen, K., Ed.; Applied Science Publishers: London, 1980; p 185.
17. Broutman, L. J.; Panizza, G. *Int. J. Polym. Mater.* **1971**, 1, 95.
18. Kunz-Douglass, S.; Beaumont, P. W. R.; Ashby, M. F. *J. Mater. Sci.* **1980**, 15, 1109.
19. Kunz, S. C.; Beaumont, P. W. R.; Ashby, M. F.; In *Proceedings of the PRI International Conference on Toughening of Plastics*; Plastics and Rubbers Institute: London, 1978; p 15-1.
20. Levita, G.; Butta, E.; Marchetti, A. Lazzeri, A. *Polym. Eng. Sci.* **1986**, 26, 63.
21. Bascom, W. D.; Hunston, D. L. In *Proceedings of the PRI International Conference on Toughening of Plastics*; Plastics and Rubbers Institute: London, 1978; p 22-1.
22. Sultan, J. N.; McGarry, F. J. *J. Polym. Sci.* **1973**, 3, 29.

Received for review March 6, 1991. Accepted revised manuscript September 23, 1991.

# Toughening Epoxies Using Rigid Thermoplastic Particles

## A Review

Raymond A. Pearson

**Department of Materials Science and Engineering, Lehigh University, Bethlehem, PA 18015-3195**

*Rubber-toughened epoxies have been widely used and studied for well over 20 years. In contrast to rubber-toughened epoxies, thermoplastic-modified epoxies are new materials whose potential as toughened epoxies has yet to be realized. Reports of successfully toughened thermoplastic-modified epoxies have been sporadic. Even more important, the underlying principals governing the toughening effect remain unclear. This chapter reviews the need for this new approach for toughening epoxies, describes the conventional synthetic methods, summarizes the progress achieved to date, discusses the importance of the various proposed toughening mechanisms, and provides directions for future research on these novel materials.*

THERMOSETTING POLYMER–MATRIX COMPOSITES are often used in high-performance, high-temperature aerospace applications. The aerospace industry is expected to purchase more than 100 million pounds of these advanced composites by the year 2000, and epoxy-based composites will account for about 80% of the total market share (*1*). However, the actual amount of market share captured by epoxy-based composites depends on whether the technical problem of damage tolerance can be solved.

Damage tolerance of composite materials is controlled by the fracture toughness of the epoxy matrix. The fracture toughness of epoxies is extremely low: typically below 1.0 MPa-m$^{1/2}$ (*1–7*). Fortunately, the fracture toughness of epoxies can be modified by the addition of either elastomeric (*2–4*) or

0065–2393/93/0233–0405$06.00/0

ceramic (5–7) particles. However, the use of either type of toughening agent often compromises other mechanical properties and does not always provide a significant toughening effect.

Recently, a novel approach to toughening of epoxies has emerged: the use of rigid, thermoplastic particles. Our discussion of thermoplastic-modified epoxies will begin by analyzing the need for this new approach to toughening. Next, we will describe the conventional synthetic schemes used to produce thermoplastic-modified epoxies and will provide an overview of the progress made in the toughening of epoxies using thermoplastic resins. Then, we will discuss the type of toughening mechanisms proposed to explain these improvements in toughness and present some directions in future research.

## The Need for a New Approach to Toughening

There are two reasons why a new approach to the toughening of epoxies is needed. First, although the rubber-toughening of epoxy polymers has received a significant amount of attention in the literature (2–4, 8–20), the addition of soft inclusions into a stiff matrix results in a reduction of standard mechanical properties. For example, both tensile modulus and yield strength are significantly reduced (see Figure 1). Second, the addition of soft inclusions into highly cross-linked epoxies (2, 10, 12) does not provide significant improvements in toughness. Figure 2 clearly shows the inefficiency of rubber-toughening in highly cross-linked epoxies.

In contrast, to rubber modification, the use of rigid thermoplastic resins to toughen epoxies has shown great promise (1, 21–33). However, this new approach for toughening epoxies is not always successful and is not well understood. Understanding the type of toughening mechanisms in these novel epoxy alloys is paramount for the development of this crucial technology. Therefore, it is the aim of this paper to critically review the current literature regarding the toughening mechanisms in thermoplastic-modified epoxies.

## Conventional Synthetic Method To Prepare Thermoplastic-Modified Epoxies

Before discussing toughening mechanisms, we must first understand their microstructure and the synthetic method used to produce these materials. The microstructure of tough, thermoplastic-modified epoxies always consists of two phases: one phase is predominately epoxy, whereas the other is rich in thermoplastic polymer. The most common synthetic route to prepare thermoplastic-modified epoxies consists of dissolving the thermoplastic resin of

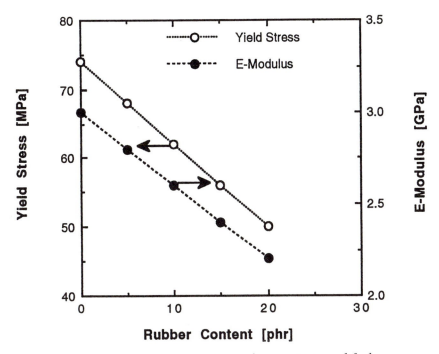

*Figure 1. Mechanical properties of a CTBN-modified epoxy (DGEBA–piperidine). Rubber modification has its drawbacks: both yield strength and modulus are reduced.*

interest into the epoxy resin (a cosolvent can also be present) and precipitating the thermoplastic particles during cure. The more common types of epoxies, curatives, and thermoplastic modifiers cited in the literature are illustrated in Charts I–III. Note that the thermoplastic polymers have similar chemical structures to the epoxy polymers. In thermoplastic-modified epoxies, the solubility parameter match is important to enable the thermoplastic polymer to dissolve. The synthetic route to prepare thermoplastic-modified epoxies is similar to that used in epoxies modified by carboxyl-terminated butadiene–acrylonitrile (CTBN): the second phase precipitates during the cure cycle. However, unlike CTBN-modified epoxies, the morphology of two-phase thermoplastic-modified epoxies can either be particulate or co-continuous structures.

The shape, size, and size distribution of the second phase in thermoplastic-modified epoxies depend on the kinetics of cure and the mechanism of phase separation. Two types of phase-separation mechanisms can occur: either binodal or spinodal decomposition. Binodal decomposition often re-

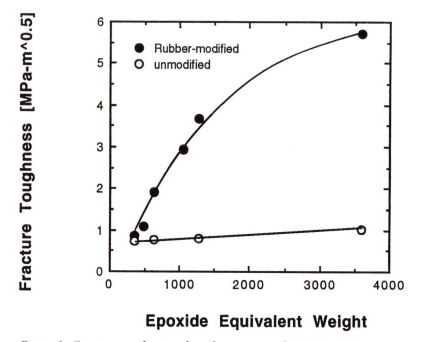

Figure 2. Fracture toughness values for a series of DGEBA–DDS epoxies. Highly cross-linked epoxies (low EEW) are difficult to toughen with CTBN rubber.

sults in a spherical morphology with very sharp interfaces. Spinodal decomposition often yields co-continuous domains with more diffuse interfaces. Unfortunately, the mechanisms of phase separation in thermoplastic-modified epoxies are not as well understood as those for CTBN-modified epoxies (31–33). Thus, the size and shape of the thermoplastic domains remain difficult to control. Schematic diagrams of morphologies resulting from binodal and spinodal decomposition are shown in Figure 3. Figure 3 also contains scanning electron microscopy (SEM) micrographs of the fracture surfaces of thermoplastic-modified epoxies with these morphologies. The three-dimensional nature of the co-continuous domains can be seen best using SEM on a surface where the thermoplastic has been removed by dissolving it in a suitable solvent. Now that we have identified the synthesis method and the observed morphologies in thermoplastic-modified epoxies, let us examine the various thermoplastic-modified epoxy systems developed to date.

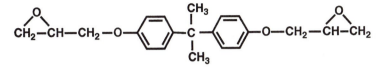

**DGEBA: Diglycidyl Ether of Bisphenol A (a diepoxide resin)**

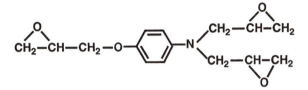

**TGAP: Triglycidyl epoxide based on aminophenol**

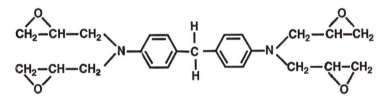

**TGDDM: Tetraglycidyl epoxide based on DDM**

*Chart I. Typical epoxy resins used in the thermoplastic-modification studies.*

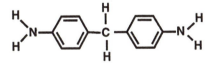

**DDM: 4,4'-diamino diphenyl methane**

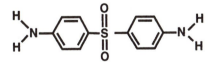

**DDS: 4,4'-diamino diphenyl sulfone**

*Chart II. Typical curing agents used with the epoxy resins in Chart I.*

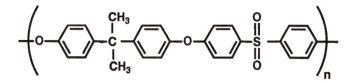

**PSU: poly(sulfone)**

**PESU: poly(ether sulfone)**

*Chart III. Typical thermoplastic tougheners evaluated.*

## Summary of Progress Made Using Conventional Approach

The concept for this new class of modified epoxies appeared during the early 1980s. Initially, the improvements in fracture toughness were not very significant. In terms of a change in the critical stress intensity factor, $\Delta K_{Ic}$, only modest increases of 0.1–0.2 MPa-m$^{1/2}$ were observed. Recently, the development of tougher thermoplastic-modified epoxies has led to significant increases in $\Delta K_{Ic}$ to 1.0 MPa-m$^{1/2}$. The evolution of thermoplastic-modified epoxies is shown in Table I and is discussed in chronological order in the following text.

**1983.** Bucknall and Partridge (21) appear to be the first to have published a study on thermoplastic-toughened epoxies. In their study, a trifunctional and a tetrafunctional epoxy, cured with 4,4'-diaminodiphenyl-sulfone (DDS), were mixed with a commercial grade of poly(ether sulfone) (PES) to improve fracture toughness. Bucknall and Partridge reported that the PES phase-separated only in the trifunctional epoxy to give PES particles, which were roughly 0.5 μm in diameter. Unfortunately, the resulting improvement in fracture toughness was not impressive: $\Delta K_{Ic} = 0.1$–0.2 MPa-m$^{1/2}$. During this same period, Raghava (22) reported that two commercial grades of PES would phase-separate in a difunctional epoxy and a tetrafunc-

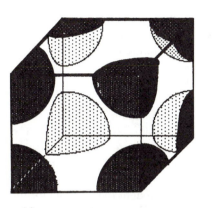

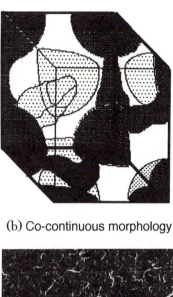

(a) Particulate morphology                 (b) Co-continuous morphology

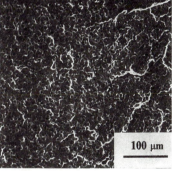

(c) Particulate morphology                 (d) Co-continuous morphology

*Figure 3. Particulate and co-continuous morphologies are observed for thermoplastic-modified epoxies: schematic diagrams (a), (b); SEM micrographs (c), (d).*

tional epoxy cured with an anhydride. Again, the improvements in toughness were not impressive and in some cases the fracture toughness decreased.

**1985.** Hendrick et al. (23) used two low-molecular-weight oligomers of hydroxyl-terminated bisphenol-A-based polysulfones (PSF) to modify a diglycidyl ether of bisphenol-A-based epoxy (DGEBA) cured with DDS. In addition to molecular weight, end-group concentration on the thermoplastic oligomers also influenced fracture toughness. The PSF-modified epoxies reported by Hendrick et al. exhibited a particulate morphology in which the average particle size was roughly 0.2 $\mu$m in diameter. These two-phase PSF-modified epoxies showed significant improvements in toughness, which were strongly influenced by the molecular weight of the PSF resin. The toughest system in the study exhibited a $\Delta K_{Ic}$ increase of 0.7 MPa-m$^{1/2}$. Table II contains a summary from Hendrick et al. (23). The success of Hendrick et al. was attributed to the strengthening of the PSF–epoxy interface. The strengthening of this interface was thought to be crucial for the operation of a particle-bridging mechanism, and SEM micrographs provided by these authors showed signs of ductile tearing on the fracture surface of the alloys.

**1986.** Bucknall and Partridge (24) developed a solubility parameter approach to predict miscibility between various thermoplastic–epoxy combinations. Parenthetically speaking, the formation of a second phase is crucial for the toughening effect; thus, the ability to predict the phase-separation of new epoxy–thermoplastics pairs would be quite useful. Bucknall and Partridge concluded that a simple one-component solubility parameter approach can predict the miscibility of low-molecular-weight polymeric modifiers with epoxy systems. Moreover, based on the method developed by Fedors (25), Bucknall and Partridge showed the success of using calculated solubility parameters to predict miscibility of epoxy systems (resin plus curing agent) with thermoplastic resins. It is important to note that Bucknall and Partridge stressed that the final morphologies of these two-phase alloys are governed by the kinetics of cure and its effects on the mobility of phase-separating species.

During this same period, Cecere and McGrath (26) developed thermoplastic-modified epoxies using amine-terminated PSF and a poly(ether ketone) (PEK). These thermoplastic-modified epoxies were also two-phase systems whose average particle size increased with molecular weight: 0.5 to > 100 $\mu$m in diameter. The dramatic improvements in toughness using high molecular weight thermoplastic resins are shown in Table III. The use of

**Table I. Overview of the Progress Made in the Toughening of Epoxies Using Thermoplastic Resins**

| | | |
|---|---|---|
| 1983 | Bucknall and Partridge | |
| | Several epoxies–DDS–PES | $\Delta K_{Ic} = +0.2$ MPa-m$^{1/2}$ |
| 1985 | Hendrick, Yilgor, Wilkes, and McGrath | |
| | DGEBA–DDS–PSF | $\Delta K_{Ic} = +0.7$ MPa-m$^{1/2}$ |
| 1986 | Cecere and McGrath | |
| | DGEBA–DDS–PEK | $\Delta K_{Ic} = +1.4$ MPa-m$^{1/2}$ |
| 1987 | Chu, Jabloner, and Swetlin | |
| | Epoxy blends–DDS–PSF | $\Delta K_{Ic} = +0.9$ MPa-m$^{1/2}$ |
| | S. Kim and Brown | |
| | DGEPM–DDS–thermoplastic | $\Delta K_{Ic} = +0.4$ MPa-m$^{1/2}$ |
| 1988 | Raghava | |
| | TGDDM–anhydride–PES | $\Delta K_{Ic} = +0.2$ MPa-m$^{1/2}$ |
| | Fu and Sun | |
| | DGEBA–DDS–PSF | $\Delta K_{Ic} = +0.5$ MPa-m$^{1/2}$ |
| | TGDDM–DDS–PSF | $\Delta K_{Ic} = +0.5$ MPa-m$^{1/2}$ |
| | DGEBA–DDS–PES | $\Delta K_{Ic} = +0.2$ MPa-m$^{1/2}$ |
| 1989 | Bucknall and Gilbert | |
| | TGDDM–DDS–PEI | $\Delta K_{Ic} = +1.0$ MPa-m$^{1/2}$ |
| 1990 | Pearson and Yee | |
| | DGEBA–PIP–PP0 | $\Delta K_{Ic} = +0.8$ MPa-m$^{1/2}$ |
| | DGEBA–PIP–PEI–PDMS | $\Delta K_{Ic} = +0.9$ MPa-m$^{1/2}$ |
| | J. Kim and Robertson | |
| | DGEBA–DDM–PVDF | $\Delta K_{Ic} = +0.5$ MPa-m$^{1/2}$ |
| | DGEBA–DDM–PBT | $\Delta K_{Ic} = +1.1$ MPa-m$^{1/2}$ |

**Table II. The Importance of Molecular Weight of the Thermoplastic Resin First Shown by Hendrick et al. (23) for DGEBA–DDS–PSF**

| PSF (%) | $M_n$ | $K_{Ic}$ (MPa-m$^{1/2}$) |
|---|---|---|
| 0 | N/A | 0.6 |
| 15 | 5300 | 0.9 |
| 15 | 8200 | 1.3 |

high-molecular-weight resins is thought to increase the ductility of the second phase. Unfortunately, the use of high-molecular-weight resins also made control of the morphology more difficult and created the possibility of co-continuous domains that could significantly effect the type of toughening mechanism that occurred.

**1987.** Two patents obtained by Hercules Chemical Company (27, 28) were awarded for a series of tough thermoplastic-modified epoxies and for a series of polymer-based composites that employed these thermoplastic-modified epoxies as their matrices. The first patent discloses the use of aromatic oligomers as the toughening agents: Example 1 uses an amine end-capped PSF oligomer. Other PSF modifiers are also described in this patent. In these patents, toughness improved only when the thermoplastic phase became co-continuous. The use of low-molecular-weight oligomers seems unwise in light of the work by Cecere and McGrath (26), who demonstrated the positive effect of higher molecular weight materials.

During this same period, Kim and Brown (29) published a study on the toughening mechanism in a resorcinol-based epoxy [diglycidyl ether of polymethylene (DGEPM)] modified with an undisclosed thermoplastic oligomer from Hercules Chemical Company. This material had a reasonable glass transition temperature ($T_g \sim 170$ °C) and at 30-wt % oligomer had a $\Delta K_{Ic} = 0.4$ MPa-m$^{1/2}$. The source of this toughness was attributed to a particle-bridging mechanism and to the yielding of individual particles.

> It seems that the situation is similar to that in the rubber-toughened materials, the particles yield before the matrix and, as they are not greatly cross-linked and so do not strain harden, a stress concentration is caused in the epoxy which will initiate localized yielding between the particles.

The quote reveals the belief of Kim and Brown that thermoplastic particles can behave like rubber particles and provide toughness by shear banding. Kim and Brown also noted evidence of de-wetting around the particles. However, observations of deformation mechanisms in 60-$\mu$m thick films may not correspond to those made on thicker specimens where the triaxial component of stress is greater.

**1988.** Raghava (30) published more work on PES-modified tetrafunctional epoxy [tetraglycidyldiaminodiphenylmethane resin (TGDDM)]: In this instance, the use of reactive end groups and a higher molecular weight PES was employed. Unfortunately, the improvements in fracture toughness were still not large. Low elongations to failure for PES were cited as the reason for the lack of improvement.

Fu and Sun (*31*) studied the use of end-capped PES and PSF to toughen epoxies. PES-modified epoxies were not significantly toughened, despite end-capping. These results corroborate Raghava's findings. In contrast, the PSF-modified epoxies were effective toughening agents. However, the measured increases in toughness were somewhat low: $\Delta K_{\mathrm{Ic}} = 0.5$ MPa-m$^{1/2}$. Interestingly, Fu and Sun successfully toughened a ductile epoxy (DGEBA–piperidine) using PSF particles.

**1989**. Bucknall and Gilbert (*32*) from Cranfield Institute used a commercial poly(ether imide) (PET) to modify a tetrafunctional epoxy with remarkable success (*see* Figure 4). The poly(ether imide) phase-separated during cure into 0.5–3.0-$\mu$m particles. At 25 parts per hundred parts resin (phr) PEI content, the PEI-modified epoxy showed a toughness increase of $\Delta K_{\mathrm{Ic}} = 1.0$ MPa-m$^{1/2}$. However, it is uncertain whether a particulate morphology was maintained at high PEI content. Bucknall and Gilbert concluded that the toughening mechanism was a particle-bridging mechanism. It is important to mention that their use of a commercial PEI cast doubts on whether end-capping is a necessary requirement for tough, thermoplastic-modified epoxies and whether tailor-made oligomers are necessary.

**1990**. Pearson and Yee (*33*) used various amorphous thermoplastic polymers to toughen a low $T_{\mathrm{g}}$ epoxy and reported notable successes. Several

**Table III. The Importance of Molecular Weight of the Thermoplastic Resin Reemphasized by Cecere and McGrath (*26*) for DGEBA–DDS–PSF and DGEBA–DDS–PEK**

| PSF (%) | $M_n$ (g/mole) | $K_{Ic}$ (MPa-m$^{1/2}$) | PEK (%) | $M_n$ (g/mole) | $K_{Ic}$ (MPa-m$^{1/2}$) |
|---|---|---|---|---|---|
| 0 | N/A | 0.6 | 0 | N/A | 0.6 |
| 15 | 4,100 | 0.6 | 15 | 6,980 | 0.8 |
| 15 | 5,300 | 0.8 | 15 | 12,700 | 1.2 |
| 15 | 7,100 | 0.8 | 15 | 15,400 | 1.2 |
| 15 | 13,100 | 1.0 | 15 | 17,800 | 2.0 |
| 15 | 19,500 | 1.0 | | | |
| 15 | 36,500 | 1.2 | | | |

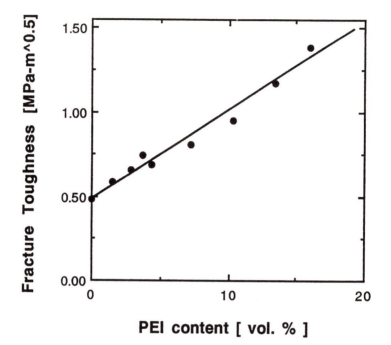

*Figure 4. Bucknall and Gilbert (32) found that a commercial poly(ether imide) (PET) could modify a tetrafunctional epoxy system with remarkable success.*

commercial polymers tried included poly(phenylene oxide) (PPO), polystyrene (PS), poly(ether imide) (PEI), poly(ether imide–dimethyl siloxane) random copolymer (PEI–PDMS), polycarbonate, and poly(carbonate–dimethyl siloxane) random copolymer (PC–PDMS). These commercial resins were injection-molded grades; thus, their average molecular weights were much higher than the reactive oligomers investigated by the research team at Virginia Polytechnic Institute (26, 28). Results from Pearson and Yee are summarized in Table IV.

Again, phase separation was a prerequisite for toughness. However, phase separation is a necessary but not sufficient condition for toughness improvement. Evidence that suggests that particle size influences toughness has been reported. For example, the PC–PDMS copolymer did not provide any improvements in toughness although it did phase-separate; the particles were large, 100 μm in diameter. The varied morphologies observed in these materials masked any trends attributable to the influence of modulus and yield strengths of the particle on toughness improvement. At moderate concentrations of PPO or PEI–PDMS, these thermoplastic-modified epoxies exhibited a toughness increase of $\Delta K_{Ic} = 0.9$ MPa-m$^{1/2}$ without significantly reducing the modulus or yield strength. Both types of particles initiated shear

**Table IV. Pearson and Yee (33) Found That Some Thermoplastic Modifiers
Worked Better Than Others in a DGEBA–Piperidine Epoxy System**

| Thermoplastic Modifier | $T_g$ (°C) | E modulus (MPa) | $\sigma_{yield}$ (MPa) | Phase Separation | max $\Delta K_{Ic}$ (MPa-m$^{1/2}$) |
|---|---|---|---|---|---|
| poly(ether imide) (PEI) | 220 | 2.8 | 105 | no | 0.0 |
| poly(phenylene oxide) (PPO) | 205 | 2.6 | 78 | yes | 0.9 |
| poly(carbonate) (PC) | 145 | 2.4 | 62 | no | 0.0 |
| poly(butylene terephthalate) (PBT) | 50 | 2.3 | 55 | no | 0.0 |
| reactive poly(styrene) (RPS) | 105 | 3.3 | ~ 35 | yes | −0.1 |
| poly(butylene terephthalate– butyl ether) (PBT–PBE) | — | 0.6 | — | yes | 0.0 |
| poly(ether imide–dimethyl siloxane) (PEI–PDMS) | — | 0.5 | < 30 | yes | 0.9 |
| poly(carbonate–dimethyl siloxane) (PC–PDMS) | − 100,120 | 0.3 | — | yes | 0.2 |
| poly(styrene–butadiene– styrene) (SBS) | − 81,87 | 0.03 | N/A | yes | N/A |

banding in the epoxy matrix when these modified epoxies were deformed in uniaxial tension. However, the toughening mechanism active in each system consisted of a zone of profuse microcracking, which was elucidated using SEM and optical microscopy techniques on crack-tip damage zones.

During this same period, Kim and Robertson (34) tried various semicrystalline thermoplastic polymers to toughen a high $T_g$ epoxy system [DGEBA–diaminodiphenylmethane (DDM)], also with notable success. Several commercial polymers tried included poly(butylene terephthalate) (PBT), polyamide-6 (PA), and poly(vinylidene fluoride) (PVDF). These particular crystalline polymers exhibit polymorphism and are likely candidates for a stress-induced solid phase transformation, a well known toughening mechanism in structural ceramics (35–39). The PBT-modified epoxies exhibited a significant increase in toughness: $\Delta K_{Ic} = 1.1$ MPa-m$^{1/2}$. This increase in toughness is thought to be provided by transformation toughening, but the current evidence is inconclusive.

In summary, many thermoplastic polymers–epoxy network combinations have been evaluated with regard to improving toughness. Not all attempts were successful. Early successes used thermoplastic modifiers with chemical structures similar in nature to the epoxy resin. More recent work has shown that other thermoplastic resins can be used as tougheners. The successful attempts appear to have one similarity: a two-phase morphology that consists of a ductile thermoplastic phase embedded in a brittle epoxy matrix. However, the specific requirements on the toughening phase have not been clearly identified yet. For example, the influences of type of morphology, particle size, particle–matrix adhesion, and the mechanical attributes of the thermoplastic phase have not been clearly established. Therefore, one may conclude that the true potential of these novel toughening agents has not yet been realized. To understand the influences of such parameters on toughness enhancement, it is useful to discuss the toughening mechanisms (models) most often cited as being responsible for the enhancement in toughness.

## Evaluation of Toughening Mechanisms: A Guide to Tougher Epoxies

Many toughening mechanisms have been proposed to explain the improvements in toughness for thermoplastic-modified epoxies: crack pinning, particle bridging, crack-path deflection, massive shear-banding, and microcracking. It is important that we have a clear understanding of each mechanism because the type of mechanism will determine the optimal particle size and requisite particle–matrix adhesion. The following paragraphs review several toughening mechanisms by examination of quantitative models used to rationalize their effect.

**Crack Pinning.**   The original concept of crack pinning developed by Lange (*40*) assumes that the crack front changes in length as it interacts with inhomogeneous particles. The propagation of the mechanism is shown in Figure 5. The crack-pinning mechanism has been modeled by Lange (*40*), Evans (*41*), and Rose (*42*). The model developed by Lange (*40*) relates the particle diameter, $d_p$, and the center-to-center particle distance, $d_s$, to the increase in fracture toughness. Evans's model accounts for the interaction of the segments of the crack front, thus predicting a nonlinear increase in fracture toughness as a function of the volume fraction of pinning particles. The model proposed by Rose is the most quantitative of all and can fit the maxima in toughness observed in glass-filled epoxies (*43*). All three models favor smaller particles as toughening agents. The models of Evans and Rose also cite the interpenetrable strength of the particles as an important

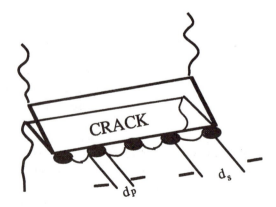

*Figure 5. A schematic diagram of the crack-pinning mechanism.*

parameter. However, only Rose's model considers the traction applied by trailing particles, which emphasizes the stiffness of the particle.

**Particle Bridging (Rigid Particles).** A schematic diagram of the particle-bridging mechanism is shown in Figure 6. The toughening of a brittle resin using rigid, ductile particles has been modeled by Sigl et al. (*44*). According to Sigl's model, the particle plays two roles in this mechanism: (1) It acts as a bridging particle that applies compressive traction in the crack wake. (2) The ductile particles plastically deform in the material surrounding the crack tip, which provides additional crack shielding. Sigl et al. (*44*) contend that shielding that results from yielded particles is negligible and that particle bridging provides most of the improvement in toughness. In contrast to crack-pinning models, the particle-bridging models favor large particles and emphasize the work-to-rupture of the ductile phase.

**Crack-Path Deflection.** A schematic diagram of the crack-path deflection mechanism is shown in Figure 7. The increase in toughness due to crack-path deflection can be explained by a stress intensity approach: The deflection of the crack path decreases the mode-I character of the crack opening and increases the mode-II character (essentially a shearing mode). Because most materials are more resistant to mode-II crack opening, they exhibit a greater apparent toughness when this mechanism is in operation. A quantitative crack-path deflection model has been developed by Faber and Evans (*45*). The model assumes no dependence on particle size, however. Toughness can increase when the model is used to consider the distribution of the spacing of the particles. Uneven spacing is thought to provide better results than uniform spacing.

**Massive Shear-Banding.** A schematic diagram of the crack-shielding effect introduced by massive shear-yielding is shown in Figure 8. Toughening due to massive shear-banding at the crack tip has been modeled by

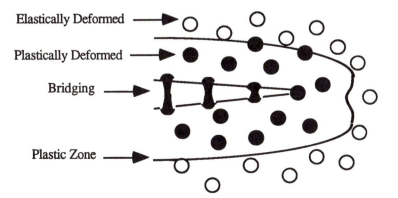

*Figure 6. A schematic diagram of the particle-bridging mechanism.*

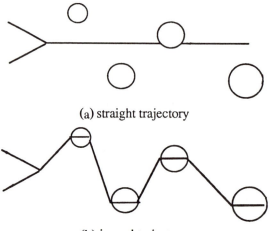

Figure 7. A schematic diagram of the crack-deflection mechanism.

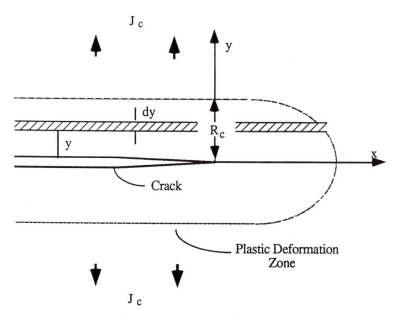

Figure 8. A schematic diagram of the shear-yielding model proposed by Argon.

Argon (46) using a J-integral approach. The model sums the energy contribution of infinitesimal elements ($dy$) as a function of the distance from the crack ($y$) up to the edge of the plastic zone ($R_c$). Argon proposed that additional crack-tip shielding in rubber-modified epoxy occurs due to the reduction in yield stress by the stress concentration of the compliant rubber particles that facilitate shear yielding. Evidence of shear-yielding in epoxies modified with thermoplastic particles has been presented by Kim and Brown (29) and by Pearson (47). However, shear-banding in these materials cannot be explained using a simple elastic analysis of the stress distribution about a rigid sphere. A nonlinear elastic analysis by Sue et al. (48) correctly predicts the occurrence of localized shear-yielding around the particle. The nonlinear analysis predicts that a 60% difference in the tangent modulus prior to failure is sufficient to cause localized shear-yielding. Therefore, Argon's model for massive shear-banding will require modification before it can be applied to thermoplastic-modified epoxies toughened by shear-banding. Other researchers (9, 49) have attempted to model shear-yielding as a toughening mechanism in rubber-modified polymers. The model by Wu (49) is particularly attractive because it assumes that the interparticle distance is a fundamental parameter in rubber-toughened polymers. The model proposes that uniformly spaced, small particles are the most efficient toughening agents for polymers that shear-band. Unfortunately, other parameters, such as cavitation resistance of the rubber particle, have a greater influence on toughening (50).

**Microcracking.** A schematic diagram of the microcracking mechanism is shown in Figure 9. Not unlike crack pinning, there have been several models proposed to quantify the toughness provided by a microcracking mechanism (51–55). However, the problem is more complicated because many different types of microcracking mechanisms exist. Because it is not the intention of this review to be comprehensive, but rather to identify possible requirements on the toughening phase, a brief discussion of one microcrack-

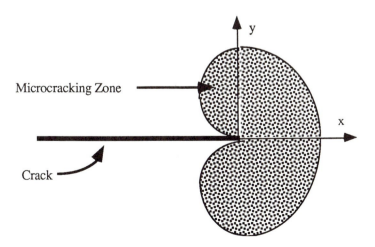

*Figure 9. A schematic diagram of the microcracking mechanism.*

ing model should suffice. Evans et al. (*51*) developed a model to predict the toughness in materials where the rigid glass spheres were not well bonded to the matrix. Debonding or microcracking effectively lowers the modulus in the frontal process zone around the crack tip (*see* Figure 9) and thus effectively reduces the stress intensity there. The model appears to be quantitative and correctly predicts the toughness of weakly bonded glass spheres in epoxy as a function of volume fraction and particle size. The particle size effect in microcracking is opposite to particle bridging, that is, small particles are better.

In summary, we have discussed six different toughening mechanisms often quoted as being responsible for the observed improvements in toughness when thermoplastic particles are embedded into an epoxy matrix. The requirements on the toughening particle vary significantly for each mechanism. Through our discussion of the toughening models four parameters have been identified as being important to toughening: (1) the size of the particles; (2) the strength of the particles; (3) the adhesion between the particle and the matrix; and (4) the distribution of the particles in the matrix. Unfortunately, the current synthetic routes used to synthesize thermoplastic epoxies cannot independently control these parameters. For example, the use of reactive oligomers increases adhesion but also reduces the particle size.

## Recommended Research: Try To Control the Important Material Parameters

Future studies of improved toughness of thermoplastic-modified epoxies should focus on the aforementioned material parameters: (1) the size of the thermoplastic particles; (2) the strength of the thermoplastic particles; (3) the adhesion between the particle and the matrix; and (4) the distribution of the particles in the matrix. Independent control of these parameters would be beneficial. Thus, a novel synthetic approach for preparing thermoplastic-modified epoxies appears necessary.

A possible approach for producing thermoplastic-modified epoxies would utilize existing emulsion polymerization techniques to produce structured core–shell particles. The use of such structured core–shell latex particles should allow independent control of particle size, ductility, and adhesion. The structured core–shell latex particles are preformed. Thus the size of particles can be independently controlled with surfactants. The ductility of the particles can be controlled by care selection of the core polymer. The shell polymer can accommodate reactive groups, which would allow independent control of adhesion.

As mentioned during the discussion of several toughening mechanisms, the determination of the role of adhesion on the work-to-fracture of ductile particles is crucial for optimizing toughness. Therefore, new techniques, such

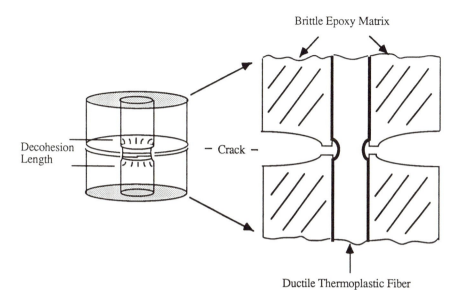

*Figure 10. Composite cylinder experiment modeled after Mataga and Ashby et al.*

as those described by Mataga (56) and Ashby et al. (57), should be applied to model the influence of adhesion on the work-to-fracture of ductile particles embedded in a brittle matrix. The two techniques are very similar and both techniques examine the influence of constraining a ductile fiber with a brittle matrix on the work-to-fracture. These techniques also provide a relative measure of adhesion; thus, the influence of adhesion on the side of the process zone could be determined. A schematic diagram of an embedded ductile fiber technique is shown in Figure 10. Note that highly constrained fibers display brittle behavior, whereas unconstrained fibers can be quite ductile. This same phenomenon also occurs in particulate composites containing spherical particles.

## Acknowledgments

I gratefully acknowledge the advice and support of Dr. Albert F. Yee, who has assumed the roles of both mentor and colleague during various periods of my career. I would also like to thank Dr. C. Keith Riew for his patience during the preparation of this manuscript.

## References

1. Kubel, E. J. *Adv. Mater. Process.* **1989**, 8, 23.
2. Pearson, R. A.; Yee, A. F. *J. Mater. Sci.* **1989**, 24, 2571.

3. Yee, A. F.; Pearson, R. A. *J. Mater. Sci.* **1986**, *21*, 2462.
4. Pearson, R. A.; Yee, A. F. *J. Mater. Sci.* **1986**, *21*, 2475.
5. Moloney, A. C.; Kausch, H. H.; Kiaser, T.; Beer, H. R. *J. Mater. Sci.* **1987**, *22*, 381.
6. Spanoudakis, J.; Young, R. J. *J. Mater. Sci.* **1984**, *19*, 473.
7. Spanoudakis, J.; Young, R. J. *J. Mater. Sci.* **1984**, *19*, 487.
8. Pearson, R. A.; Yee, A. F. *J. Mater. Sci.* **1991**, *26*, 3828.
9. Kinlock, A. J. In *Rubber-Toughened Plastics*; Riew, C. K., Ed.; Advances in Chemistry 222; American Chemical Society: Washington, DC, 1989; p 67.
10. Hwang, J. F.; Manson, J. A.; Hertzberg, R. W.; Miller, G. A.; Sperling, L. H. *Polym. Eng. Sci.* **1989**, *29*, 1466.
11. Low, I. M.; Mai, Y. W. *J. Mater. Sci.* **1989**, *24*, 1634.
12. Levita, G. *Polym. Prepr.* (*Am. Chem. Soc. Div. Polym. Chem.*) **1988**, *29*, 175.
13. Schultz, W. J.; Portelli, G. B.; Jordan, R. C.; Thompson, W. L. *Polym. Prepr.* (*Am. Chem. Soc. Div. Polym. Chem.*) **1988**, *29(1)*, 136.
14. Garg, A. C.; Mai, Y. W. *Comp. Sci. Tech.* **1988**, *31*, 179.
15. Kinlock, A. J.; Kodokian, F. A.; Jamarani, M. B. *J. Mater. Sci.* **1987**, *22*, 4111.
16. Kinlock, A. J.; Hunston, D. L. *J. Mater. Sci. Lett.* **1986**, *5*, 909.
17. Kinlock, A. J.; Gilbert, D. G.; Shaw, S. J. *J. Mater. Sci.* **1986**, *21*, 1051.
18. Butta, E.; Levita, G.; Marchetti, A.; Lazzeri, A. *Polym. Eng. Sci.* **1986**, *26*, 63.
19. Yang, T. F.; We, Y. Z. *Mater. Chem. Phys.* **1986**, *15*, 505.
20. Kinloch, A. J.; Shan, S. J.; Tod, D. A.; Hunston, D. L. *Polymer* **1983**, *24*, 1341, 1355.
21. Bucknall, C. B.; Partridge, I. K. *Polymer* **1983**, *24*, 639.
22. Raghava, R. S. *Natl. SAMPE Symp.* **1983**, *28*, 367.
23. Hendrick, J. L.; Yilgor, I.; Wilkes, G. L.; McGrath, J. E. *Polym. Bull.* **1985**, *13*, 201.
24. Bucknall, C. B.; Partridge, I. K. *Polym. Eng. Sci.* **1986**, *26*, 54.
25. Fedors, R. F. *Polym. Eng. Sci.* **1974**, *14*, 147.
26. Cecere, J. A.; McGrath, J. E. *Polym. Prepr.* (*Am. Chem. Soc. Div. Polym. Chem.*) **1986**, *27(1)*, 299.
27. Jabloner, H.; Swetlin, B. J.; Chu, S. G. U.S. Patent 4,656,207, 1987.
28. Chu, S. G.; Swetlin, B. J.; Jabloner, H. U.S. Patent 4,656,208, 1987.
29. Kim, S. C.; Brown, H. R. *J. Mater. Sci.* **1987**, *22*, 2589.
30. Raghava, R. S. *J. Polym. Sci., Polym. Phys. Ed.* **1988**, *26*, 65.
31. Fu, Z.; Sun, Y. *Polym. Prepr.* (*Am. Chem. Soc. Div. Polym. Chem.*) **1988**, *29(2)*, 177.
32. Bucknall, C. B.; Gilbert, A. H. *Polymer* **1989**, *30*, 213.
33. Pearson, R. A.; Yee, A. F. *Polym. Mater. Sci. Eng.* **1990**, *63*, 311.
34. Kim, J.; Robertson, R. *Polym. Mater. Sci. Eng.* **1990**, *63*, 301.
35. Amazigo, J. C.; Budiansky, B. *J. Mech. Phys. Solids* **1988**, *36*, 581.
36. Swain, M. V.; Rose, L. R. F. *J. Am. Ceram. Soc.* **1986**, *69*, 511.
37. Rose, L. R. F. *J. Am. Ceram. Soc.* **1986**, *69*, 208.
38. McMeeking, R. M.; Evans, A. G. *J. Am. Ceram. Soc.* **1982**, *65*, 208.
39. Evans, A. G.; Heuer, A. H. *J. Am. Ceram. Soc.* **1980**, *63*, 208.
40. Lange, F. F. *Philos. Mag.* **1970**, *22*, 983.
41. Evans, A. G. *Philos. Mag.* **1972**, *26*, 1327.
42. Rose, L. R. F. *Mech. Mater.* **1987**, *8*, 11.
43. Lange, F. F.; Radford, K. C. *J. Mater. Sci.* **1971**, *6*, 1197.
44. Sigl, L. S.; Mataga, P. A.; Dageleidsh, B. I.; McMeeking, R. M.; Evans, A. G. *Acta Metall.* **1988**, *36*, 945.
45. Faber, K. T.; Evans, A. G. *Acta Metall.* **1983**, *31*, 565.

46. Argon, A. S. In *ICF7: Advances in Fracture Research*; Samala, K.; Ravi-Chander, K.; Taplin, D. M. R.; Rama Rao, P., Eds.; Pergamon Press: New York, 1989; Vol. 4.
47. Pearson, R. A., Ph.D. Thesis, University of Michigan, 1990.
48. Sue, H. J.; Pearson, R. A.; Yee, A. F., submitted for publication in *Polym. Eng. Sci.*
49. Wu, S. *Polymer* **1985**, *24*, 643.
50. Borggreve, R. J. M.; Gaymans, R. J.; Schuijer, J. *Polymer* **1989**, *30*, 71.
51. Evans, A. G.; Williams, S.; Beaumont, P. W. R. *J. Mater. Sci.* **1985**, *20*, 3668.
52. Chambides, P. G.; McMeeking, R. M. *Mech. Mater.* **1987**, *6*, 71.
53. Evans, A. G.; Faber, K. T. *J. Am. Ceram. Soc.* **1984**, *67*, 255.
54. Hutchinson, J. W. *Acta Metall.* **1987**, *35*, 1605.
55. Ortiz, M. *J. Appl. Mech.* **1987**, *54*, 54.
56. Mataga, P. A. *Acta Metall.* **1989**, *37*, 3349.
57. Ashby, M. F.; Blunt, F. J.; Bannister, M. *Acta Metall.* **1989**, *37*, 1847.

RECEIVED for review August 2, 1991. ACCEPTED revised manuscript June 26, 1992.

# Preparation of Poly(butylene terephthalate)-Toughened Epoxies

**Junkyung Kim and Richard E. Robertson***

**Macromolecular Science and Engineering Center and Department of Materials Science and Engineering, The University of Michigan, Ann Arbor, MI 48109-2136**

*Several alternative mixing and processing procedures were explored to increase further the fracture toughness of poly(butylene terephthalate) (PBT)–epoxy blends and to simplify their manufacture. These procedures included (1) the annealing of PBT just below its melting point, both before mixing with the epoxy and after mixing and curing, and (2) heating mixtures of PBT and epoxy without curing agent to various temperatures, with and without mechanical mixing. The annealing had no effect on the fracture energy, but the exposure of PBT to hot epoxy agent did. The highest fracture energy was obtained by heating a mixture of 15-μm PBT particles in epoxy without curing agent at 10 °C/min to 193 °C and immediately cooling back to room temperature, at which point the curing agent was added. After curing, a 5-wt % PBT–epoxy blend has a fracture energy of 900 J/m² vs. 190 J/m² without the PBT.*

$\mathbf{E}$XAMINATION OF ALTERNATIVE MIXING AND PROCESSING PROCEDURES that either could increase further the fracture toughness of poly(butylene terephthalate) (PBT)–epoxy blends or simplify their manufacture is the focus of this chapter. As shown in Figure 1, particles of PBT dispersed in an aromatic amine-cured epoxy raise the fracture energy $G_{\text{Ic}}$ from 190 J/m² for unfilled epoxy to 600 and 1100 J/m² for 5 and 20% PBT additions, respectively (1). For this epoxy, the fracture energy increases were about twice the increase attainable with particulate inclusions of other polymers, including nylon 6,

---

* Corresponding author.

0065-2393/93/0233-0427$06.75/0

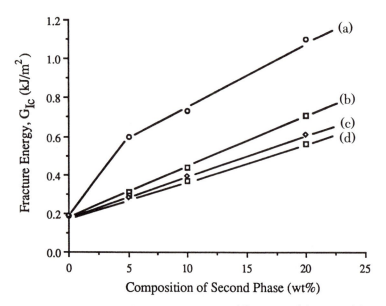

*Figure 1. Fracture energy for initiation for PBT (a); nylon 6 (b); CTBN (c), and PVDF (d). Key:* ○, *PBT–epoxy;* □, *nylon 6–epoxy and PVDF–epoxy;* ◇, *CTBN–epoxy. (Reproduced with permission from reference 1. Copyright 1992 Chapman and Hall.)*

poly(vinylidene fluoride), and carboxyl-terminated butadiene–acrylonitrile (CTBN) rubber. The PBT particles employed were obtained prior to use in epoxy by dissolving PBT in phenol, precipitating it in methanol, and grinding the resulting material to a uniform particle size.

## Experimental Procedures

**Materials.** Two diglycidyl ether epoxies, Epon 828 and Epon 1002F (Shell Chemical Company), were used for this investigation. These epoxies differ in the number of bisphenol A units linked together in a chain: Epon 828 has one and Epon 1002F has four. The curing agent (Curing Agent Z; Shell Chemical Company) was a mixture of aromatic amines with 4,4′-methylenedianiline as the main component. The optimum concentration of 20 parts per hundred parts resin (phr) of curing agent was used. The epoxy–curing agent mixture was cured at 80 °C for 2 h and postcured at 150 °C for another 2 h. The glass-transition temperature of the cured Epon 828 was 156 °C and that of the Epon 1002F was 78 °C.

Poly(butylene terephthalate) (PBT) was obtained from Aldrich Chemical Company. The inherent viscosity in phenol at 40 °C was 0.71. For some of the studies, the PBT was processed into particulate form by dissolution in phenol and precipitation in methanol (1). After further washing in methanol and drying, the PBT was milled and sieved to obtain a uniform particle size. The dilute solution viscosity of the powder was the same as the original bead, indicating that no molecular weight reduction occurred in the powder-making process. All samples were vacuum dried before mixing.

**Characterization.**   To test for changes in molecular weight of PBT, solution viscometry was used. The inherent viscosity ($\eta_{inh}$) of the PBT was measured in 5-g/L phenol solution at 40 °C.

A differential scanning calorimeter (DSC; Perkin-Elmer DSC 7) was used to measure the melting ($T_m$) and glass-transition ($T_g$) temperatures and the heat of fusion. Specimens weighing approximately 10 mg were scanned in an inert environment and at a heating rate of 10 °C/min.

Wide angle X-ray scattering was used to investigate crystal structure changes. An X-ray diffractometer (Rigaku) with nickel-filtered CuK$_\alpha$ radiation ($\lambda = 1.541$ Å) was used.

**Mechanical Tests.**   The double-torsion (DT) test, which was first reported by Kies and Clark (2), was used to measure toughness. The DT method was initially designed for brittle ceramic materials (3–8), but has been used widely since for brittle polymers such as epoxies (9–14).

The double-torsion specimen is a pair of elastic torsion bars each having a rectangular cross section, that are laid side-by-side and joined together along a portion. A diagram of the DT specimen and loading configuration is shown in Figure 2. For small deflections, the fracture toughness ($K_{Ic}$) is given by

$$K_{Ic} = P_{crit}w_n\left[3(1 + \nu)/Wt^3t_n\right]^{1/2} \tag{1}$$

where $P_{crit}$ is the load at which crack-growth initiation or arrest occurs, $w_n$ is the moment arm of each torsion, $\nu$ is Poisson's ratio, $W$ is the specimen width, $t$ is the specimen thickness, and $t_n$ is the specimen thickness where fracture is occurring. As can be seen from the equation, the stress-intensity factor is independent of crack length and is a function only of the applied load, the specimen dimensions, and Poisson's ratio. This feature makes the double torsion test extremely useful for subcritical crack-growth studies.

The fracture energy $G_{Ic}$ is related to the fracture toughness by

$$K_{Ic} = (E\,G_{Ic})^{1/2} \tag{2}$$

where $E$ is Young's modulus.

The dimensions of the rectangular specimens are 100 mm long, 30 mm wide, and 6 mm thick. A sharp crack was inserted at one end to prevent the formation of a plastic zone in the vicinity of the crack tip. The crack was inserted by tapping with a hammer on a razor blade held against the end of a

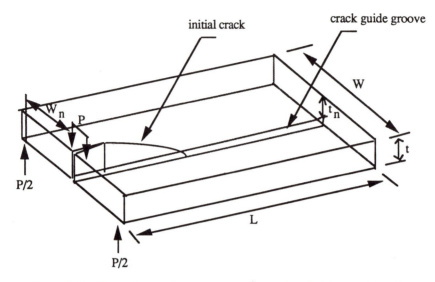

*Figure 2. Double torsion specimen geometry. (Reproduced with permission from reference 1. Copyright 1992 Chapman and Hall.)*

saw cut. Initial crack lengths were made longer than 0.55 times the specimen width to avoid end effects (6). A shallow center groove was cut with a table saw along one face of the specimen to guide the propagating crack. The width of the groove was less than 0.5 mm and the depth was nearly one half the specimen thickness.

The load was applied with a screw-driven mechanical tester (Instron) at 0.04 mm/s crosshead speed. The load against time was recorded with a strip chart recorder.

To examine the yielding behavior of the materials, specimens were deformed in uniaxial compression. The studied epoxies deformed plastically in uniaxial compression but were too brittle to do so in uniaxial tension. Rectangular rod-shaped specimens were used with a height-to-width ratio of 2:1. The specimens were deformed at a constant displacement rate of 0.01 mm/s between polished ceramic plates lubricated with molybdenum disulfide. The load from the measured load-displacement curve was converted to stress using the initial cross-sectional area of the specimen. The compressive modulus ($E$) and the compressive yield stress ($\sigma_y$) were also determined.

**Fractography.** Scanning electron microscopy (SEM) was used to study and record the fractured surfaces of the pure and modified epoxies. The fractured sample surfaces were coated with a thin layer (10–30 nm) of gold–palladium. The coating was carried out by placing the specimen in a high vacuum evaporator and vaporizing the metal held in a heated tungsten basket.

Polarized transmission optical microscopy was used to investigate the structure and deformation below the fracture surface. The specimens used for optical microscopy were $30 \times 10 \times 6$ mm, with a 3-mm-deep notch in one side that was inserted with a 0.2-mm-thick saw blade. The specimens were broken by three-point bending, and thin sections perpendicular to the fracture surface were obtained by metallographic thinning techniques (15). Under polarized light, the plastic deformation zone was detected by induced birefringence. Thin sections were taken from the middle region of the three-point bending specimen. The section plane was parallel to the crack propagation direction and normal to the fracture surface.

## Results

**Effect of Annealing.** Annealing a polymer near the crystalline melting point can perfect the crystallinity that exists, thicken the crystal lamellae, and add to the crystallinity. Short periods of annealing are easily performed. In the present case, however, an especially long period of annealing was undertaken to ascertain what could be expected from annealing. Poly(butylene terephthalate) powder and a fully cured PBT–epoxy blend were each annealed for 24 h at 200 °C.

The effect of annealing on the melting temperature and heat of fusion of the PBT and on the glass-transition temperature of the epoxy are given in Table I. The melting of PBT has been split into two events. One melting occurred at 210 °C and the other occurred at the original $T_m \simeq 224$ °C. The most probable cause of observed heat-of-fusion increase with annealing is an increase in crystallinity, possibly by as much as 15–20%. All of the new, as well as some of the original, crystallinity melted at the lower temperature.

The effect of annealing on the fracture energy for crack initiation is shown in Table II. Even the long 24-h annealing at 200 °C had little effect on the toughness.

### Table I. Properties of PBT and Epoxy Separately and in Blends

| Material | Epoxy $T_g$ (°C) | $T_m$ of PBT (°C) | PBT Heat of Fusion[a] (J/g) |
|---|---|---|---|
| Epoxy | 155 | | |
| Unannealed PBT | — | 224 | 65 |
| Annealed PBT[b] | — | 210, 224[c] | 74 |
| Unannealed PBT–epoxy | 155 | 223 | 63 |
| Annealed PBT—epoxy[b] | 154 | 209, 223[c] | 72 |

[a] Based on mass of PBT.
[b] Annealing 24 h at 200 °C.
[c] Temperatures of pairs of endotherm maxima that resulted from annealing.

**Table II. Effect of Annealing on PBT–Epoxy Blends**

| PBT (wt %) | Heat Treatment | PBT Heat of Fusion[a] (J/g) | $G_{Ic}$ (kJ/m$^2$) |
|---|---|---|---|
| 5 | Unannealed | 63 | 0.60 |
| 20 | Unannealed | 64 | 1.10 |
| 5 | Annealed[b] | 74 | 0.72 |
| 20 | Annealed | 75 | 1.08 |

NOTE: Average particle size 15 μm.
[a] Based on the mass of PBT.
[b] 24 h at 200 °C.

**Effect of PBT Particle Size and Epoxy Molecular Weight on Mixing Behavior.** Near and above its melting point, PBT is soluble in uncured epoxy resins. The temperature at which the dissolution process begins and reaches completion are of interest. These dissolution temperatures are equivalent to the lowering of the PBT melting point by the epoxy. For processing, the effect of the average PBT particle size on the dissolution process is of interest and can readily be examined with differential scanning calorimetry (DSC). In Figure 3, thermograms of Epon 828 (without curing agent) containing 20-wt % PBT particles of three different sizes (Figures 3b–3d) are compared with the thermogram of dry PBT powder (Figure 3a). For PBT particles as large as 3 mm, the shape of the DSC thermogram changed little from that for dry PBT except that the melting point was slightly reduced. As the particle size decreased to 15 μm, the thermogram became increasingly complex, and finally exhibited three apparent endothermic peaks. The melting temperature associated with the highest-temperature endothermic maxima also decreased with decreasing particle size. Prior to the second DSC scans, the specimens were quickly returned to room temperature from the maximum temperature (250 °C) of the first scans. The resultant thermograms (Figure 4b) exhibited a different shape, even for the smallest particles. The three apparent endothermic peaks remain, but the lower two are shifted to lower temperatures. For the mixture containing 15-μm-diameter PBT particles, there was no further change in the thermogram during subsequent scans (Figure 4c). The thermogram in Figure 4c indicates that the solid PBT begins dissolving in the epoxy at 170–175 °C, and the process is essentially complete at 210–215 °C.

The dissolution of PBT also depends on the molecular weight of the initial epoxy (i.e., Epon 828 vs. Epon 1002F). The effect of the molecular weight for 15-μm-diameter PBT particles is shown in Figure 5. (Figure 5a is for unmixed PBT particles.) The thermogram is less complex with the higher molecular weight epoxy (Figure 5b). Only two apparent endotherms are seen, and there is less lowering of the melting temperature.

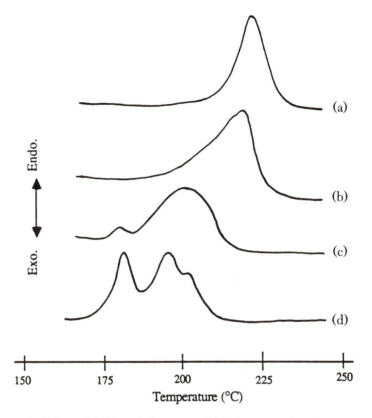

*Figure 3. Effect of PBT particle size on DSC thermographs of mixtures of 20-wt % PBT and low molecular weight epoxy (Epon 828): pure PBT (a); 3 mm diameter (b); 0.5 mm diameter (c); 0.015 mm diameter (d).*

Above 220 °C, PBT is completely miscible with both epoxies. Cooling causes the PBT to precipitate from these solutions. The crystalline morphology of the PBT that develops depends strongly on the epoxy molecular weight, as evidenced by Figures 6 and 7. From the lower molecular weight epoxy (Epon 828), the PBT forms spherulites as seen in the micrograph of a thin section in Figure 6a, which was obtained with transmitted light. Although these spherulites do not exhibit Maltese crosses when viewed with crossed polarizers (as in this micrograph), they do have the circular and fibrous characteristics of spherulites. In SEM micrographs of the fracture surfaces (Figure 7a), the PBT appears as spherical particles. For the higher molecular weight epoxy blend (Figure 7b), no crystalline phase separation is apparent on the fracture surface despite the relatively high heat of fusion ($\sim$ 65 J/g) after curing. Tiny crystals dominate the field of view in optical micrographs of thin sections, as seen in Figure 6b.

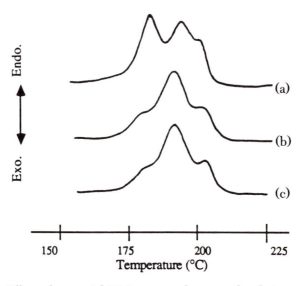

*Figure 4. Effects of sequential DSC scans on thermographs of mixtures of PBT (20 wt %; 0.015 mm) and low molecular weight epoxy (Epon 828): first scan (a); second scan (b); third scan (c).*

**Effect of "Mixing Temperature".**    In the following experiments, 15-$\mu$m-diameter PBT particles were dispersed in the lower molecular weight epoxy (Epon 828). Different portions of this mixture were heated at the rate of ~ 10 °C/min, without stirring, to specific temperatures, referred to here as the "mixing temperature" ($T_{mix}$). When $T_{mix}$ was reached, the mixture was rapidly cooled to room temperature, again without stirring. At room temperature, the curing agent was added (with stirring) and the mixture was cured for 2 h at 80 °C and 2 h at 150 °C.

The effect of $T_{mix}$ on the thermal properties of the blend is summarized in Table III. A pair of endotherms for the melting of PBT was found in the 170–200 °C mixing temperature range. The temperature of the upper endotherm remained relatively constant, whereas the temperature of the lower endotherm, as well as the glass-transition temperature of the epoxy ($T_g$), decreased with increasing mixing temperature. Observation of Figure 8 indicates that both decreases are almost linear with increasing mixing temperature. When the mixing temperature was 220 °C, only a single PBT melting endotherm was found: the endotherm that continued the linear decrease of the lower endotherms with increasing mixing temperature. The higher temperature endotherm disappeared. The thermograms in Figures 3–5 for the 15-$\mu$m-diameter particles show that complete melting–dissolution of the PBT in the epoxy occurs at 210–215 °C at a heating rate of 10 °C/min. Thus,

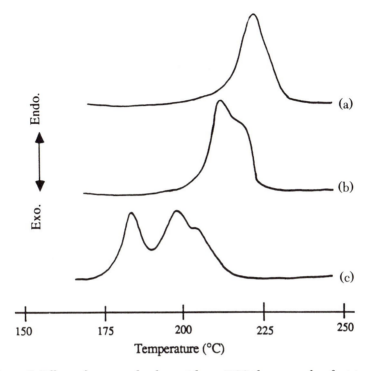

*Figure 5. Effects of epoxy molecular weight on DSC thermographs of mixtures of PBT (20 wt %; 0.015 mm) and epoxy: pure PBT (a); high molecular weight epoxy (Epon 1002F) (b); low molecular weight epoxy (Epon 828) (c).*

the pair of endotherms that occurs when the mixing temperature is in the 170–200 °C range seems to arise from the partial melting–dissolution of the PBT. The lowering of the epoxy $T_g$ with mixing temperature suggests that not all of the PBT dissolving in the epoxy is reprecipitated when cooled. However the fraction of crystalline PBT, as indicated by the heat of fusion given in Table III, increased with increasing mixing temperature. In addition, X-ray diffraction analysis indicated the occurrence of a general perfecting of the crystals with no change in crystal structure as mixing temperature increased. The diffraction peaks became sharper and narrower without displacement.

The viscosity of the mixtures after cooling to room temperature also was affected by the mixing temperature. The viscosity appeared to go through a maximum with increasing mixing temperature. For mixing temperatures in the 185–200 °C range, the mixture at room temperature behaved like a thick grease. Even with only 5-wt % PBT, uniform mixing of the curing agent was very difficult, and at 10-wt % PBT and above, further processing could not

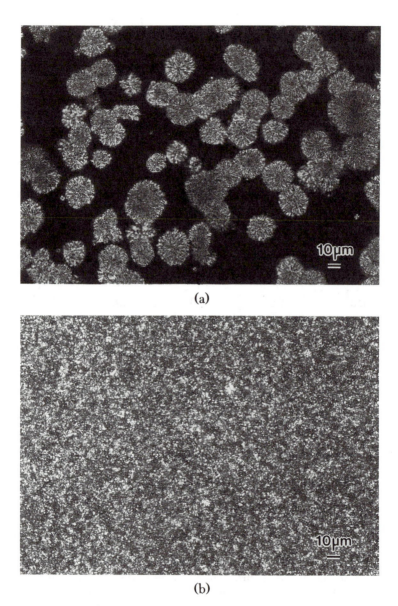

(a)

(b)

*Figure 6. Optimal micrographs of thin sections of PBT–epoxy blends with polarizer and analyzer at 45° to each other: low molecular weight epoxy (Epon 828) blend (a); high molecular weight epoxy (Epon 1002F) blend (b).*

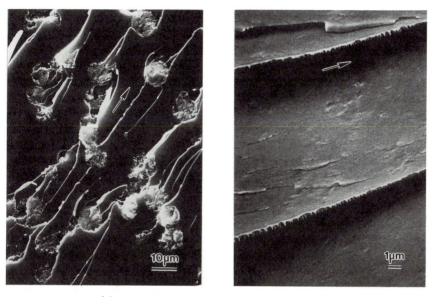

(a)                                              (b)

*Figure 7. SEM micrographs of fracture surfaces of PBT–epoxy blends: low molecular weight epoxy (Epon 828) blend (a); high molecular weight epoxy (Epon 1002F) blend (b). Arrows indicate the crack-propagation direction.*

**Table III. Effect of Mixing Temperature on Thermal Properties of PBT–Epoxy Blends**

| $T_{mix}$ (°C) | Epoxy $T_g$ (°C) | $T_m{}^a$ of PBT (°C) | PBT Heat of Fusion[b] (J/g) |
|---|---|---|---|
| 25  | 157.3 | 224        | 62 |
| 170 | 157.3 | 217.1, 224 | 63 |
| 185 | 157.1 | 215.8, 223 | 68 |
| 193 | 154.8 | 215.7, 223 | 69 |
| 200 | 154.3 | 213.7, 222 | 71 |
| 220 | 151.8 | 209.1      | 72 |

NOTE: 5-wt % PBT; average particle size 15 μm.
[a] Temperatures of endotherm maxima.
[b] Based on the mass of PBT.

be done. When the mixing temperature was below 185 °C or at 220 °C and above, the room-temperature viscosity of the mixture differed little from the viscosity of the unheated mixture of PBT particles in epoxy.

The effect of mixing temperature on the mechanical properties of cured PBT–epoxy blends, including the fracture toughness $K_{Ic}$ and fracture energy

$G_{\mathrm{Ic}}$, is summarized in Table IV. The effect of mixing temperature on the fracture energy of the 5-wt % PBT–epoxy blend is plotted in Figure 9. The fracture energy increased rapidly when the mixing temperature increased from 185 to 193 °C. For higher mixing temperatures, the fracture energy slowly decreased. The fracture energy resulting from the 193 °C mixing temperature was 900 J/m$^2$. The room-temperature Young's modulus ($E$) and yield stress ($\sigma_y$) given in Table IV are unaffected by the mixing temperature or the process conditions, and remain essentially the values of the epoxy without PBT addition.

The morphologies of the blends after cure reflect the mixing and recrystallization behavior. For mixing temperatures below 175 °C, the mor-

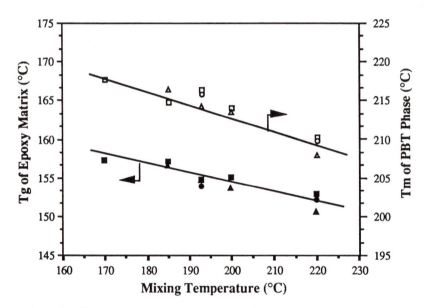

*Figure 8. Effect of mixing temperature on the thermal properties of PBT–epoxy blends. Different symbols refer to different specimens. At a given annealing temperature, the same symbol (filled and unfilled) refers to the same specimen.*

**Table IV. Effect of Mixing Temperature on Mechanical Properties**

| $T_{mix}$ (°C) | E (GPa) | $\sigma_y$ (MPa) | $K_{Ic}$ (MPa · m$^{1/2}$) | $G_{Ic}$ (kJ/m$^2$) |
|---|---|---|---|---|
| 25 | 3.06 | 118 | 1.34 | 0.60 |
| 170 | 3.02 | 115 | 1.28 | 0.55 |
| 185 | 3.10 | 114 | 1.35 | 0.60 |
| 193 | 3.00 | 114 | 1.62 | 0.87 |
| 200 | 2.98 | 113 | 1.60 | 0.85 |
| 220 | 3.00 | 114 | 1.50 | 0.75 |

NOTE: 5-wt % PBT; average particle size 15 μm.

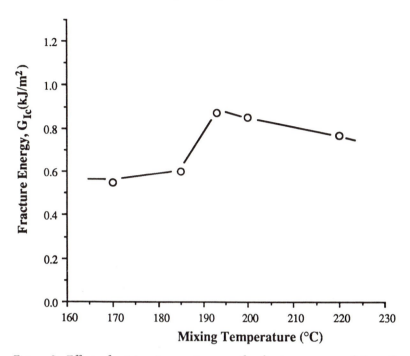

*Figure 9. Effect of mixing temperature on the fracture energy of 5-wt % PBT–epoxy blends. The open circles denote experimental data.*

phology was unchanged from that of the blend mixed at room temperature (Figure 10a). As the mixing temperature increased above 175 °C, the interface became less distinct, the size of the PBT phase decreased, and the fracture surfaces became rough and complex (Figures 10b and 10c). For mixing temperatures above 220 °C, however, there was a sudden change in the morphology of the PBT–epoxy blend (Figure 10d). The PBT phase assumed a spherical form with a uniform size of about 8–10 μm diameter and a distinct interface.

The decrease in viscosity of PBT–epoxy mixtures when the mixing temperature is 220 °C or greater allowed study of the effect of PBT content. The effect of the 220 °C mixing temperature on the epoxy matrix $T_g$ and the PBT $T_m$ for 5-, 10-, and 20-wt % PBT is shown in Table V. The preparation of these materials, referred to as Process 2, was the same preparation used for the preceding materials: Without curing agent, 15-μm-diameter PBT particles were mixed with the epoxy at room temperature and the mixture was heated at a rate of 10 °C/min to 220 °C and then immediately cooled to room temperature. Subsequently, the curing agent was added and the mixture was cured at 80 °C and postcured at 150 °C. Process 1 refers to the original procedure in which 15-μm-diameter PBT particles, curing agent, and

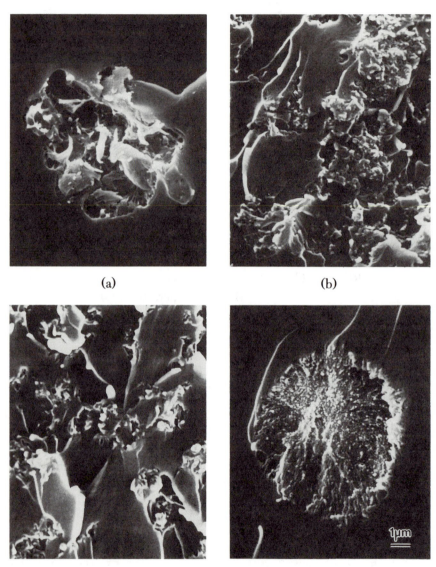

(a)                                    (b)

(c)                                    (d)

*Figure 10. SEM micrographs of fracture surfaces of PBT–epoxy blends made by mixing at 25 °C (a); 193 °C (b); 200 °C (c); 220 °C (d). The scale is the same for all micrographs.*

epoxy were mixed together at room temperature, degassed cured at 80 °C, and postcured at 150 °C.

The $T_g$s and $T_m$s of the Process 2 blends were lowered to around 152 and 212 °C, respectively, independent of PBT content. By contrast, the $T_g$s and $T_m$s of the Process 1 blends remained at 157 °C and 223 °C, respectively, the same values as the unmixed materials. The Young's modulus and yield stress at room temperature of the PBT–epoxy blends given in Table VI remain unaffected by the different processing conditions, again independent of PBT content, despite the resulting variety of morphologies and thermal properties. The fracture energy was affected by the mixing temperature and by the PBT content, as can be seen in Table VI and Figure 11. For the Process 1 blend, the fracture energy increased rapidly up to 5-wt % PBT and then less rapidly

### Table V. Effect of Preparation on the Thermal Properties of PBT–Epoxy Blends

| Process | PBT (wt %) | Epoxy $T_g$ (°C) | $T_m$ of PBT (°C) | PBT heat of Fusion[a] (J/g) |
|---------|-----------|------------------|-------------------|------------------------------|
| 1 | 5 | 157.2 | 223.5 | 65 |
| 1 | 10 | 157.0 | 223.8 | 63 |
| 1 | 20 | 156.3 | 223.3 | 64 |
| 2 | 5 | 152.2 | 209.8 | 74 |
| 2 | 10 | 151.2 | 211.9 | 72 |
| 2 | 20 | 152.6 | 214.0 | 71 |
| 3 | 5 | 148.5 | 212.3 | 72 |
| 3 | 10 | 151.7 | 212.2 | 71 |
| 3 | 20 | 149.7 | 212.7 | 67 |

[a] Based on the mass of PBT.

### Table VI. Effect of Preparation on the Mechanical Properties of PBT–Epoxy Blends

| Process | PBT (wt %) | E (GPa) | $\sigma_y$ (MPa) | $K_{Ic}$ (MPa · $m^{1/2}$) | $G_{Ic}$ (kJ/$m^2$) |
|---------|-----------|---------|------------------|----------------------------|----------------------|
| 1 | 5 | 2.99 | 115 | 1.34 | 0.60 |
| 1 | 10 | 3.04 | 117 | 1.48 | 0.73 |
| 1 | 20 | 2.98 | 115 | 1.82 | 1.10 |
| 2 | 5 | 3.00 | 114 | 1.50 | 0.75 |
| 2 | 10 | 2.97 | 112 | 1.47 | 0.72 |
| 2 | 20 | 3.03 | 112 | 1.49 | 0.74 |
| 3 | 5 | 3.03 (2.69[a]) | 116 | 0.97 | 0.31 |
| 3 | 10 | 3.10 (2.60[a]) | 116 | 0.85 | 0.24 |
| 3 | 20 | 2.98 (2.74[a]) | 114 | 0.87 | 0.25 |

[a] Measured by uniaxial tension.

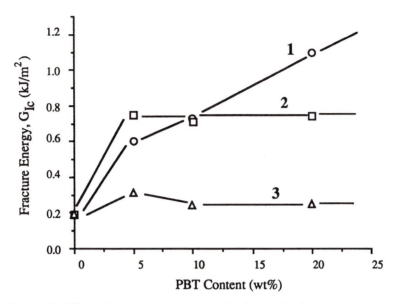

*Figure 11. Effect of three preparation methods on the fracture energy of PBT–epoxy blends. Key:* △ *, Process 3;* ○*, Process 1;* □ *, Process 2.*

but linearly with increasing amounts of PBT. For the Process 2 blends, the fracture energy increased even more rapidly up to 5-wt % PBT, but remained relatively constant at around 750 $J/m^2$ above 5-wt % PBT.

**Effect of Mechanical Mixing and Time at Temperature.**  In the following experiments, the preparation of particulate PBT as a separate step was eliminated. Instead, pellets were mixed directly with the epoxy. In this process, referred to as Process 3, $\frac{1}{8}$-in.-diameter PBT pellets (without curing agent) were mixed with the epoxy at room temperature and the mixture was heated at a rate of 10 °C/min to 230 °C. After stirring for 10 min at this temperature, the PBT had dissolved in the epoxy. The mixture was cooled to room temperature, the curing agent was added, and the mixture was cured (at 80 °C) and postcured (at 150 °C) as in Processes 1 and 2.

The effect of this mixing procedure on the epoxy matrix $T_g$ and the $T_m$ of the PBT is shown in Table V. The $T_g$s and $T_m$s of the Process 3 blends were lowered to around 150 and 212 °C, respectively, which are about the same as results from Process 2. Apparently neither the degree of agitation during mixing nor the amount of PBT in the blend had an effect on the thermal properties. The Young's modulus and yield stress at room temperature of the PBT–epoxy blends also remained unaffected by the different processing conditions, as can be seen in Table VI. The fracture energy, however, was affected by the mixing temperature, the degree of agitation during mixing, and the PBT composition, as can be seen in Table VI and Figure 11. The fracture-energy enhancement by the PBT was negligible over the whole composition range tested and remained almost the same as that of the unfilled epoxy.

## Discussion

A variety of mixing and processing procedures has been described. The purpose of these procedures was to increase the fracture toughness and simplify the manufacture of PBT–epoxy blends for "nontoughenable" epoxies beyond the previously described (*1*) constraints of dispersing PBT particles with curing agent in epoxy at room temperature. Although none was singularly successful, a number of aspects of toughening were revealed.

**Effect of Annealing on Toughness**. Annealing is a relatively simple process that was expected to increase the perfection and amount of crystallinity. Although this expectation was fulfilled, after 24 h of annealing at 200 °C, the crystallinity increased only 10–15%. No net effect within experimental error (good or bad) on the fracture energy of the blends was observed, perhaps because the experimental error of the fracture-energy measurements was so large that small changes were difficult to detect. Also, to the extent that toughening arose from phase-transformation toughening, it is possible that conformational changes similar to the changes occurring in the crystalline phase and absorbing a similar amount of energy are induced in the amorphous phase and obscure the effect of the small 10–15% increase in crystallinity. The observed transformation during annealing of higher melting crystals to lower melting crystals entails thermodynamic restrictions that couple the transformation to the overall increase in crystallinity. A description of this transformation mechanism is described elsewhere (*16*).

**Effect of PBT Particle Size and Epoxy Molecular Weight on Mixing Behavior**. Poly(butylene terephthalate) becomes soluble in bisphenol A-type epoxies at temperatures below the melting point of the PBT. The effect of PBT particle size on the DSC thermogram of PBT–epoxy mixtures reflects the kinetics of interdiffusion of PBT and epoxy in comparison with the heating rate of 10 °C/min. The kinetics of PBT and epoxy interdiffusion is slow enough that very little PBT in the 3-mm-diameter pellets becomes molecularly mixed with the epoxy in the 40–50 °C temperature range (4–5 min duration) during which the PBT is melting. However, the kinetics are fast enough that a significant amount of the 0.5-mm PBT particles and most of the 0.015-mm particles become molecularly mixed with the epoxy in this temperature–time interval. The preceding interdiffusion mixing phenomena are probably true for both of the epoxies studied. The difference between the two epoxies is that the higher molecular weight epoxy has less solvating power for PBT. Hence, the melting temperature is not lowered as much, which narrows the melting-temperature range.

The complexity of the thermograms in Figures 3–5 can arise from several sources:

1. The epoxy has different access to the PBT crystals. This difference certainly contributes to the complexity of the thermograms in Figures 3 and 5, where the PBT particles and epoxy are initially mixed, and probably continues to a small extent for the thermograms in Figure 4, where the PBT has recrystallized from the epoxy solution, because of the finite size of the crystals.

2. There probably are crystals or parts of crystals of different degrees of perfection. The crystals developing from the melt earlier and at higher temperature during cooling will have greater perfection than the crystals developing later. The crystals developing later and at lower temperature (higher supercooling) often incorporate less regular material, such as molecular chains with jogs from improper conformations, chains that are constrained (from parts of the chain being incorporated into crystals that formed earlier), or defects like chain ends.

3. A single endotherm may appear to be multiple because of the superposition of a crystallization or recrystallization exotherm. The superposition of a small exotherm on a broader endotherm can make the broader endotherm appear as two endotherms (17), and the superposition of a broad exotherm on a similarly broad endotherm, depending on their shapes, can give the appearance of three endotherms (18). An exotherm would result from the crystallization of remaining uncrystallized PBT and from the recrystallization after melting of the less perfect crystals. The three apparent peaks in the DSC thermograms in Figure 4 that occur on the second and subsequent scans probably arise from the superposition of an endotherm with either one or two maxima and an exotherm. The general melting–dissolution of the less perfect and more epoxy-accessible crystals seems to be followed and accompanied by recrystallization into more perfect crystals. This superposition causes the melting to appear in the form of the two left-hand peaks, if not all three peaks. The melting–dissolution of the most perfect crystals, slightly smeared by the different accessibility of the epoxy, then occurs at the right edge of the right-hand peak.

The change in the thermogram for the 15 μm PBT particles from the first to the second scan indicates that the 15-μm spherulites that form contain some epoxy.

The fibrous texture and the absence of Maltese crosses in the spherulites seen in the thin section shown in Figure 6a indicate the presence of a small number of coarse crystals within the spherulites. This could be another indication of the incorporation of epoxy within the spherulites. When PBT is fully dissolved in the lower molecular weight epoxy, the crystallization on cooling exhibits the common spherulitic morphology, which involves fairly rapid diffusion of the PBT molecules to the sites of primary nucleation. When PBT is dissolved in the higher molecular weight epoxy, the morphology is an irregular packing of tiny crystals, as visible in the micrograph in Figure 6b, which suggests that diffusion is sufficiently inhibited in the solution that none of the nuclei formed at higher temperature accumulates significant amounts of polymer. Instead, the major part of the PBT crystallizes by short-range diffusion to an enormous number of nuclei that form at large undercoolings. Alternatively, though less likely, the massive numbers of crystals may have formed by spinodal decomposition rather than by nucleation and growth (*19*).

**Effect of Mixing Temperature**. The thermal properties, fracture behavior, and morphology of PBT–epoxy blends after cure are strongly dependent on the mixing temperature. As the mixing temperature increases above 170 °C, the $T_g$ of the epoxy decreases almost linearly. The lowering of $T_g$ indicates a miscibility between PBT and the epoxy (*20*). Such miscibility was anticipated from the lowering of melting–dissolution temperature of PBT in epoxy seen in the DSC scans. As indicated in Table III, "double melting" is exhibited in PBT–epoxy blends mixed at temperatures ranging from 170 to 200 °C. The double melting of the PBT is because of the partial dissolution of PBT in the epoxy. Complete PBT dissolution in even the lower molecular weight epoxy is not obtained below 210–215 °C, as seen in Figure 3d. The higher melting at 224 °C corresponds to the melting of undissolved PBT. With mixing temperature increase, the temperature of the lower endotherm decreases almost linearly. The apparent lowering of the $T_m$ is caused by the dissolution–melting of the less perfect PBT crystals (*17*). Another possible component of the lower endotherm is a reduction of lamellar thickness caused by a depolymerization or transesterification between the epoxy and PBT during mixing. Depolymerization reportedly occurs with PBT in the presence of moisture (*21, 22*), and transesterification occurs in the presence of polycarbonate (*23*). The α structure of the PBT remains unchanged with changes in mixing temperature, however. The increase in the heat of fusion and the sharpening of the X-ray diffraction pattern with increasing mixing temperature suggest that the melting or dissolution of an increasingly greater fraction of the PBT is occurring in the process. Recrystallization from solution that follows yields a greater fraction of more perfect crystals being formed than occurs during crystallization from the melt.

The fracture energy of the blends is also affected by the mixing temperature. The fracture energy rapidly increases as the mixing temperature increases from 185 to 193 °C; at higher mixing temperatures, the fracture energy slowly decreases. At its maximum at a mixing temperature of 193 °C, the fracture energy is 900 J/m². The maximum fracture energy is associated with a very complex fracture surface, which is indicative of large deformation of both the epoxy and PBT. The large viscosity increase shown by these mixtures before curing suggests that a PBT network interpenetrating the epoxy may have formed, and it may be this interpenetrating network that is contributing to the large increase in toughness.

The mixing behavior of PBT and epoxy is exhibited in the fracture surface morphology of the blends. The PBT particles made by dissolution–reprecipitation appear as agglomerates of smaller particles. When the PBT is mixed with the epoxy below the onset temperature of the melting–dissolution endotherm (175 °C) the fracture morphology is the same as that from room-temperature mixing. When the PBT is mixed with the epoxy at temperatures ranging from 185 to 200 °C, however, the fracture morphology becomes more complex because of the partial dissolution of the PBT particles. For mixing temperatures above 220 °C, however a sudden change occurs in the morphology of the PBT–epoxy blend, and it reverts to a simpler mixture. The probable cause of the change is the complete dissolution of the PBT and recrystallization, which results in homogeneous resin with approximately spherical PBT particles that are nearly all the same size (8–10 μm).

A final characteristic of mixing temperature is the viscosity of the PBT–unreacted epoxy mixture after cooling to room temperature. The rapid viscosity increase with increasing mixing temperature above 185 °C and the sudden viscosity fall off when the mixing temperature reaches 220 °C is indicative of melting, dissolution, and recrystallization processes. As the mixing temperature is increased above 170 °C, an increasing fraction of the PBT melts and dissolves in the epoxy. At temperatures below 210–215 °C, recrystallization of the dissolved PBT can occur and, as long as the mixing temperature has remained below 210–215 °C, a massive number of crystals or crystal nuclei remain upon which crystallization can occur. When the density of crystal nuclei is high enough, a single PBT molecule can be incorporated into more than one crystal, thus creating a PBT network interpenetrated by the epoxy. The presence of this PBT network is the likely cause of the large increase in viscosity noted, particularly when the mixing temperature is in the range from 185 to 200 °C. When the mixing temperature exceeds 210–215 °C, most, if not all, of the PBT crystals melt. As the mixture cools, only a modest number of crystal nuclei reappear, and they are far enough apart that a PBT network is unable to form. Hence, the viscosity of the mixture at room temperature is little more than that of the epoxy alone.

**Effect of Mechanical Mixing and Time at Temperature**. Because of the high melt viscosity of PBT, a uniform mixture of PBT and the epoxy cannot be obtained even at high temperature without stirring. It is disappointing, therefore, that the fracture energy of PBT–epoxy blends made at 230 °C with stirring (Process 3) is about the same as that of the epoxy alone. As discussed previously (*1*), the fracture energy of the blends arises from a combination of mechanisms. Phase-transformation toughening in blends made by Process 3 is thought to be suppressed by the isolation of the crystals in the spherulites by a surrounding epoxy matrix (*24*). This isolation prevents the crystals from rotating and allowing the maximum tensile stress to be applied along the chain axis, which is requisite for all the α- to-β phase transition to occur. Other mechanisms that are suppressed are primary crack bridging, crack bifurcation, secondary crack bridging, and ductile fracture of the particles. This suppression may result from a general weakening of the spherulite caused by epoxy incorporation. Primary crack bridging, crack bifurcation, and secondary crack bridging are interdependent and rely on the strength of the toughening particle, which in the present case is the PBT spherulite. Ductile fracture depends on the general ductility of the particle. PBT is ductile, but a spherulitic collection of PBT crystals in an epoxy matrix is not nearly as ductile.

The Young's modulus and yield stress remained unaffected by any of the variables of blending with PBT, including agitation during mixing and the PBT content.

Like the Young's modulus and yield stress, although the $T_g$ of the epoxy and the $T_m$ of the PBT exhibit a strong dependency on mixing temperature, they do not depend on the degree of agitation during mixing or on the PBT content. The PBT composition affects the precipitation temperature, which for mixtures with 5-, 10-, and 20-wt % PBT were around 165, 185, and 200 °C, respectively, but did not reflect in the melting temperatures. The lowering of the $T_g$ with increasing mixing temperature indicates miscibility between PBT and epoxy, but the relative independence of the epoxy $T_g$ from PBT content suggests that the epoxy is saturated with PBT even at 5-wt % PBT.

## Summary

Particles of poly(butylene terephthalate) (PBT) dispersed in an aromatic amine-cured epoxy were found to increase the fracture energy $G_{Ic}$ at about twice the rate attainable with particulate inclusions of other polymers, including nylon 6, poly(vinylidene fluoride), and CTBN rubber. The PBT particles employed were obtained prior to use by dissolving PBT in phenol, precipitating it in methanol, and grinding the resulting dried polymer to a uniform particle size. To increase further the fracture toughness of PBT–epoxy blends or simplify their manufacture, several alternative mixing and processing procedures were explored. These procedures included (1) the annealing of PBT just below its melting point, either before mixing with the epoxy or after mixing with and curing the epoxy, and (2) exposing mixtures of PBT and epoxy without curing agent to various temperatures, with and without mechanical mixing.

The annealing essentially had no effect (beneficial or otherwise) on the fracture energy of the blend. The degree of crystallinity was increased by annealing by only perhaps 15–20%.

The fracture energy was generally sensitive to the exposure of the PBT to the epoxy at elevated temperatures. The highest fracture energy was obtained by heating a mixture of 15-$\mu$m PBT particles in epoxy without curing agent at 10 °C/min to 193 °C and immediately cooling to room temperature to add the curing agent. After curing, a 5-wt % PBT–epoxy blend had a fracture energy of 900 J/m$^2$ vs. 190 J/m$^2$ without the PBT. A PBT network interpenetrating the epoxy may contribute to this large increase in toughness. Compositions with higher amounts of PBT were too viscous to be processed in this manner. When the mixture of PBT particles in epoxy without curing agent was heated to 220 °C, the viscosity problems were avoided and the fracture energy of the cured blend was 750 J/m$^2$ at the 5-wt % PBT level, although higher fracture energies were not obtained with greater PBT content.

The glass-transition temperature of the epoxy and the melting temperature of the PBT were generally lowered by forming these blends. For the PBT–epoxy mixture heated to 200 °C, the $T_g$ was lowered from 157 to 152 °C, and the $T_m$ was lowered from 223 to 212 °C. Alternatively, the Young's modulus and the yield strength of all blends remained the same as those of the unblended epoxy.

## Acknowledgments

We acknowledge useful discussions with our colleagues, Professor I-Wei Chen, Professor Ray Pearson, and Professor Albert Yee. This work was supported by the National Science Foundation through Materials Research Group Grant DMR–87–08405.

## References

1. Kim, J. K.; Robertson, R. E. *J. Mater. Sci.* **1992**, 27, 161.
2. Kies, J. A.; Clark, B. J. In *Proceedings of the Second International Conference on Fracture*; Pratt, P. L., Ed.; Chapman and Hall: London, 1969; p 83.

3. Williams, D. P.; Evans, A. G. *J. Test. Eval.* **1972**, *1*, 264.
4. Evans, A. G. *J. Mater. Sci.* **1972**, *7*, 1137.
5. Evans, A. G.; Wiederhorn, S. M. *J. Mater. Sci.* **1974**, *9*, 270.
6. Trantina, G. G. *J. Amer. Ceram. Soc.* **1977**, *60*, 7.
7. Evans, A. G.; Russel, L. R.; Richerdson, D. W. *Metall. Trans. A* **1975**, *6*, 707.
8. Govila, R. K. *J. Amer. Ceram. Soc.* **1980**, *63*, 319.
9. Beaumont, P. W. R.; Young, R. J. *J. Mater. Sci.* **1975**, *10*, 1334.
10. Young, R. J.; Beaumont, P. W. R. *J. Mater. Sci.* **1977**, *12*, 684.
11. Phillips, D. C.; Scott, J. M.; Jones, M. *J. Mater. Sci.* **1978**, *13*, 311.
12. Yamini, S.; Young R. J. *J. Mater. Sci.* **1979**, *14*, 1609.
13. Leevers, P. S.; Williams, J. G. *J. Mater. Sci.* **1987**, *22*, 1097.
14. Cantwell, W. J.; Roulin-Moloney, A. C.; Kausch, K. H. *J. Mater. Sci. Lett.* **1988**, 7, 976.
15. Holik, A. S.; Kambour, R. P.; Hobbs, S. Y.; Fink. D. G. *Microstruct. Sci.* **1979**, *7*, 357.
16. Kim, J. K.; Nichols, M. E.; Robertson, R. E., submitted for publication in *J. Phys. Chem.*
17. Todoki, M.; Kawaguchi, T. *J. Polym. Sci. Polym. Phys. Ed.* **1977**, *15*, 1067.
18. Nichols, M. E.; Robertson, R. E. *J. Polym. Sci. Polym. Phys. Ed.* **1992**, *30*, 305, 755.
19. Kwei, T. K.; Wang, T. T. In *Polymer Blends*; Paul, D. R.; Newman, S., Eds.; Academic: New York, 1978; Vol. 1, pp 141–184.
20. MacKnight, W. J.; Karasz, F. E.; Fried, J. R. In *Polymer Blends*; Paul, D. R.; Newman, S., Eds.; Academic: New York, 1978; Vol. 1, pp 185–200.
21. Borman, W. F. H. *Polym. Eng. Sci.* **1982**, *22*, 883.
22. Kishore, K.; Sankaralingam, S. *Polym. Eng. Sci.* **1984**, *24*, 1043.
23. Devaux, J.; Godard, P.; Mercier, J. P. *Polym. Eng. Sci.* **1982**, *22*, 229.
24. Kim, J. K.; Robertson, R. E. *J. Mater. Sci.* **1992**, *27*, 3000.

RECEIVED for review March 6, 1991. ACCEPTED revised manuscript January 23, 1992.

# Toughening of Epoxy Resins by Epoxidized Soybean Oil

Isabelle Frischinger[1] and Stoil Dirlikov

Coatings Research Institute, Eastern Michigan University, Ypsilanti, MI 48197

*Homogeneous mixtures of a liquid rubber based on prepolymers of epoxidized soybean oil with amines, diglycidyl ether of bisphenol A epoxy resins, and commercial diamines form, under certain conditions, two-phase thermosetting materials that consist of a rigid epoxy matrix and randomly distributed small rubbery soybean particles (0.1–5 μm). These two-phase thermosets have improved toughness, similar to that of other rubber-modified epoxies, low water absorption, and low sodium content. In comparison to the unmodified thermosets, the two-phase thermosets exhibit slightly lower glass-transition temperatures and Young's moduli, but their dielectric properties do not change. The epoxidized soybean oil is available at a price below that of commercial epoxy resins and appears very attractive for epoxy toughening on an industrial scale.*

TOUGHENING OF EPOXY RESINS BY AN ELASTOMERIC SECOND PHASE was first demonstrated by Sultan et al. (*1*, *2*). It is now well established that a small amount of discrete rubbery particles that have an average size of several micrometers and are randomly distributed in a glassy brittle epoxy thermoset dissipates part of the impact energy and, thus, improves crack and impact resistance without major deterioration of other properties of the unmodified epoxy thermosets (*3*, *4*). The main role of the rubber phase in toughened epoxies is to relieve the constraints in the matrix through the principal mechanisms of cavitation and formation of shear bands (*5*, *6*). This phenomenon is used for industrial-scale toughening of epoxy resins.

---

[1] Current Address: Polymer Division, Ciba-Geigy A.G., CH–4002 Basel, Switzerland

0065–2393/93/0233–0451$10.75/0
© 1993 American Chemical Society

Epoxy toughness is usually achieved by separation of a rubbery phase that has a unimodal size distribution from the matrix during the curing process. Different reactive liquid rubbers, based on low molecular weight carboxy- or amino-terminated oligomers of butadiene and acrylonitrile [carboxyl-terminated butadiene–acrylonitrile (CTBN) and amine-terminated butadiene–acrylonitrile (ATBN)], are usually used for the formation of the rubbery phase. Low molecular weight amino-terminated (methyl) siloxanes offer other alternatives, but some of these oligomers are quite expensive.

Epoxidized vegetable oils, such as vernonia, epoxidized soybean (ESO), and linseed oils, open new opportunities. Vegetable oil epoxy resins with commercial diamines are elastomers at room temperature that have low glass-transition temperatures in the range of $-70$ to $0$ °C, which depend on the nature of the amine used for curing.

Epoxidized vegetable oils, however, are characterized by good compatibility with polymers, and it appears unlikely that they form a second phase with epoxy resins. Qureshi et al. (7) observed small-scale heterogeneity in epoxy thermosets that contained epoxidized vegetable oils. The broadening of the glass-transition region in dynamic mechanical analysis (DMA) with the appearance of small additional loss peaks indicates that these formulations form semimiscible thermosets.

In our previous paper (8), we described initial results on toughening of epoxy thermosets with vernonia oil, a natural epoxidized vegetable oil that contains three epoxy rings and three carbon–carbon double bonds:

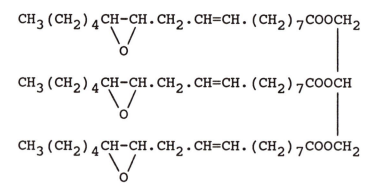

We obtained two-phase thermosets that consisted of a rigid matrix of commercial epoxy resin with randomly distributed small rubbery, spherical vernonia particles. The vernonia particles separate during the curing process from the initial homogeneous mixture of diglycidyl ether of bisphenol A epoxy [Epon resin 825; Shell Chemical Company; 4,4'-diaminodiphenyl methane (DDM) or 4,4'-diaminodiphenyl sulfone (DDS)] and vernonia liquid rubber, which is a B-staged material (soluble prepolymer) of vernonia oil with DDM or 1,12-dodecanediamine.

Although vernonia oil is at a developmental stage, it is unavailable for industrial applications. Therefore we are evaluating other epoxidized vegetable oils for toughening epoxy resins. The industrially produced epoxidized soybean and linseed oils are available at a low price (in the range of $0.50 to $0.65 per pound), which makes them very attractive. In this chapter, our initial unoptimized results on toughening commercial epoxy resins by epoxidized soybean oil are described. Epoxidized soybean oil consists of a mixture of different triglycerides. The triglyceride structures are schematically illustrated by the following formula:

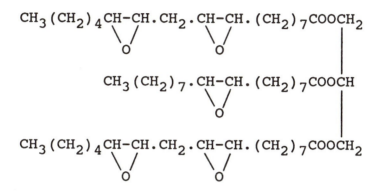

## Experimental Details

**Materials.**   The epoxidized soybean oil is produced under the commercial name Flexol Plasticizer EPO (Union Carbide Corporation). The average epoxy functionality of this oil is approximately 4.5. Epon epoxy resin 825, a solid diglycidyl ether of bisphenol A (DGEBA), is a Shell Chemical Company product. All amine curing agents were purchased from Aldrich Chemical Company.

**Epoxidized Soybean Liquid Rubber.**   The epoxidized soybean liquid rubber (ESR) was prepared by B-staging (prepolymerization) of epoxidized soybean oil with a stoichiometric amount of DDM. For this purpose, a mixture of 100 g of epoxidized soybean oil and 22.28 g of DDM was heated in a nitrogen atmosphere at 135 °C for 37–40 h until a highly viscous liquid, which was still soluble in the commercial DGEBA epoxy resin at 70 °C, was obtained. The slight B-staging time variation from batch to batch was dependent on the quality of the epoxidized soybean oil used.

ESR density (1.05 g/cm$^3$ at 20 °C) was determined with a pycnometer.

**Cure Procedure.**   The epoxy formulations are based on stoichiometric mixtures of DGEBA and amine curing agent (DDM, DDS, etc.) containing 10, 15, 20, and 30 wt % of epoxidized soybean liquid rubber or a stoichiometric mixture of the initial epoxidized soybean oil and amine curing agent. The cure was carried out according to the following procedure: First, the mixture of DGEBA and epoxidized soybean oil or rubber was heated at 75 °C under vacuum with stirring for 15–30 min. Then the stoichiometric amount of the diamine (DDS, DDM, etc.) was added while stirring. The mixture was degassed and stirred under a vacuum for an additional 15 min at 150 °C for the DDS formulations, or for 30 min at 75 °C for the DDM formulations. Then it was poured into a mold that was preheated at 150 °C for DDS and at 75 °C for DDM formulations, and again left under a vacuum for 15 min. For the DDS formulations, the final cure was carried out in an air-circulating oven at 150 °C for 2 h. DDM formulations were cured first at 75 °C for 4 h and then at 150 °C for 2 h.

**Morphology.**    The morphology of the neat and rubber-modified epoxy resins was examined by scanning electron microscopy (SEM) of the fracture surfaces using an electron microscope (Amray model 1000B).

**Fracture Toughness.**    Single-edge-notched (SEN) specimens with approximate dimensions of 100 mm long × 12.7 mm wide × 6.7 mm thick (Figure 1) were machined from castings. A sharp crack was introduced into the specimen by the strike of a razor blade (previously chilled in liquid nitrogen) with a rubber mallet. The tests were carried out with a three-point bending assembly, which was monitored by a servohydraulic materials testing machine (Instron 1331) with a span of 50.8 mm and a piston rate of 2.54 mm/s. A computer interface controlled the machine and recorded the data. A computer (Hewlett-Packard model 310) was programmed to calculate the

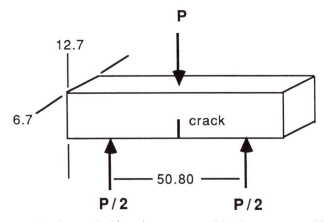

*Figure 1. Single-edge-notched (SEN) specimen used for determination of fracture toughness by the three-point bending test. The dimensions are in millimeters.*

critical stress intensity factor, $K_{Ic}$, using the relation (9)

$$K_{Ic} = Y \frac{2 \, PS\sqrt{a}}{3tw^2}$$

where $P$ is the critical load for crack propagation in Newtons ($N$), $S$ is the length of the span (mm), $a$ is the crack length (mm), $w$ is the width (mm), $t$ is the thickness (mm), and $Y$ is the nondimensional shape factor given by

$$Y = 1.9 - 3.07(a/w) + 14.53(a/w)^2 - 25.11(a/w)^3 + 25.80(a/w)^4$$

The following simple relationship, which holds in our case of linear-elastic-fracture mechanics (LEFM) under plane strain conditions, was used for determination of the fracture energy, $G_{IC}$:

$$G_{Ic} = \frac{(1 - v^2)(K_{Ic})^2}{E}$$

where $v$ is the Poisson's ratio and $E$ is the elastic or Young's modulus. At least eight specimens of each formulation were used for determination of average fracture energy $G_{Ic}$.

**Uniaxial Tensile Test**.   Dog-bone bars with dimensions of 6 in. × 0.5 in. × 1/8 in. (152 mm × 12.7 mm × 3.5 mm) were cut with a high speed router and their external surface was polished with very fine aluminum oxide sandpaper 220 (3M Corporation). A screw-driven tensile tester (Instron model 1185), equipped with an extensometer for determination of the longitudinal strain and a computer interface type 4500 series, was used at a stroke rate of 60 mm/min. At least 10 specimens of each formulation were used for determination of the average tensile properties at room temperature.

**Dynamic Mechanical Analysis**.   A thermal analyzer (DuPont 2100) instrument with dynamic mechanical analysis (DMA) model 983, based on a flexural bending deformation measurement, and rectangular bar specimens with approximate dimensions of 45 mm × 7.8 mm × 3.5 mm were used to study the dynamic mechanical properties over the temperature range from −130 to 200 °C at a heating rate of 2 °C/min.

**Differential Scanning Calorimetry**.   Differential scanning calorimetry (DSC) measurements were carried out on a thermal analyzer (DuPont 2100) instrument with DSC model 2910 over the temperature range from −130 to 200 °C at a heating rate of 2 °C/min.

**Water Absorption.** Maximum water absorption was determined on rectangular specimens with approximate dimensions of 45 mm × 7.8 mm × 3.5 mm, which were predried in a vacuum oven at 60 °C to constant weight and then kept in boiling water for 3–4 weeks until saturation (e.g., to constant weight).

**Dielectric Properties.** The dielectric constant and dissipation factor were measured at 1 MHz and room temperature on 3 in. × 3 in. × 1/8 in. specimens on a dielectric analyzer Genrad 1687 B Digibridge equipped with an LD–3 cell and using the two-fluid method (air and DC–200; 1 cs).

**Gel Permeation Chromatography.** Gel permeation chromatography (GPC) was measured on a GPC instrument (Hewlett–Packard) with a differential refractometer (Waters Associates model R401) and gel columns (Polymer Laboratory).

**Infrared Spectroscopy.** Infrared spectra were taken with an infrared spectrometer (IBM 44).

**Sodium Content.** The epoxidized soybean oil and rubber samples were first dissolved by acid digestion. Sodium content was determined with an atomic absorption spectrophotometer (Perkin–Elmer 2380). Our results were confirmed by sodium analysis at Galbraith Analytical Laboratory.

## Results and Discussion

**Morphology.** The morphology of the two-phase DGEBA–soybean epoxy thermosets was evaluated by scanning electron microscopy as an average of several micrographs taken at different fracture surfaces for each formulation. Our primary interest was to elucidate the evolution process of the second (liquid or rubbery) phase and particle size determination. Therefore, we studied the dependence of morphology on the weight percentage of soybean fraction and on the compatibility of the formulations: polarity of the diamine curing agent, degree of B-staging of the epoxidized soybean oil, and so forth.

The commercial epoxy resin used in the present study is Epon resin 825, which is practically pure DGEBA. Most of the research was done with stoichiometric DDM or DDS formulations of DGEBA. Only initial screening evaluation was carried out on DGEBA formulations with other cycloaliphatic (isophorone diamine) and aliphatic diamines and polyamines (1,10-decanediamine, 1,12-dodecanediamine, diethylenetriamine, triethylenetetramine, etc.). These epoxy–amine formulations were toughened by the addition of 10, 15, 20 and 30% of epoxidized soybean oil or liquid rubber (prepolymer) as described in the "Experimental Details" section.

Initial attempts to toughen epoxy resins were carried out directly with epoxidized soybean oil. For this purpose, homogeneous mixtures of DGEBA and diamine (DDM, isophorone diamine, different aliphatic di- and polyamines, etc.) that contained 10–30% epoxidized soybean oil were cured according to a conventional curing procedure for epoxy resin at 70 °C. Because the diamines have a much higher reactivity with DGEBA than with the epoxy groups of the epoxidized soybean oil, they form a rigid matrix at 70 °C in which the epoxidized soybean oil separates as a second phase of small liquid droplets. The electromicrograph of such a thermoset based on DGEBA, isophorone diamine, and 20% epoxidized soybean oil shows a rigid matrix with random distribution of liquid soybean droplets with a diameter in the range of 1 μm (Figure 2). Unfortunately, at the higher temperature of 150 °C these liquid droplets do not cure fast enough with the remaining diamine to form rubbery particles, and we have been unable to prepare two-phase thermosets. Instead, soybean oil dissolves and plasticizes the rigid DGEBA matrix. One-phase homogeneous thermosets with single lower glass-transition temperatures are obtained as observed in SEM and DSC, respectively.

The direct toughening of the corresponding more polar (less miscible) DDS formulations by epoxidized soybean oil is under investigation. Similar DGEBA–DDS formulations toughened by vernonia oil undergo (macroscopic) phase separation at 150 °C.

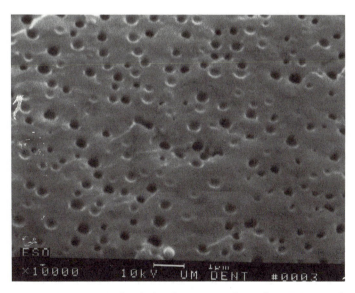

*Figure 2. SEM micrograph of the fracture surface of DGEBA–isophorone diamine–ESO-20 specimen.*

We were able to obtain two-phase thermosets with rubbery soybean particles under the same curing conditions by using epoxidized soybean liquid rubber (ESR) instead of the initial pure epoxidized soybean oil. These liquid rubbers were prepared by B-staging (prepolymerization) of a stoichiometric mixture of epoxidized soybean oil and DDM at 135 °C for about 40 h as described in the "Experimental Details" section. A typical gel permeation chromatogram (GPC) of these soybean liquid rubbers is given in Figure 3. The soybean liquid rubbers contain oligomers with a broad molecular weight distribution in addition to the unreacted initial epoxidized soybean oil that appears as a sharp peak at about 25 min.

The soybean particles again separated from the initial homogeneous mixture of DGEBA, diamine, and soybean liquid rubbers. The higher molecular weight of soybean liquid rubber (in comparison to that of the soybean oil) increases its incompatibility with the epoxy matrix at higher temperature. As a result, the final thermoset, after being cured at 150 °C, consists of a rigid epoxy matrix with randomly distributed small rubbery spherical soybean particles (Figures 4–10). The diamine molecules on the interface are expected to react with both the epoxy groups of the commercial epoxy resin and the unreacted epoxy groups of the epoxidized soybean rubber, and are expected to form chemical bonds between the rigid matrix and the rubbery particles.

A plastification phenomenon at 150 °C, similar to the foregoing description for the pure epoxidized soybean oil, occurred under one-stage curing conditions for certain formulations based on soybean liquid rubbers as well. This phenomenon was observed for formulations whose epoxy matrix and soybean rubbery phase had similar solubility parameters. Such formulations are based on less polar diamines like DDM; for example, the formulations based on DGEBA, DDM, and epoxidized soybean liquid rubber, which are abbreviated here as DGEBA–DDM–ESR. Complete miscibility and homogeneous one-phase thermosets were obtained for these formulations at any soybean liquid rubber content if one-stage curing was carried out directly at high temperatures (for instance at 150 °C). Under the higher temperature conditions, the remaining epoxy groups of the soybean liquid rubber had higher reactivity and probably copolymerized with DGEBA without phase separation.

We have avoided the plastification of the DGEBA–DDM–ESR formulations and have prepared their two-phase final thermosets, which contain 15, 20, and 30% soybean liquid rubber (DGEBA–DDM–ESR-15, DGEBA–DDM–ESR-20, and DGEBA–DDM–ESR-30), by a two-stage curing procedure (see the Experimental Details section). The curing is carried out initially at a low temperature (70 °C) at which the soybean epoxy groups have very low reactivity. The low reactivity occurs because the soybean epoxy groups require higher curing temperatures. Practically, at 70 °C only DGEBA reacts with the diamine and free hydroxyl groups, which increase the matrix

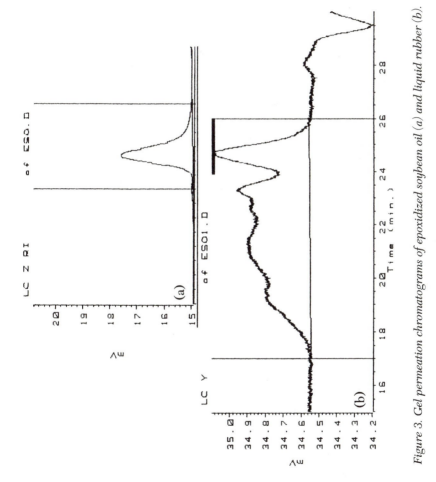

*Figure 3. Gel permeation chromatograms of epoxidized soybean oil (a) and liquid rubber (b).*

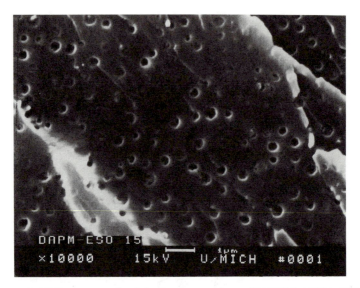

*Figure 4. SEM micrograph of the fracture surface of DGEBA–DDM–ESR-15 specimen.*

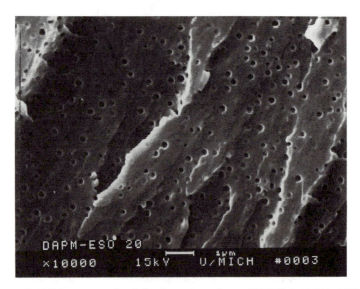

*Figure 5. SEM micrograph of the fracture surface of DGEBA–DDM–ESR-20 specimen.*

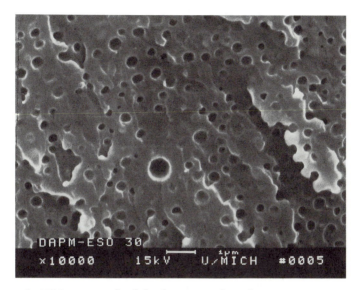

*Figure 6. SEM micrograph of the fracture surface of DGEBA–DDM–ESR-30 specimen.*

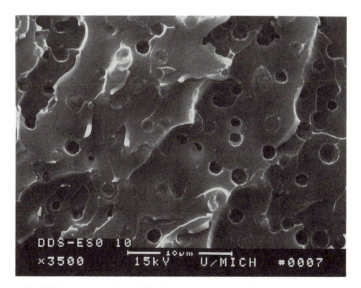

*Figure 7. SEM micrograph ˙of the fracture surface of DGEBA–DDS–ESR-10 specimen.*

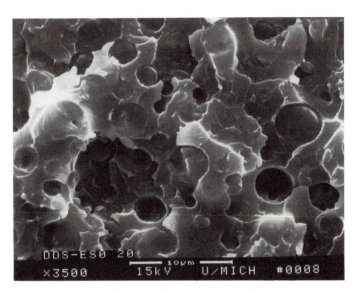

*Figure 8. SEM micrograph of the fracture surface of DGEBA–DDS–ESR-20 specimen.*

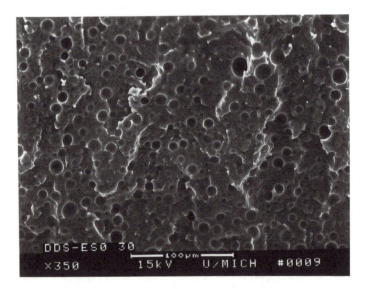

*Figure 9. SEM micrograph of the fracture surface of DGEBA–DDS–ESR-30 specimen used for tensile measurements.*

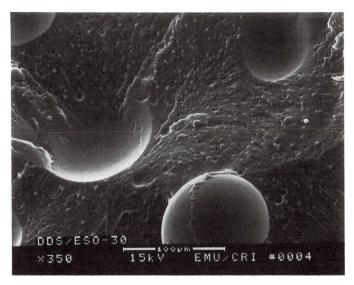

*Figure 10. SEM micrograph of the fracture surface of DGEBA–DDS–ESR-30 specimen used for fracture toughness measurements.*

polarity, are formed. The molecular weight of the DGEBA gradually increases and the rigid cross-linked matrix is formed at 70 °C. Higher molecular weight, cross-linking, and increased polarity in the DGEBA result in higher incompatibility with the less polar hydrophobic epoxidized soybean liquid rubber, which separates into small rubbery particles at this stage. During the second curing stage, the temperature is increased to 150 °C and the less reactive soybean epoxy groups cure with the remaining diamine and form rubbery particles; the matrix post-cures simultaneously. Thus, the two-stage curing procedure allows thermodynamic equilibrium (phase separation) to be achieved kinetically.

The formulations containing only 10% soybean liquid rubber (DGEBA–DDM–ESR-10), however, produce a completely transparent one-phase material even following the two-stage curing procedure. Obviously, only plastification of the DGEBA epoxy rigid matrix without phase separation occurs at low soybean liquid rubber concentration. We have not been able to obtain two-phase thermosets in this case. The solubility of the soybean liquid rubber in the DGEBA–DDM resin, however, is limited. The formulations with 10% soybean rubber are closed to saturation and we observe phase separation at slightly higher soybean content (15%).

Comparison of the DGEBA–DDM formulations toughened by epoxidized soybean oil (ESO) and rubber (ESR) at a 10% level shows that DGEBA–DDM–ESO-10 forms a two-phase thermoset with small liquid soybean oil droplets at 70 °C. However, the droplets dissolve in the

DGEBA–DDM matrix and form a homogeneous thermoset at 150 °C. In contrast, the DGEBA–DDM–ESR-10 formulation does not phase-separate, but forms a homogeneous casting at both 70 and 150 °C. The epoxidized soybean rubber is more polar than the epoxidized soybean oil due to its free hydroxyl and amine groups, and obviously it has better solubility in the DGEBA–DDM matrix at 70 °C despite its higher molecular weight.

This plastification phenomenon was not observed for formulations based on diamines more polar than DDM, for instance, those based on DDS. In the DDS formulations (abbreviated, analogously, DGEBA–DDS–ESR) the solubility parameters of the polar epoxy matrix and nonpolar soybean phase were quite different and the phase separation occurred at a lower conversion. DDS-based formulations, therefore, do not require a two-stage curing procedure to build the molecular weight of the matrix at lower temperature and increase the incompatibility between DGEBA and soybean phases. DDS formulations form two-phase thermosets directly in a one-stage curing procedure at a high temperature (150 °C) (*see* the "Experimental Details" section).

Several factors control phase separation: Miscibility of the initial formulation of epoxy resin, amine, and epoxidized soybean oil or its liquid rubber is required. Incompatibility and phase separation should start during the curing process before gelation for spherical morphology formation. Simultaneously, the viscosity at this stage and the rate of cure should be high enough to prevent coalescence and macroscopic phase separation. The effective rubbery phase depends on the volume fraction of epoxidized soybean rubber; the degree of its B-staging; the nature (polarity, reactivity, molecular weight, etc.) of the epoxy resin and curing amine; and the time and temperature of the curing regime (kinetic factor).

The introduction of the epoxidized soybean liquid rubber into the DGEBA epoxy resin has obvious advantages for preparation of two-phase thermosets over the pure epoxidized soybean oil. The soybean liquid rubber was prepared by B-staging (prepolymerization) the epoxidized soybean oil with DDM (at 135 °C for ~ 40 h) under conditions at which an advanced degree of B-staging was achieved and the conversion was just below the gel point. Therefore, the soybean liquid rubber required a very short time at 150 °C to reach its gel point and cross-link. The cross-linked soybean liquid rubber was not able to dissolve into the matrix. Obviously, the monomeric epoxidized soybean oil requires a much longer time under the same conditions to reach the equivalent gel point. Thus, the lower level of matrix plastification in the presence of soybean liquid rubber, in comparison to the soybean oil, was probably due to two factors: higher molecular weight and incompatibility at higher temperature (thermodynamic factor) and rapid cross-linking at 150 °C (kinetic factor).

As mentioned earlier, the epoxidized soybean liquid rubber is a heterogeneous mixture of liquid oligomers and it contains a certain amount of the unreacted starting epoxidized soybean oil (Figure 3). No doubt, the unreacted soybean oil has a stronger plastification effect on the matrix than the higher molecular weight oligomers of ESR.

**Particle Size.**   The average particle size of the rubbery soybean phase was determined by scanning electron microscopy simultaneously with the morphology as described in the preceding section. The average range of particle size distribution for the DGEBA–DDM–ESR and DGEBA–DDS–ESR formulations is given in Table I. Both the minimum and the maximum values are listed.

As previously mentioned, DGEBA–DDM–ESR formulations with 10% soybean liquid rubber content form homogeneous castings without separation of a second rubbery phase. The particle size of the DGEBA–DDM–ESR formulations with 15 and 20% soybean liquid rubber is in the range from 0.1 to 0.5 μm and it increases slightly to 0.8 μm at 30% rubber content (Table I and Figures 4–6). DDM thermosets at any soybean rubber content are characterized by a unimodal particle size distribution.

The DGEBA–DDS–ESR formulations with 10% soybean liquid rubber form two-phase thermosets with ~ 1–2-μm particle size (Table I). With 20% soybean rubber, the particle size increases and ranges from 1–5 μm. DGEBA–DDS–ESR-10 and DGEBA–DDS–ESR-20 formulations are characterized by relatively unimodal size distributions as well (Figures 7 and 8).

The average particle size and unimodal distribution in both DGEBA–DDM–ESR and DGEBA–DDS–ESR formulations are quite reproducible with the exception of DGEBA–DDS–ESR-30 at a 30% soybean content. SEM micrographs show that the average size, range, and mode of distribution of the rubbery particles in this DGEBA–DDS–ESR-30 formulation are not reproducible and vary from specimen to specimen. Some specimens, for instance those used for tensile measurements, are characterized by a very

**Table I. Average Range of Soybean Particle Size Distribution of the Rubber-Modified DGEBA–DDM–ESR and DGEBA–DDS–ESR Thermosets**

| Thermoset Formulation | Particle Size (μm) |
|---|---|
| DGEBA–DDM–ESR-10 | No phase separation |
| DGEBA–DDM–ESR-15 | 0.1–0.5 |
| DGEBA–DDM–ESR-20 | 0.1–0.4 |
| DGEBA–DDM–ESR-30 | 0.1–0.8 |
| DGEBA–DDS–ESR-10 | 1–2 |
| DGEBA–DDS–ESR-20 | 1–5 |
| DGEBA–DDS–ESR-30[a] | 1–17 |
| DGEBA–DDS–ESR-30[b] | 5–10, 100–200 |

[a] Tensile specimens.
[b] Fracture toughness specimens.

broad unimodal distribution with particle sizes ranging from 1–17 μm (Table I and Figure 9). A bimodal particle distribution, however, is often observed for this formulation as well. For example, the specimens used for determination of fracture toughness show a bimodal size distribution with smaller particles in the range of 5–10 μm and larger particles in the range of 100–200 μm (Table I and Figure 10). SEM micrographs of the formulations with such bimodal particle distributions show that not only are the average size and range of the particles not reproducible, but also the apparent volume fraction varies from specimen to specimen without change in the overall morphology. As a matter of fact, the apparent volume fraction of the rubbery phase, observed on the SEM micrographs of the DGEBA–DDS–ESR-30 formulations with both unimodal and bimodal particle size distribution, is greater than the actual volume soybean fraction. The increased volume fraction indicates (1) that the larger particles in the DGEBA–DDS–ESR-30 formulations with bimodal distribution are the result of occlusion of the DGEBA epoxy component into the soybean rubbery particles and (2) are in close proximity to the phase-inversion point at 30% soybean rubber. Such phase-inversion phenomena were observed by Lee et al. (10) for a similar formulation, DGEBA–DDS–CTBN at a high 30% CTBN liquid rubber content.

Recently, we experimentally confirmed the phase-inversion phenomena and the occlusion of DGEBA resin in the rubbery particles in a similar formulation, DGEBA–DDM–vernonia rubber at 20% vernonia rubber, by SEM micrographs of specimens treated with osmium tetraoxide (unpublished data). In contrast to the epoxidized soybean oil, vernonia oil contains carbon–carbon double bonds, stains with osmium tetraoxide, and its rubbery phase is easily distinguished from the DGEBA resin in SEM micrographs.

The miscibility and phase separation in close proximity to the phase-inversion point are highly dependent on the kinetic factors. A small deviation in temperature or time of the curing process causes a shift in the phase-inversion point, broad variation of morphology for different specimens, and fluctuation of physicomechanical properties, as will be discussed later.

It is not clear, however, why DGEBA–DDS–ESR-30 results in a bimodal distribution (in some specimens) and appears to be close to the phase-inversion point, whereas the corresponding DGEBA–DDM–ESR-30 with the same soybean rubber content does not exhibit these phenomena. The differences are probably due to two factors.

First, different curing procedures were used for the DGEBA–DDM–ESR and DGEBA–DDS–ESR formulations. The DGEBA–DDS–ESR-30 formulation was cured directly for 2 h at 150 °C. The DGEBA–DDM–ESR-30 was cured according to a two-stage curing procedure: first at 75 °C for 4 h and then at 150 °C for 2 h. As previously discussed, the two-stage procedure allows the matrix formation at a lower temperature and results in better separation of the soybean rubber (perhaps without DGEBA occlusion).

Direct comparison of the morphology and rubbery particle size and distribution of DGEBA–DDM–ESR and DGEBA–DDS–ESR formulations prepared under the same curing conditions is difficult. DDS has a high melting point (mp = 177 °C) and requires a higher temperature (> 130 °C) to dissolve in the initial DGEBA–ESR formulation and form homogeneous mixtures. DGEBA–DDS–ESR formulations, therefore, cannot be cured by the two-stage procedure, but were cured directly in one stage at a higher temperature (150 °C). In contrast, DDM has a lower melting point (mp = 91 °C) and its initial DGEBA–DDM–ESR formulations are homogeneous mixtures at much lower temperature. DDM formulations, however, are more compatible and form only one-phase thermosets without phase separation by the direct one-stage curing procedure at a high temperature (150 °C).

Second, the formation of larger rubbery particles and DGEBA occlusion at the phase-inversion point depends on the mutual miscibility of DGEBA resin and soybean rubber, and especially depends on the solubility partition constants of the DGEBA monomer in the matrix and in the rubbery phase at different stages of curing. The DGEBA–DDM–ESR and DGEBA–DDS–ESR formulations are characterized by different miscibility. At an earlier stage of curing, the initial unreacted (nonpolar) DGEBA monomer of the DGEBA–DDS–ESR formulations, especially those with a higher ESR content, has a relatively higher solubility in the nonpolar ESR rubber phase than in the highly polar DDS-dominated DGEBA–DDS "matrix" phase, which is characterized by strong hydrogen bondings between amine, hydroxyl, and sulfone groups. [As a matter of fact, the DGEBA monomer is miscible at any ratio at room or elevated (70 °C) temperature with the epoxidized soybean rubber, but is completely insoluble in DDS.] The DGEBA monomer, dissolved in the rubbery phase, reacts with the diamine over progressing cure time and phase-separates at a later stage with the formation of small rigid particles within the large rubbery particles. Therefore, we observe much larger particles (100–200 μm), greater apparent rubbery volume fraction (than the actual soybean volume fraction), and the phase-inversion phenomenon at a relatively lower soybean content (at ~ 30%) for the DGEBA–DDS–ESR formulations. In contrast, the DGEBA monomer of the DGEBA–DDM–ESR formulations has relatively lower solubility under the same conditions in the rubbery phase because the DGEBA–DDM "matrix" phase is much less polar than the DGEBA–DDS phase. As a matter of fact, the DGEBA monomer is miscible with DDM at a much lower temperature. As a result, the rubbery particles of the DGEBA–DDM–ESR-30 formulation do not contain smaller rigid DGEBA particles even at 30% soybean rubber and their apparent volume fraction corresponds to the actual soybean content. The formation of larger particles and the phase-inversion phenomenon for the DDM formulations is expected to occur at a higher soybean content. Phase diagrams of both formulations and further SEM evaluation on the fracture surface of these thermosets are under investigation. Similar miscibility phenomena have been reported by Romanchick et al. (*11*) for CTBN-modified DGEBA epoxy resins.

The SEM micrographs of all DDM- and DDS-based formulations with 10, 15, 20, and 30% soybean content display rough fracture surfaces. Evidently, the soybean rubbery particles cavitate under shear (fracture) deformation with the formation of deep voids (Figures 4–9). A comparison of both formulations shows rougher fracture surface and less pronounced cavitation phenomena in the DGEBA–DDS–ESR formulations. Some of the DGEBA–DDS–ESR particles are clearly observed in SEM, but only a portion of the particles have cavitated with the formation of some deep voids. The cavitation process is much more efficient in the smaller rubbery particles of DGEBA–DDM–ESR formulations where complete cavitation and a smoother fracture surface are observed.

The results also show that DGEBA–DDS–ESR and DGEBA–DDM–ESR formulations are characterized by different particle sizes at the same soybean rubber content (Table I), which indicates that the particle size depends on the compatibility of the formulations.

The DGEBA–DDM–ESR formulations are expected to have better compatibility than the DGEBA–DDS–ESR formulations because both the DGEBA epoxy matrix and the soybean rubber phase are based on the same DDM diamine. In addition, DDM is less polar than DDS. The solubility parameter of the DDM matrix in the DGEBA–DDM–ESR formulations, therefore, matches the solubility parameter of the nonpolar soybean liquid rubber much better than the more polar DDS matrix in the DGEBA–DDS–ESR formulations. As a result of better compatibility, the DGEBA–DDM–ESR formulations with a 10% soybean content form homogeneous thermosetting material without phase separation. The rubbery particles observed in the formulations with 15, 20, and 30% soybean rubber are formed at a later stage in curing and have a smaller diameter.

The DGEBA–DDS–ESR formulations are less compatible because (1) they are based on two different diamines—DDS in the rigid matrix and DDM in the rubbery phase—and (2) their DDS matrix is more polar (*see* preceding text). The DGEBA–DDS–ESR formulations, therefore, are characterized by bigger rubbery particles that separate at earlier stages of curing from the initial liquid homogeneous mixture. In comparison with DGEBA–DDM–ESR formulations, the particle size of DGEBA–DDS–ESR formulations increases much more rapidly with increasing soybean rubbery content, especially in close proximity to the phase-inversion point, due to DGEBA occlusion (Table I).

The foregoing results show that under the same curing conditions, but using different diamines, the particle size of the rubbery phase in the range of desirable particle size for toughening epoxy thermosets (e.g., ranging from 0.1 to 10 $\mu$m) can be varied easily. In general, larger particles (and even macroscopic total phase separation) were observed for less compatible formu-

lations with more polar diamines used for curing the rigid matrix, for instance
DDS. In addition, a lower viscosity of the formulations (at higher temperature) allows coalescence of the liquid soybean droplets and results in the
formation of bigger rubbery particles. In contrast, higher curing rates result
in rapid cross-linking, prevent coalescence, and thus produce smaller rubbery
particles.

In summary, the particle size depends on the nature of the diamines
used for curing the matrix and for preparation of the soybean liquid rubber;
the volume fraction of the epoxidized soybean rubber; the degree of B-staging of the soybean rubber; the viscosity of the formulations and rate of cure
(e.g., temperature and time of B-staging); the cure of the final thermoset
(cure kinetics); and so forth.

**Glass-Transition Temperature**.  The temperature dependence of
the storage modulus and tangent delta of the two "pure" matrices—
DGEBA–DDM and DGEBA–DDS—were determined by dynamic mechan-
ical analysis in the temperature range of $-130$ to $+200$ °C (Figures 11–14).
Both thermosets were prepared without soybean rubber under the same
curing conditions as the corresponding two-phase materials in the presence of
soybean rubber.

The glass-transition temperatures, $T_g$, of these one-phase DGEBA–DDM
and DGEBA–DDS thermosets were observed at 190 and 185 °C, respectively. In addition, the DGEBA–DDM and DGEBA–DDS thermosets exhibited beta relaxations at lower temperatures: $-30$ and $-35$ °C, respectively
(Table II).

We were unable to measure the DMA of the pure epoxidized soybean
rubber (ESR) cured under the same conditions in the absence of the
DGEBA–diamine component. The ESR glass-transition temperature ($T_g =
-25$ °C) was determined by differential scanning calorimetry (Figure 15 and
Table II). Obviously, the ESR thermoset exists in a rubbery state at room
temperature, and it appears to be suitable for toughening commercial (brittle)
epoxy resins.

The temperature dependence of the storage modulus and tangent delta
of both DGEBA–DDM–ESR and DGEBA–DDS–ESR formulations at different contents of soybean liquid rubber was determined by DMA (Figures
12–15). All formulations exhibited transitions at high and low temperatures
(Table II).

The higher temperature transitions of the DGEBA–DDM–ESR and
DGEBA–DDS–ESR formulations occur in close proximity to the glass-
transition temperature of the pure DGEBA–DDM and DGEBA–DDS
thermosets: in the range of 145–190 °C. Therefore, these transitions were
assigned to the glass-transition temperatures of their DGEBA matrices.
These glass-transition temperatures gradually decrease in the corresponding

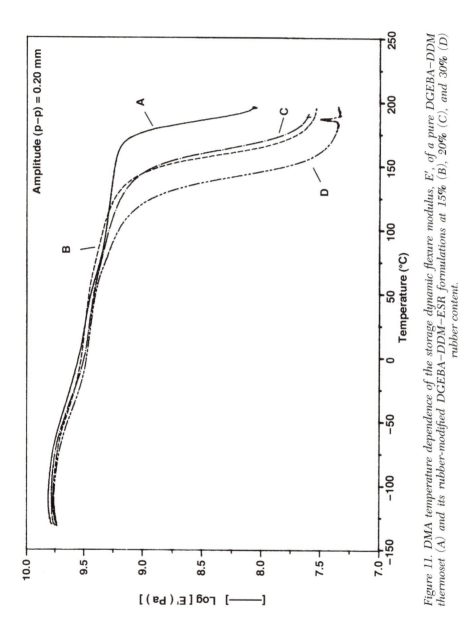

Figure 11. DMA temperature dependence of the storage dynamic flexure modulus, E', of a pure DGEBA–DDM thermoset (A) and its rubber-modified DGEBA–DDM–ESR formulations at 15% (B), 20% (C), and 30% (D) rubber content.

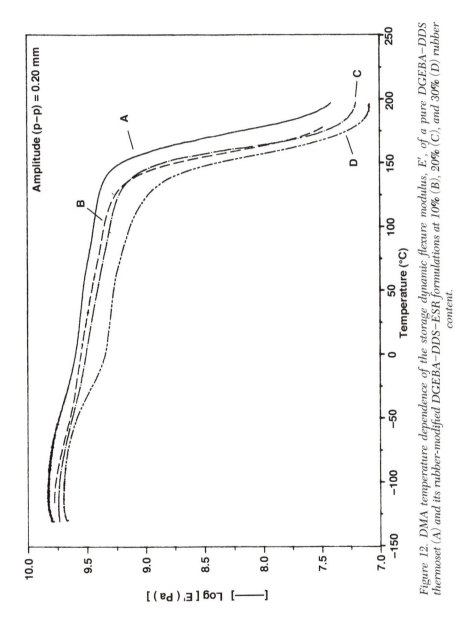

*Figure 12. DMA temperature dependence of the storage dynamic flexure modulus, E', of a pure DGEBA–DDS thermoset (A) and its rubber-modified DGEBA–DDS–ESR formulations at 10% (B), 20% (C), and 30% (D) rubber content.*

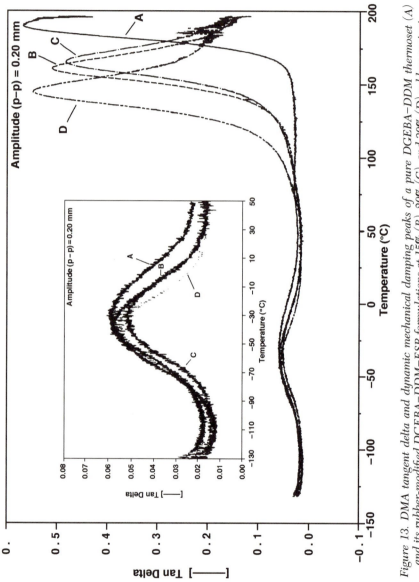

*Figure 13. DMA tangent delta and dynamic mechanical damping peaks of a pure DGEBA–DDM thermoset (A) and its rubber-modified DGEBA–DDM–ESR formulations at 15% (B), 20% (C), and 30% (D) rubber content.*

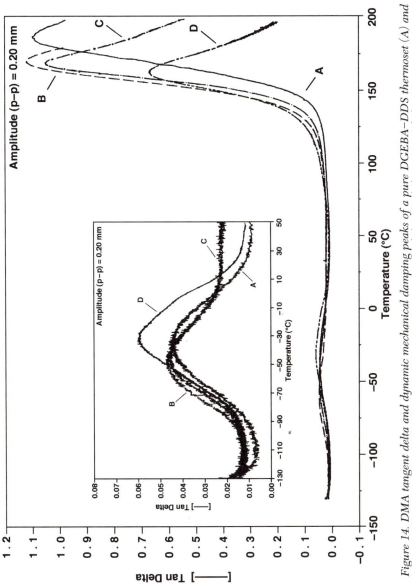

*Figure 14. DMA tangent delta and dynamic mechanical damping peaks of a pure DGEBA–DDS thermoset (A) and its rubber-modified DGEBA–DDS–ESR formulations at 10% (B), 20% (C), and 30% (D) rubber content.*

**Table II. DMA Lower and Higher Temperature Transitions of Pure
DGEBA–DDM and DGEBA–DDS Thermosets and Their Rubber-Modified
DGEBA–DDM–ESR and DGEBA–DDS–ESR Formulations**

| Thermoset | Transition (°C) | |
| Formulation | Low[a] | High[b] |
| --- | --- | --- |
| DGEBA–DDM | − 30 | 190 |
| DGEBA–DDM–ESR-15 | − 35 | 161 |
| DGEBA–DDM–ESR-20 | − 33 | 165 |
| DGEBA–DDM–ESR-30 | − 39 | 145 |
| DGEBA–DDS | − 35 | 185 |
| DGEBA–DDS–ESR-10 | − 45 | 168 |
| DGEBA–DDS–ESR-20 | − 43 | 166 |
| DGEBA–DDS–ESR-30 | − 30 | 161 |
| ESR[c] | − 25 | |

[a] The lower transition temperature corresponds to an overlap in the glass-transition temperature of the rubbery particles and beta transition of the DGEBA matrix.
[b] Glass-transition temperature of the DGEBA matrix.
[c] Determined by DSC.

DGEBA–DDM–ESR and DGEBA–DDS–ESR formulations with an increasing amount of soybean liquid rubber. Obviously, a part of the soybean rubber, and especially its lower molecular weight oligomers and its unreacted epoxidized soybean oil, dissolves and plasticizes the DGEBA rigid matrix.

This plastification phenomenon is more pronounced for the DGEBA–DDM–ESR formulations. The depression of the glass-transition temperature of the DGEBA–DDM–ESR matrices is about 25 °C at 15–20% of soybean rubber content and almost 45 °C at 30% loading. As discussed in the preceding section, the DGEBA matrix and soybean rubber are characterized by better compatibility in the DDM formulations than in those based on DDS. This compatibility results in better plastification and high concentration of soybean rubber in the DGEBA phase, both of which are confirmed here by greater depressions observed for the glass-transition temperature of the DDM matrices.

Smaller depressions of the matrix glass-transition temperature are observed for the less compatible DGEBA–DDS–ESR formulations. The DDS formulation $T_g$ decreases from about 15 °C at 10% soybean liquid rubber to ~ 25 °C at 30% loading.

The lower temperature transitions of the DGEBA–DDM–ESR and DGEBA–DDS–ESR formulations are observed in the range − 30 to − 45 °C. No other transitions for these formulations are detected, either at a higher temperature in the range of − 30 to + 130 °C or at a lower temperature in the range of − 45 to − 130 °C. Therefore, we believe that the glass-transition temperature of the ESR rubbery particles and the beta transition of the matrix of the two-phase formulations overlap in the range of

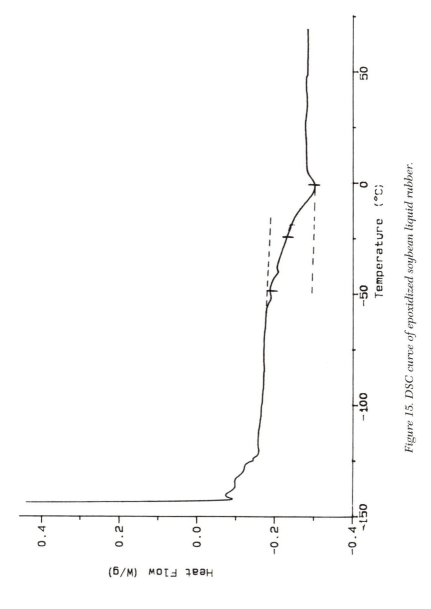

*Figure 15. DSC curve of epoxidized soybean liquid rubber.*

$-30$ to $-45$ °C and are observed as a single transition at a lower temperature.

The glass-transition temperature of the rubbery particles in the two-phase thermosets was expected to correspond to that of the pure ESR rubber or to appear at a slightly higher temperature if the rubbery particles dissolved a certain amount of DGEBA epoxy component that had a higher $T_g$. Surprisingly, the low transition temperatures of the two-phase thermosets are observed at a slightly lower temperature than the glass-transition temperature of the pure ESR rubber. This discrepancy is probably due to the different methods used for determination of these transitions: DMA is used for the two-phase thermosets and DSC is used for the pure ESR rubber. The discrepancy might also result as an artifact of the overlapping of the glass transition of the rubbery particles and the beta transition of the matrix or from stresses induced by thermal shrinkage as discussed by Manzionne and Gillham (12). The slight increase of the lower temperature transition observed for the DGEBA–DDS–ESR-30 formulation with bimodal particle size distribution could be attributed to the occlusion of epoxy DGEBA resin into the larger rubbery particles. In any case, the glass-transition temperature of the pure soybean rubber (ESR) does not change much in the presence of the DGEBA matrix in both DGEBA–DDM–ESR and DGEBA–DDS–ESR formulations. The relatively constant $T_g$ indicates that the rubber phase in these two-phase thermosets does not contain a significant amount of the DGEBA component below the phase-inversion point and the thermosets are formed practically by pure soybean rubber.

**Fracture Toughness.** The fracture toughness, in terms of the stress intensity factor, $K_{Ic}$, and fracture energy, $G_{Ic}$, is given in Table III for both DGEBA–DDM–ESR and DGEBA–DDS–ESR formulations.

**Table III. Stress Intensity Factor, $K_{Ic}$, and Fracture Energy, $G_{Ic}$, of Pure DGEBA–DDM and DGEBA–DDS Thermosets and Their Rubber-Modified DGEBA–DDM–ESR and DGEBA–DDS–ESR Formulations**

| Thermoset Formulation | $K_{Ic}$ $(MPa \cdot m^{1/2})$ | $G_{Ic}$ $(J/m^2)$ |
|---|---|---|
| DGEBA–DDM | 0.75 | 175 |
| DGEBA–DDM–ESR-10[a] | | |
| DGEBA–DDM–ESR-15 | 1.24 | 564 |
| DGEBA–DDM–ESR-20 | 1.33 | 802 |
| DGEBA–DDM–ESR-30 | 1.40 | 1008 |
| DGEBA–DDS | 0.75 | 145 |
| DGEBA–DDS–ESR-10 | 1.09 | 374 |
| DGEBA–DDS–ESR-20 | 1.19 | 596 |
| DGEBA–DDS–ESR-30 | 0.54 | 145 |

[a] This formulation does not phase-separate and its fracture toughness has not been determined.

A better toughening effect, which gradually increases at a higher soybean content, is observed for the DGEBA–DDM–ESR formulations. The plastification phenomenon that occurs at a larger scale for these formulations (*see* the two preceding sections) probably affects the improved toughening via the soybean component, which acts as a matrix plasticizer. These formulations also are characterized by smaller particle size (Table I; 0.1–0.8 μm), and their fracture surfaces show extensive particle cavitation and formation of deep voids (*see* the "Particle Size" section and Figures 4–6). Pearson and Yee (*13*) have shown that the toughening mechanism of elastomer-modified epoxies is governed by the phenomena of internal cavitation of rubber particles and formation of shear bands. The efficiency of these mechanisms depends on rubber particle size. Pearson and Yee also demonstrated that smaller particles ranging from 0.2–2 μm are more efficient and produce a better toughening effect than larger particles. The higher values for fracture energy observed for the DGEBA–DDM–ESR formulations are probably due to both better compatibility (plastification) and smaller particles (in the range from 0.1–0.8 μm; Table I) with a more efficient toughening mechanism.

A smaller improvement of fracture toughness is observed for the DGEBA–DDS–ESR formulations. Plastification plays a smaller role here and the formulations are characterized by larger rubbery particles. In agreement with the results of Pearson and Yee, the larger particles act as "bridging" particles with only a modest increase in fracture toughness. In addition, DGEBA–DDS–ESR formulations exhibit optimum toughening at 20% soybean component. The DGEBA–DDS–ESR-30 specimens at 30% soybean liquid rubber, which were used here for determination of fracture toughness, are characterized by bimodal particle size distribution and are close to their phase-inversion point (*see* the "Particle Size" section). Physicomechanical properties of epoxy resins start to deteriorate rapidly at the phase-inversion point and, therefore, we observe lower fracture toughness for the DGEBA–DDS–ESR-30 thermoset.

In summary, our measurements show that the introduction of soybean liquid rubber remarkably improves the fracture toughness of commercial highly cross-linked brittle epoxy resins, such as DGEBA–DDS and DGEBA–DDM, which exhibit low ductility and, as shown by other authors (*14, 15*), are very difficult to toughen.

**Tensile Properties**.   Young's moduli were determined from the corresponding tensile stress–strain curves. The Young's moduli are reported here as an average value of several independent tensile measurements to minimize the fluctuations observed in our tensile data as a result of the high stiffness of the specimens (Table IV).

The dependence of Young's modulus on the soybean content is plotted in Figure 16 for both DGEBA–DDM–ESR and DGEBA–DDS–ESR formulations. As expected, the elastic moduli gradually decrease with increasing

**Table IV. Young's modulus, *E*, of pure DGEBA–DDM and DGEBA–DDS Thermosets and Their Rubber-Modified DGEBA–DDM–ESR and DGEBA–DDS–ESR Formulations**

| Thermoset Formulation | E (MPa) |
|---|---|
| DGEBA–DDM | 2840 |
| DGEBA–DDM–ESR-10[a] | |
| DGEBA–DDM–ESR-15 | 2410 |
| DGEBA–DDM–ESR-20 | 1950 |
| DGEBA–DDM–ESR-30 | 1720 |
| DGEBA–DDS | 3420 |
| DGEBA–DDS–ESR-10 | 2810 |
| DGEBA–DDS–ESR-20 | 2100 |
| DGEBA–DDS–ESR-30 | 1780 |

[a] This formulation does not phase-separate and its tensile properties have not been determined.

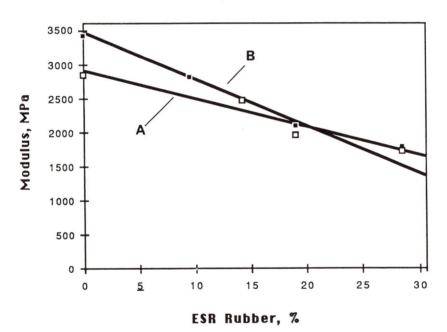

**ESR Rubber, %**

*Figure 16. Dependence of Young's modulus on the ESR rubber content of the DGEBA–DDM–ESR (A) and DGEBA–DDS–ESR formulations (B).*

soybean fraction. Linear relationships can approximately fit the different points for both formulations, and a slight deviation is observed only with a higher content of soybean rubber. The slope of the DGEBA–DDS–ESR formulations, however, is steeper than that of DGEBA–DDM–ESR. The tensile properties of DGEBA–DDS–ESR decrease faster with increasing soybean content. This result may be attributed to the larger rubbery particles of the DGEBA–DDS–ESR formulations.

Surprisingly, the linear tensile dependence of the DGEBA–DDS–ESR formulations on the soybean rubber content does not correlate well with the corresponding fracture toughness relationship, which shows an optimum at 20% and a drop in fracture toughness at 30% soybean content. As discussed in the "Particle Size" section, the DGEBA–DDS–ESR formulation at 30% soybean content is close to its phase-inversion point and its different specimens have different morphology. Indeed, direct comparison of the SEM micrographs on the fracture surface of the DGEBA–DDS–ESR-30 formulation shows that the tensile specimens used in this study are characterized by a unimodal particle size distribution ranging from 1–17 μm (Table I and Figure 9). In contrast, the specimens used for fracture-toughness measurements show a bimodal distribution with smaller particles ranging from 5–10 μm and larger particles ranging from 100–200 μm (Table I and Figure 10). The different morphology of the DGEBA–DDS–ESR-30 specimens used for tensile and fracture-toughness measurements obviously results in fluctuation of their physicomechanical properties.

**Water Absorption.**    High water absorption is another major disadvantage of the commercial epoxy resins in addition to their poor fracture toughness. The maximum water absorption of the DGEBA–DDS–ESR and DGEBA–DDM–ESR formulations was determined on predried samples immersed in boiling water until saturation (constant weight) as described in the "Experimental Details" section. Although our results are somewhat lower than expected, the water absorption of both formulations decreases gradually and linearly with increasing soybean content (Table V). Obviously, the highly hydrophobic long fatty chains of the epoxidized soybean oil reduce the water

**Table V. Maximum Water Absorption of Pure DGEBA–DDM and DGEBA–DDS Thermosets and Their Rubber-Modified DGEBA–DDM–ESR and DGEBA–DDS–ESR Formulations**

| Thermoset Formulation | Water Absorption (%) |
|---|---|
| DGEBA–DDM | 2.42 |
| DGEBA–DDM–ESR-15 | 2.22 |
| DGEBA–DDM–ESR-20 | 2.00 |
| DGEBA–DDM–ESR-30 | 1.62 |
| DGEBA–DDS | 3.61 |
| DGEBA–DDS–ESR-10 | 3.33 |
| DGEBA–DDS–ESR-20 | 3.13 |
| DGEBA–DDS–ESR-30 | 2.80 |

absorption of the epoxy thermosets. As expected, the DGEBA–DDS–ESR formulations at different soybean liquid rubber content have a higher water absorption than the corresponding DGEBA–DDM–ESR formulations, due to the more polar character of DDS in comparison to DDM.

**Dielectric Properties.** The dielectric constants and dissipation factors of the DGEBA–DDM–ESR and DGEBA–DDS–ESR formulations are given in Table VI. The introduction of epoxidized soybean oil, with its long aliphatic chains, was expected to lower the dielectric constant and dissipation factor of the commercial epoxy resins simultaneously, despite the formation of free hydroxyl groups by the opening of the epoxy rings. Surprisingly, the dielectric properties of both types of formulations do not change (decrease) much with increasing soybean liquid rubber content. Although we do not have an explanation for this observation at the present moment, D. Shimp (Rhone-Poulenc, personal communication) has suggested that, in addition to the traditional epoxy curing reaction, the epoxidized soybean oil (and rubber) might undergo amidolysis with the aromatic DDS or DDM diamines under our curing conditions. As demonstrated in Scheme I, the amidolysis proceeds via the formation of additional glycerol free hydroxyl groups and fatty acid amide groups, both with very high dielectric properties. These by-products increase the dielectric constant and dissipation factor of their two-phase thermosets and balance the reduction effect of the long aliphatic chains of the epoxidized soybean oil.

To prove that such amidolysis takes place, we measured consecutive infrared spectra of a stoichiometric mixture of epoxidized soybean oil and DDM with progressing B-staging time (Figure 17). B-staging (or prepolymerization) was carried out at 135 °C for 48 h under conditions similar to those at which the castings used for determination of the dielectric properties were prepared. The infrared spectrum of the initial pure epoxidized soybean oil exhibited a single absorption band at about 1730 cm$^{-1}$ that corresponds to

**Table VI. Dielectric Properties of Pure DGEBA–DDM and DGEBA–DDS Thermosets and Their Rubber-Modified DGEBA–DDM–ESR and DGEBA–DDS–ESR Formulations**

| Thermoset Formulation | Dissipation Factor ($\times 10^{-2}$) | Dielectric Constant |
|---|---|---|
| DGEBA–DDM | 3.3 | 3.78 |
| DGEBA–DDM–ESR-15 | 3.5 | 3.78 |
| DGEBA–DDM–ESR-20 | 3.6 | 3.77 |
| DGEBA–DDM–ESR-30 | 3.7 | 3.77 |
| DGEBA–DDS | 2.5 | 3.95 |
| DGEBA–DDS–ESR-20 | 3.3 | 3.88 |

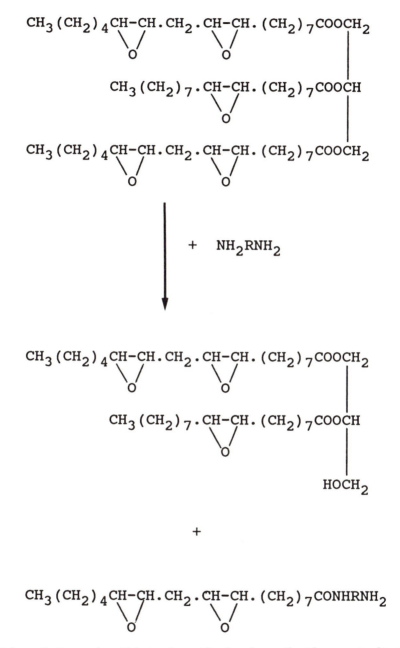

Scheme I. *Proposed amidolysis of epoxidized soybean oil with aromatic diamines* $(NH_2 \cdot R \cdot NH_2)$ *under curing conditions (D. Shimp, Rhone-Poulenc, personal commu - nication).*

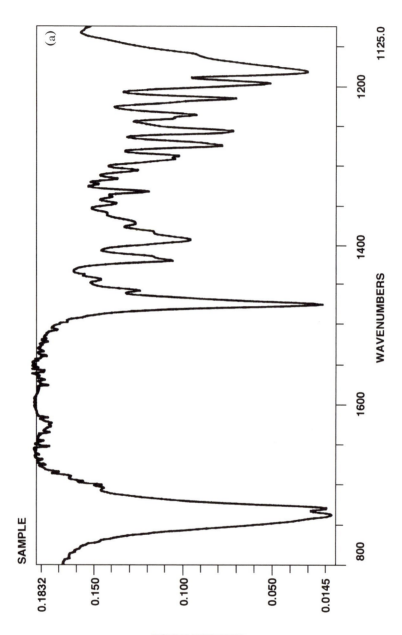

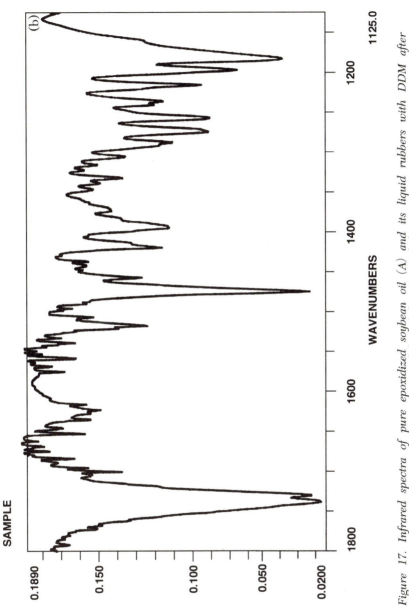

*Figure 17. Infrared spectra of pure epoxidized soybean oil (A) and its liquid rubbers with DDM after prepolymerization at 135 °C for 0 (B), 24 (C), and 48 h (D). Continued on next page.*

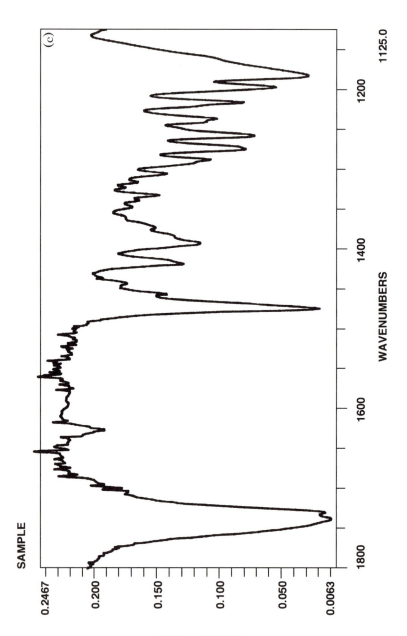

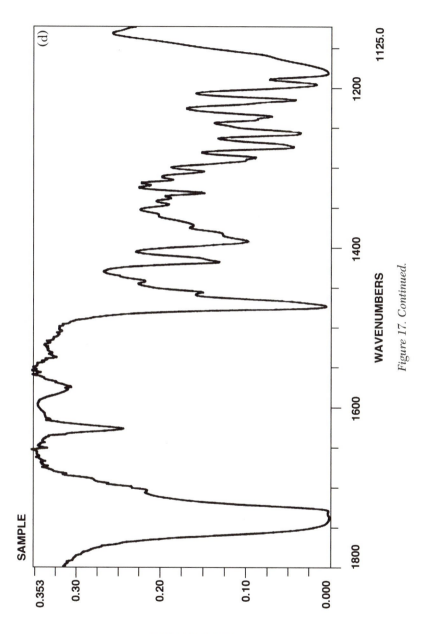

Figure 17. Continued.

the stretching $> C = O$ vibrations of the ESO carbonyl $-COO$ groups. A comparison of this infrared spectrum with the spectrum of the B-staged materials with progressing B-staging time reveals no new absorption bands for the $> C = O$ vibrations of the amide $-COONH$ groups of the by-product nor any other changes in the region of $1800-1650$ cm$^{-1}$. The comparison results indicate that if the amidolysis reaction indeed proceeds simultaneously with the traditional epoxidation reaction, it does so only to a limited extent.

**Sodium Content.** Low ionic content, often below 10 ppm (especially for sodium and chlorine ions) is a requirement for the application of epoxy resins in the electronics industry. Electronics-grade epoxy resins are significantly more expensive because an additional purification procedure is needed to lower ionic content.

Epoxidized soybean oil and liquid rubber, without any purification, is characterized by very low sodium content (16.5 and 12.5 ppm, respectively). The content of other ionic species is probably also low. The hydrophobic nonpolar molecular structure of the epoxidized soybean oil, which lacks free hydroxyl or carboxyl groups, is probably responsible for low ionic content. Thus, epoxidized soybean rubber appears especially suitable for toughening epoxy resins for electronics applications.

**Epoxidized Soybean Oil vs. Vernonia Oil.** In general, the morphology, particle size, toughening, and so forth of DGEBA–DDM and DGEBA–DDS epoxy formulations modified by epoxidized soybean oil or rubber resemble the characteristics of the corresponding two-phase thermosets based on vernonia oil and rubber (8) prepared under similar conditions. A comparison of the effect of the two oils also shows several differences. Although vernonia oil has a structure similar to that of the epoxidized soybean oil, it is characterized by a lower epoxy functionality. Vernonia oil contains an average of 2.4 epoxy groups per triglyceride molecule, which is only half of the average 4.5 epoxy groups found in epoxidized soybean oil. Vernonia oil and rubber, therefore, is expected to have less polar character than the epoxidized soybean oil and rubber, to have lower compatibility with commercial epoxy resins, and to produce rubbery particles with lower cross-linking density. Our results (8) agree with the projections and show the formation of bigger vernonia particles whose separation proceeds at an earlier stage of B-staging and curing of the vernonia formulations of commercial epoxy resins. For example, the DGEBA–DDM formulation containing 10% vernonia rubber forms two-phase thermosets, whereas the corresponding DGEBA–DDM–ESR-10 is homogeneous. In addition, vernonia rubbery particles have a lower glass-transition temperature, which reflects lower cross-linked density.

These results show that the morphology, particle size, and, presumably, the fracture toughness (together with the other physicomechanical properties) of the commercial epoxy resins can be varied not only by changing the nature and polarity of the diamine curing agent as discussed previously, but also can be altered by variation of the nature and degree of epoxidation of the vegetable oils.

**Epoxidized Vegetable Oils vs. CTBN Liquid Rubber**. It is also interesting to compare the effect of epoxidized vegetable oils with the effect of CTBN liquid rubber, which presently is commonly used for epoxy toughening. For this purpose, we compared our results on DGEBA–DDS–ESR-10 with published data (*14*) on the DGEBA–DDS thermoset modified with 10% CTBN (DGEBA–DDS–CTBN-10). Although both thermosets were prepared by different procedures, they had similar morphology. CTBN was incorporated into the epoxy resin without prepolymerization.

According to recent studies by Pearson and Yee (*14*) on the toughening of epoxy resins, the introduction of 10% CTBN liquid rubber into a DGEBA–DDS epoxy resin produces two-phase thermosets with CTBN rubbery particles with a diameter of $\sim 5$ $\mu$m. The fracture energy of the unmodified DGEBA–DDS thermoset increases from $G_{Ic} = 162$ J/m$^2$ to $G_{Ic} = 242$ J/m$^2$ in the presence of 10% CTBN (e.g., with $\Delta G_{Ic} = 80$ J/m$^2$), whereas its Young's modulus simultaneously decreases from $E = 3360$ MPa to $E = 3000$ MPa (*14*). Our unmodified DGEBA–DDS thermoset exhibits fracture toughness ($G_{Ic} = 145$ J/m$^2$) and Young's modulus ($E = 3420$ MPa) (Tables III and IV) similar to those reported by Pearson and Yee (*14*). The corresponding soybean-modified DGEBA–DDS–ESR-10 thermoset at the same rubber content, however, has smaller rubbery particles ranging from 1–2 $\mu$m (Table I). The introduction of 10% soybean liquid rubber more than doubles the fracture energy of the unmodified DGEBA–DDS thermoset from $G_{Ic} = 145$ J/m$^2$ to $G_{Ic} = 374$ J/m$^2$ (Table III) (e.g., with $\Delta G_{Ic} = 229$ J/m$^2$), whereas the Young's modulus decreases from $E = 3420$ MPa to $E = 2810$ MPa (Table IV), a slightly lower value than that observed for the CTBN-modified thermoset by Pearson and Yee (*14*). Comparison of the toughening effect at 10% content level indicates that the soybean liquid rubber ($\Delta G_{Ic} = 229$ J/m$^2$) is more effective than CTBN ($\Delta G_{Ic} = 87$ J/m$^2$).

Comparison of the SEM micrographs shows a relatively smooth fracture surface for the DGEBA–DDS–CTBN-10 thermoset (*14*). In contrast, the corresponding soybean-modified DGEBA–DDS–ESR-10 fracture surfaces show extensive deep voids and more pronounced particle cavitation. These phenomena are characteristic for smaller rubbery particles and result in the improved toughening effect observed for the soybean liquid rubber.

Another advantage of the epoxidized soybean rubber is its low sodium content (12.5 ppm) in contrast to the higher sodium content of the CTBN (in the range of 300 ppm; R. Drake, BFGoodrich, personal communication). As a matter of fact, the sodium content of Hycar CTBN 1300 $\times$ 16 (BFGoodrich) is 380 ppm (Galbraith Analytical Laboratory). Low ionic content is an important characteristic for electronics applications of liquid rubbers for the toughening epoxy resins (*see* the "Sodium Content" section). Reduction of the sodium content of CTBNs requires additional washing (purification).

Finally, epoxidized soybean oil is industrially produced and available at about $0.50 per pound, a price below that of CTBN (in the range of $2.00–$2.50 per pound). Price alone makes the toughening of commercial epoxy resins by epoxidized soybean oil very attractive for large-scale commercial applications.

The epoxidized soybean oil appears to have three potential advantages over CTBN for toughening epoxy resins: better toughening effect, lower sodium content, and lower price.

## *Summary*

Homogeneous mixtures of epoxidized soybean rubber (and probably rubber of any other epoxidized vegetable oil), commercial epoxy resins based on diglycidyl ether of bisphenol A (Epon resin 825, etc.), and commercial diamines (DDS, DDM, etc.) form, under certain conditions, two-phase thermosetting materials that consist of a rigid epoxy matrix and randomly distributed small rubbery particles of cured epoxidized soybean oil. Particle size varies from 0.1–5 μm and depends on the nature of the diamine, the B-staging conditions, the kinetics of curing, and so forth.

Although further research is required, our initial unoptimized results show that these two-phase thermosetting materials have toughness comparable to (if not better than) the CTBN-modified epoxies, as well as lower water absorption and low sodium content; they may even be more environmentally friendly. The ESR formulations exhibit slightly lower glass-transition temperatures and Young's moduli in comparison with the unmodified thermosets, but their dielectric properties do not change. Optimization of the preparation conditions is expected to further improve the physicomechanical properties of ESR formulations.

Several options for direct epoxy toughening by epoxidized soybean oil, especially of DDS formulations, are under investigation.

Epoxidized soybean oil is industrially produced and available at about $0.50 per pound, a price below the least expensive epoxy resins based on DGEBA ($1.25–$1.35 per pound) and much lower than the price of CTBN (in the range of $2.00–$2.50 per pound), which is commonly used for epoxy toughening. The toughening of commercial epoxy resins by epoxidized soybean oils and rubbers is, therefore, very attractive for large-scale commercial applications.

## Acknowledgments

The authors gratefully acknowledge A. Yee (University of Michigan) for help in obtaining the fracture toughness and tensile data; D. Shimp (Rhone-Poulenc) for measurements of dielectric properties; R. Schweizer (Paint Research Associates) for sodium content determination; and R. Pearson and S. Jin (University of Michigan) for many fruitful discussions. We gratefully acknowledge the financial support of the South Coast Air Quality Management District, the U.S. Agency for International Development, Paint Research Associates, and the State of Michigan. We thank TA Instruments (formerly DuPont Instruments) for donating DSC and DMA to our Institute.

## References

1. Sultan, J. N.; Liable, R. C.; McGarry, F. J. *Polym. Sym.* **1971**, *16*, 127.
2. Sultan, J. N.; McGarry, F. J. *Polym. Eng. Sci.* **1973**, *13*, 29.
3. *Rubber-Toughened Plastics*; Riew, K. C., Ed.; Advances in Chemistry 222; American Chemical Society: Washington, DC, 1989.
4. *Rubber-Modified Thermoset Resins*; Riew, K. C.; Gillham, J. K., Eds.; Advances in Chemistry 208; American Chemical Society: Washington, DC, 1984.
5. Yee, A. F.; Pearson, R. A. *J. Mater. Sci.* **1986**, *21*, 2462.
6. Pearson, R. A.; Yee, A. F. *J. Mater. Sci.* **1986**, *21*, 2475.
7. Qureshi, S.; Manson, J. A.; Michel, J. C.; Hertzberg, R. W.; Sperling, L. In *Characterization of Highly Cross-Linked Polymers*; Labana, S. S.; Dickie, R. A., Eds.; ACS Symposium Series 243; American Chemical Society: Washington, DC, 1984; pp 109–124.
8. Dirlikov, S. K.; Frischinger, I.; Islam, M. S.; Lepkowski, T. J. In *Biotechnology and Polymers*; Gebelein, C. G., Ed.; Plenum Press: New York, 1991.
9. Brown, W. F.; Srawley, J. E. *ASTM Spec. Tech. Publ.* **1965**, *381*, 13.
10. Lee, W. H.; Hodd, K. A.; Wright, W. W. In *Rubber-Toughened Plastics*; Riew, K. C., Ed.; Advances in Chemistry 222; American Chemical Society: Washington, DC, 1989; pp 263–287.
11. Romanchick, W. A.; Sohn, J. E.; Geibel, J. F. In *Epoxy Resin Chemistry II*; Bauer, R. S., Ed.; ACS Symposium Series 221; American Chemical Society: Washington, DC, 1985; pp 85–118.
12. Manzionne, L. T.; Gillham, J. K. *J. Appl. Polym. Sci.* **1981**, *26*, 889.
13. Pearson, R. A.; Yee, A. F. *J. Mater. Sci.* **1991**, *26*, 3828.
14. Pearson, R. A.; Yee, A. F. *J. Mater. Sci.* **1989**, *24*, 2571.
15. Levita, G. In *Rubber-Toughened Plastics*; Riew, K. C., Ed.; Advances in Chemistry 222; American Chemical Society: Washington, DC, 1989; pp 93–118.

RECEIVED for review March 6, 1991. ACCEPTED revised manuscript June 23, 1992.

# APPLICATIONS
# OF TOUGHENED PLASTICS

# Structure–Property Relations of High-Temperature Composite Polymer Matrices

R. J. Morgan[1], R. J. Jurek[1], D. E. Larive[1], C. M. Tung[2], and T. Donnellan[3,4]

[1] Michigan Molecular Institute, Midland, MI 48640
[2] Northrop Corporation, Hawthorne, CA 90250–3277
[3] Naval Air Development Center, Warminster, PA 18974

*The structure-deformation–failure mode-mechanical property relations of high-temperature thermoplastic polyimide and thermoset bis-maleimide (BMI) polymeric matrices and their composites will be discussed. In the case of polyimides, the effects of test temperature, thermal history, strain rate, type of filler, and filler volume fraction on structure–property relations will be discussed. For BMIs we report systematic Fourier transform infrared spectroscopy and differential scanning calorimetry studies of the cure reactions as a function of chemical composition and time–temperature cure conditions and then describe the resultant cross-linked network structure based on our understanding of the cure reactions. The optimization of the BMI matrix toughness will be considered in terms of network structure and process-induced matrix microcracking. We also describe optimization of composite prepreg, lamination and postcure conditions based on cure kinetics, and their relationship to the BMI viscosity–time–temperature profiles. The critical processing-performance limitations of high-temperature polymer matrices will be critically discussed, and toughening approaches to address these limitations, such as toughness over a wide temperature range, will be presented.*

P︎OLYMER MATRIX FIBROUS COMPOSITES that exhibit good composite "hot–wet" properties, processibility, and toughness (residual strength after impact)

[4] Current address: Grumman Corporation, Department 584, M.S. 802–26, Beth Page, Long Island, NY 11714–3580.

0065–2393/93/0233–0493$06.00/0

will be required for future advanced military aircraft. High-temperature composite matrices such as thermoplastic polyimides exhibit both processing and toughness limitations, whereas thermoset bismaleimides are more processible but exhibit a more brittle mechanical response.

The two principal mechanisms for toughening carbon fiber–polymer matrix fibrous composites involve polymer matrix viscoelastic flow and crack deflection by the incorporation of microscopic second-phase additives into the polymer matrix.

For high glass-transition temperature ($T_g$) polymer matrices, high yield stresses at service environment conditions limit viscoelastic flow because of increasing yield stress with decreasing temperature below $T_g$. Optimization of viscoelastic flow in a cross-linked thermoset matrix additionally requires a fully reacted defect-free network (1). Networks that contain defects in the form of unreacted groups initiate crack propagation early in the deformation process and these cracks truncate significant viscoelastic flow. Hence, there is a need to characterize the cure reactions and kinetics involved in the formation of high-temperature thermosets such as bismaleimides (BMIs) to optimize network structure for high toughness. A ductile, second-phase additive that deforms viscoelastically by crack bridging also can circumvent the yield stress–temperature dependence dilemma of the polymer matrix continuous phase (2).

Small, submicrometer additives, with dimensions well below the critical matrix flaw size, can cause a tortuous crack propagation path in the polymer matrix that results in increased matrix fracture energy. Appropriately oriented organic microwhiskers that readily split longitudinally and are incorporated in the composite interfiber tow regions seem most promising from a high-fracture-surface viewpoint.

In this chapter we report on the following studies:

1. The deformation, failure modes, and toughness of poly-4,4'-oxydiphenylene pyromellitimide (POPPI: Structure I) and its

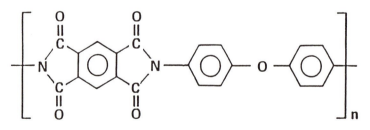

*Structure* I. The chemical structure of poly-4,4'-oxydiphenylene pyromellitimide (POPPI).

composites as a function of test temperature (23–300 °C), thermal history (300–500 °C), strain rate, type of filler (short glass fibers or graphite flakes), and filler volume fraction (15–40 wt % graphite flakes).

2. The cure reaction characteristics: optimal prepregging, lamination, and postcure conditions of thermoset 4,4'-bismaleimido-diphenylmethane–0,0'-diallyl bisphenol A (BMPM–DABPA; 1:1 molar) BMI (Matrimid 5292, Ciba-Geigy)–carbon fiber (AS4) composites. The chemical structure of the BMPM and DABPA monomers are illustrated in Structure II. Toughening procedures by network structure optimization or the incorporation of second-phase crack deflectors or viscoelastic bridgers will be described.

**4,4'-Bismaleimidodiphenylmethane**

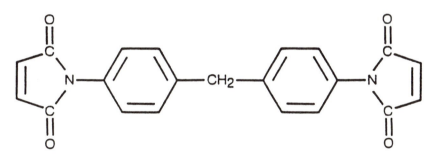

**0,0'-Diallyl Bisphenol A**

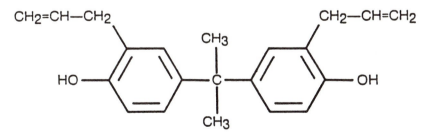

*Structure* II. The chemical structures of BMPM and DABPA monomers.

## Results and Discussion

**Thermoplastic Polyimides**.    The POPPI polyimide fails in tension by massive crazing via cavitation and fibrillation rather than by the shear-banding mode, which would involve a larger volume of matrix viscoelastic flow.

From tensile mechanical property tests of unfilled and filled POPPI matrix we determined the following effects on polyimide toughness (toughness in these studies is defined as the area under a tensile stress–strain curve):

1. In Figure 1 we illustrate the toughness of the POPPI matrix and its composites as a function of temperature. The toughness of all the polyimide materials decrease with increasing temperature. This decrease is a consequence of the lower strength of the craze fibrils with increasing temperature that results in the

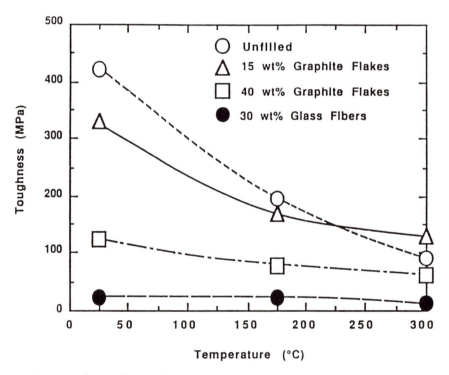

*Figure 1. The toughness of POPPI matrix and its composites as a function of temperature.*

growth of larger craze cavities and produces associated higher stress concentrations. In the case of matrices that deform by shear banding rather than crazing, the polymer matrix ductility increases with increasing temperature (3). Toughness decreases with increasing filler volume fraction of the 1–5-$\mu$m dimensional graphite flakes (Figure 1). Fracture topography observations indicate that the fillers inhibit the craze growth process. For larger dimensional fillers (glass fibers of ~ 10-$\mu$m diameter), the filler decreases the composite toughness to a greater extent than the smaller graphite flakes (Figure 1) as a result of higher fabrication stress concentrations at the fiber–matrix interfacial region induced by the fiber–matrix thermal coefficient of expansion mismatches.

2. The toughness of POPPI and its composites decreased with increasing strain rate because of the inhibition of the viscoelastic flow processes at faster testing speeds.

3. Thermal history also can modify the toughness of the polyimide composites. In the temperature regime near the POPPI $T_g$ (300–375 °C), the toughness increased after 24-h anneal as a result of relaxation of composite fabrication strains and any associated microvoids. However, exposure to temperatures above 400 °C in air results in oxidation-induced matrix cross-linking or chain scission and a corresponding decrease in ambient craze fibrillation capability and toughness. For example, air exposure at 450 °C for 24 h results in a > 50% loss in ambient composite toughness.

**Thermoset Bismaleimides**. *BMI Cure Reaction Studies.* Systematic Fourier transform infrared spectroscopy (FTIR) and differential scanning calorimetry (DSC) studies were carried out on the BMPM–DABPA BMI system as a function of chemical composition and temperature–time cure conditions. From the FTIR and DSC studies, together with published data by Ciba-Geigy researchers (4–6), we have arrived at the following conclusions: There are three principal temperature regimes associated with the cure reactions, namely: 1, 100–200 °C range; 2, 200–300 °C range; and 3, 300–350 °C range. In Figure 2 a DSC plot of a BMPM–DABPA(1:1 molar)–AS4 carbon fiber composite prepreg exhibits exotherm peaks in each of the three regimes.

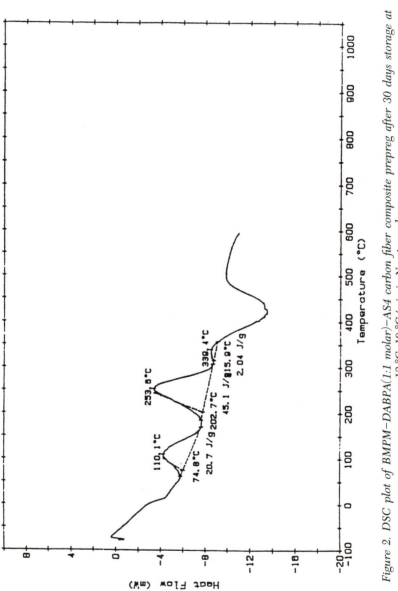

*Figure 2. DSC plot of BMPM–DABPA(1:1 molar)–AS4 carbon fiber composite prepreg after 30 days storage at −13 °C, 10 °C/min in N₂ atmosphere.*

In regime 1 (the 100–200 °C range), the "ene" reaction

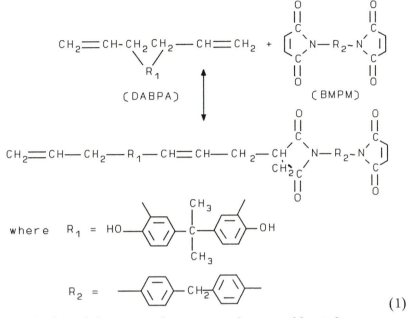

(1)

occurs slowly and there is evidence it may be reversible. Gel permeation chromatography indicates that only a small molecular weight increase occurs from 520 to 680 for the BMPM–DABPA system after 16 h at 100 °C. Consistent with our viscosity–time–temperature observations, this reaction would only occur at significant rates above 180 °C. (In this lower temperature regime the presence of propenyl isomers rather than allyl isomers in the diallyl bisphenol A will enhance chemical reactivity due to conjugation of the double bond with the phenyl ring.)

In regime 2 (the 200–300 °C range), a number of chemical reactions occur in the following sequence: The "ene" reaction 1 occurs at a significant rate; then the BMPM homopolymerization, reaction 2 occurs:

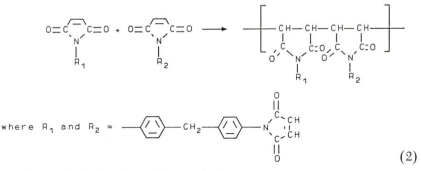

(2)

or a longer chain "ene'" reaction product.

A DSC plot of the BMPM monomer alone exhibits a reaction peak centered in the 205–210 °C range.

At higher temperatures both "ene" homopolymerization, reaction 3, and "ene"–BMI double-bond cross-linking polymerization (Diels–Alder reaction), reaction 4, occur:

$$\eta R_1 \; CH{=\!=}CHR_2 \;\longrightarrow\; \underbrace{-\!\!\left(-R_1CH\!-\!CHR_2\!-\right)\!-}_{n} \qquad (3)$$

where $R_1$ and $R_2$ are chain continuations of the "prepolymer" formed in the "ene" reaction;

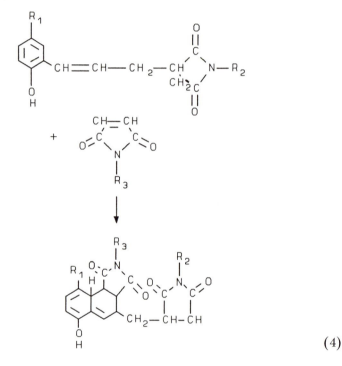

$$(4)$$

where $R_1$ and $R_2$ are chain continuations of the "prepolymer" formed in the "ene" reaction and

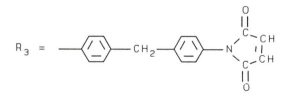

or a longer chain "ene" reaction product.

Reactions 3 and 4 are indicated by the DSC peak centered at 254 °C for the prepreg in Figure 2. A low-temperature shoulder exhibited by this peak is associated with BMPM homopolymerization.

Dehydration of the hydroxyl groups of the DABPA initiates above the 240 °C, enabling the formation of ether cross-links:

$$HO \text{---} R \text{---} OH \ + \ HO \text{---} R \text{---} OH \longrightarrow HO \text{---} R \text{---} O \text{---} R \text{---} OH$$

$$(5)$$

where $R$ represents polymer chains formed from reactions 1–4.

A DSC plot of the DABPA monomer alone exhibits thermal activity above 240 °C that is associated with dehydration. Thermogravimetric analysis weight loss and FTIR studies confirm that $\sim 75\%$ of the hydroxyl groups undergo dehydration with an associated 2-wt % $H_2O$ loss in this temperature region.

In regime 3 (the 300–350 °C range), further cure occurs as the specimen test temperature exceeds the $T_g$ of the resin. It is important from the network-toughness viewpoint that the reaction be allowed to proceed to completion, thus eliminating network defects in the form of unreacted groups. DSC studies as a function of postcure conditions indicate a postcure schedule of 2 h at 200 °C, 6 h at 250 °C, and 1 h at 300 °C would be optimal for full cure and minimization of composite fabrication stresses. A resultant $T_g$ near 340 °C results from this postcure schedule. All postcure conditions in our studies at 250 °C and above do result in inherent microcracks in composite 0°, 90° laminate layers.

***BMI Network Modeling.*** Based on our understanding of the BMPM–DABPA cure reactions, network modeling studies have been carried out. The chemical structure of the BMPM–DABPA "ene" adduct is illustrated in Structure III. The three double bonds capable of chain extension and cross-linking are designated A, B, and C in Structure III. Based on the cure reaction characteristics, geometric controlled sequential reactions for the trifunctional "ene" molecule, and symmetry conditions that allow the network to fully react, we show a proposed ring chain structure in Structure IV that would allow sequential polymerization of all three double bonds present in the "ene" molecule. The ring chains cross-link via A, B, or C "ene" double bonds (ring formation, therefore, would not occur at such cross-links) or interchain etherification by hydroxyl dehydration.

***BMI Prepreg Development.*** Isothermal viscosity profiles in the 85–95 °C prepregging temperature range indicate that initial viscosities of the BMPM–DABPA (1:1 molar) resin are in the 700–1200-cps range and only increase slowly with time. For example, the viscosity of the BMPM–DABPA (1:1 molar) system only increases to 5000 cps after 1200 min at 91°, thus

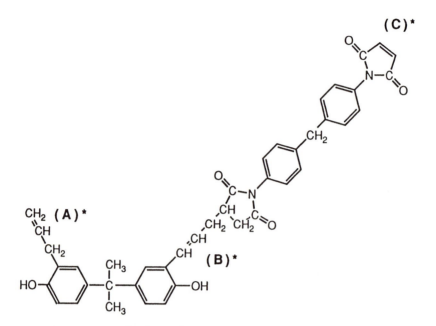

*Structure* III. *The chemical structure of BMPM–DABPA "ene" adduct.*

indicating adequate viscosity–time profiles for prepregging in the 85–95 °C
temperature range.

The BMPM–DABPA system was prepregged with adjustable wedge slit
dies developed for 3000 and 12,000 filament-tow AS4 carbon fibers. The
carbon fibers contain epoxide sizings that chemically degrade during postcure
conditions of the BMI composite and cause poor fiber–matrix interface
integrity (7). Hence, during the prepreg process the sizing is burned off the
fiber at 400 °C in air.

BMPM–DABPA(1:1 molar)–AS4 carbon fiber prepregs were consistently fabricated to meet government specifications.

**BMI Composite Lamination.** The optimum conditions to produce
BMI composites from the prepreg using a programmable press (Tetrahedron)
and miniclave were investigated.

Dynamic mechanical analyses (DMA) plots exhibited broad minima in
damping in the 160–190 °C temperature range that suggest a relatively
constant viscosity for adequate time periods to achieve laminate consolidation
and associated acceptable composite resin and void content.

A variety of lamination conditions for BMPM–DABPA(1:1 molar)–AS4
carbon fiber composite prepregs were studied and optimized. The resultant
resin weights in the composite ranged from 15 to 31 wt %. The BMI resin

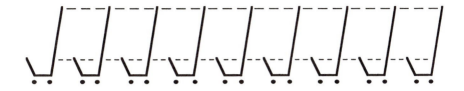

— — — (C)  BMI Double Bond Homopolymerization

- - (A) and (B) "ENE" Double Bond Polymerization

An "ENE" Molecule

• Hydroxyl Group

*Structure* IV. *Ring chain structure from "ene" double-bond polymerization.*

exhibited good flow characteristics at 177 °C and the resultant composites exhibited good surface appearance.

**BMI Composite Characterization.**    After postcure, the BMI composites were characterized to form a standard database to evaluate against toughening procedures. Scanning electron micrographs of the postcured fractured composite revealed a thoroughly and uniformly impregnated composite. Short beam shear, tensile, and toughness ($G_{Ic}$) mechanical properties, together with ultrasonic nondestructive evaluation of the composites, were monitored.

**BMI Toughening Procedures.**    The optimization of the postcure conditions based on conflicting requirements of the desirability to attain a fully reacted BMI network and to minimize composite fabrication stresses associated with the mismatch of the fiber and matrix thermal coefficients of expansion were explored.

Techniques to incorporate crack deflectors, such as organic microwhiskers of polyamide fibers (Kevlar 49; K–49) or finely divided fused silica, into the composite via the prepreg process were carried out. Our experience with the fluidized-bed approach led to variable, inconsistent third-phase

## Table I. Average Mode-I $G_{Ic}$ Toughness Values
## for BMPM–DABPA(1:1 molar)–AS4 Carbon Fiber Composites

| Additive in Resin (%) | Average Thickness (mm) | Postcure Condition (N or A)[a] | Fiber Volume Fraction (%) | Average $G_{Ic}$ (J/m²) | Standard Deviation in $G_{Ic}$ (J/M²) |
|---|---|---|---|---|---|
| 0 | 2.50 | N | 65.7 | 259.9 | 43.2 |
| 0 | 2.71 | N | 63.5 | 301.6 | 32.1 |
| 0 | 2.75 | N | — | 324.8 | 34.3 |
| 0 | 2.65 | N | 63.8 | 326.9 | 16.2 |
| 0 | 2.65 | A | 63.8 | 331.1 | 6.0 |
| PEI (3.7) | 2.48 | N | 64.0 | 462.2 | 225.7 |
| PEI (8–10) | 3.34 | A | 66.7 | 397.3 | 47.8 |
| PEI (7–8) | 2.63 | A | 71.0 | 338.2 | 15.6 |
| PEI (7–8) | 2.95 | A | 65.0 | 437.3 | 63.7 |
| K-49 (8.6) | 2.72 | N | 66.5 | 381.3 | 32.1 |
| K-49 (8.6) | 3.09 | A | 70.4 | 313.4 | 52.8 |

[a] N indicates normal postcure 2 h at 200 °C and 6 h at 250 °C. A indicates additional postcure 2 h at 200 °C, 6 h at 250 °C, and 1 h at 300 °C.

additive concentrations as a result of time-dominated instabilities in the fluidized-bed apparatus. Hence, we concentrated our efforts on the incorporation of crack deflectors into the prepreg via formation of initial slurries of the additive in the BMPM–DABPA resin. Hence, the crack deflectors were incorporated into the BMPM–DABPA mixture prior to prepregging to produce a slurry. The additives increased the resin viscosity by an order of magnitude, but prepregs were prepared successfully and the crack deflectors were uniformly distributed in the interfiber tow regions.

For the introduction of viscoelastic crack bridgers into the composite we used polyetherimide (PEI) additive and the following procedure: The PEI is dissolved in methylene chloride with the appropriate amount of DABPA (BMPM is not soluble in methylene chloride) and the methylene chloride is evaporated at 40 °C. Then the DABPA–PEI mixture is mixed with the BMPM at 130 °C and prepregged.

In Table I we summarize the composite toughness ($G_{Ic}$) values as a function of polyetherimide (PEI) additive, polyamide (Kevlar 49) powder, composite thickness, postcure conditions, and fiber volume fraction. Our principal conclusions from these data are as follows:

- The average $G_{Ic}$ values of 330 J/m² for the unmodified composites were similar for both a 250 and 300 °C final postcure temperature. However, the 300 °C postcured composite had a higher $T_g$ (near 350 °C) with no decrease in the 23 °C $G_{Ic}$ value. The higher 300 °C postcure produced two counteracting effects on the $G_{Ic}$ composite toughness values: Higher composite fabrication stresses, and probably associated microcrack formation, as a result of higher processing temperatures caused a decrease in $G_{Ic}$ composite toughness, whereas the higher postcure temperature increased the BMI network perfection and increased the $G_{Ic}$ composite toughness.

- For fiber volume fractions of approximately 65%, the incorporation of PEI increased the $G_{Ic}$ values 32% to 440 J/m$^2$ for a 300 °C postcured composite. Individual data points from the $G_{Ic}$ tests were recorded as high as 900 J/m$^2$, which suggests excellent potential for this toughening procedure, but also suggests the probability of inhomogeneous distribution of the PEI within the composite test specimens.

- For a fiber volume fraction of 66.5%, a 15% increase in $G_{Ic}$ was achieved by the incorporation of polyamide (Kevlar 49) powder.

- There is a general downward trend in $G_{Ic}$ values with increasing fiber volume fraction. From the available data illustrated in Table I for similar additive concentrations and postcure conditions, a 1% increase in fiber volume fraction results in a 5% decrease in $G_{Ic}$ composite fracture toughness. This observation identifies the need to optimize composite toughness at the expense of stiffness as a function of fiber volume fraction.

## *Summary*

Thermoplastic polyimide composite matrices are relatively tough as a result of crazing. However, their toughness is decreased with increasing temperature, strain rate, and fillers. BMIs, however, are more brittle and require network structure optimization and the incorporation of additive toughening agents.

The cure reactions of BMPM–DABPA bismaleimide are complex. In the 100–200 °C range, the BMPM and DABPA react to form an "ene" prepolymer. In the 200–300 °C range, the "ene" molecule polymerizes and cross-links by (1) cooperative reactions between the three double bonds in the "ene" molecule and (2) dehydration through hydroxyl groups, with associated moisture evolution. In the 200–250 °C range, full cure is not attained because of diffusion restrictions associated with the glassy state. Full cure is nearly attained after a 1-h 300 °C postcure that results in a final $T_g$ of near 350 °C. The cross-linked network structure is complex as a result of the pentafunctional nature of the "ene" prepolymer.

Optimum prepreg, lamination, and postcure conditions were ascertained for processing BMPM–DABPA(1: molar)–AS4 carbon fiber composites and for the introduction of toughening agents, namely, polyetherimide (PEI) particulates and polyamide (Kevlar 49) whiskers.

$G_{Ic}$ composite toughness values are enhanced by incorporation of PEI particulates and polyamide (Kevlar 49) powder into the BMI composite matrix. $G_{Ic}$ composite values decrease with increasing fiber volume fraction, which requires composite toughness to be optimized at the expense of stiffness.

## Acknowledgment

This work was partially supported by Northrop/NADC Contract No. N62269–88–C–0254.

## References

1. Morgan, R. J.; Kong, F. M.; Walkup, C. M. *Polymer* **1984**, 25, 375.
2. Morgan, R. J. *Proc. Annual ESD/ASM Advanced Composites Conf.*; ASM International: Metals Park, OH, 1986; p 179.
3. Morgan, R. J.; Walkup, C. M. *J. Appl. Polym. Sci.* **1987**, 34, 37.
4. Chaudhari, M.; Galvin, T.; King, J. *SAMPE J.* **July / August 1985**, 17.
5. Lee, B. H.; Chaudhari, M. A.; Blyakhman, V. *Polymer News* **1988**, 13, 297.
6. Zahir, S.; Chaudhari, M. A.; King, J. *Makromol. Chem., Makromol. Symp.* **1989**, 25, 141.
7. DePruneda, J. H. S.; Morgan, R. J. *J. Mater. Sci.* **1990**, 25, 4776.

RECEIVED for review March 6, 1991. ACCEPTED revised manuscript June 8, 1992.

# Model Multilayer Toughened Thermosetting Advanced Composites

**Mark A. Hoisington and James C. Seferis***

**Polymeric Composites Laboratory, Department of Chemical Engineering, University of Washington, Seattle, WA 98195**

*A model multilayer toughened high-performance thermosetting matrix composite was developed and the interlaminar fracture toughness was measured in modes I and II. Processing was accomplished using two impregnation steps followed by an autoclave curing process to consoli-date the final laminate. The final multilayer laminate structure con-tained layers of matrix resin with reinforcing carbon fibers separated by thin layers of matrix resin with modifier particles. Characterization of the processing was examined in terms of preimpregnation and final laminate morphologies. Structure–property relationships were evalu-ated by measuring the mode-I and -II interlaminar fracture toughness of the multilayer structure in comparison to the fracture toughness of a conventional composite with the same resin content but without the interlayer structure. The mode-II fracture toughness increased from 530 to 1135 J/m². No changes were observed in the mode-I fracture toughness.*

INCREASED TOUGHNESS OF THERMOSETTING MATRICES IS THE KEY in the development of primary commercial aircraft structures with advanced com-posites. Initial efforts concentrated on toughening the entire thermosetting matrix with elastomer or thermoplastic modifications. Although large tough-ness improvements were observed in neat-resin samples, only a small fraction of the neat-resin toughness improvements translated into increased laminate toughness (*1*). This lack of toughness translation from neat resin to laminate

---

* Corresponding author.

0065–2393/93/0233–0507$06.00/0

redirected many efforts from neat-resin modifications toward examining laminate toughness from the viewpoint of the composite structure.

During impact, the major type of failure that occurs within the laminate structure is ply delamination (2, 3). These delaminations create significant reductions in resulting composite mechanical properties after impact. To prevent delamination during impact, interleaving concepts have been incorporated into commercial advanced-composite systems (2–14). Interleaving is an engineering solution that minimizes the delamination problem by creating a multilayer composite structure with a thin layer of a tough resin matrix between each ply. The thin interlayers of resin are designed to absorb the large interlaminar stresses developed during impact, and can be employed as a homogeneous or a heterogeneous modification as shown schematically in Figure 1.

Homogeneous interleafs were the first modifications to commercial advanced-composite systems (2–8). In the homogeneous interleaved systems, a base-resin system that produced good composite mechanical properties was used within each ply, and a second tough homogeneous resin system was used between plies. The interlayer resin was either thermosetting or thermoplastic and created dramatic increases in the composite impact properties. The interleaf also led to reductions in the important composite compressive and hot–wet properties (4).

To maintain the improvements in composite toughness and prevent the reduction in composite compressive and hot–wet properties, second generation commercial interleaving systems were developed that utilize a heterogeneous resin layer between plies (9–14). The heterogeneous interleaved systems use the same base-resin matrix within each ply and between plies but

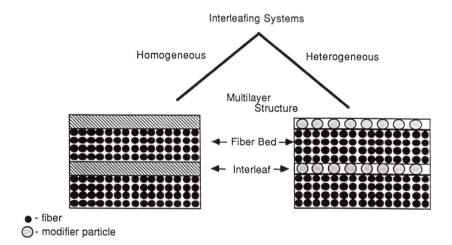

Figure 1. Two types of interleaving systems: homogeneous and heterogeneous.

the interlayer resin is mixed with either soft or rigid modifier particles approximately 10–50 μm in diameter. Modifier particles that are used in these systems possess mechanical properties similar to those of the base-resin matrix in order to maintain the overall required composite mechanical properties. The diameter of the modifier particle must be larger than the fiber diameter so that the particles can be trapped between plies to create a heterogeneous interleaf. For the interleaved composite to have the same resin content as a conventional composite without an interlayer structure, the fibers must be packed closer together to allow for some of the resin to be used between plies. Because the same base-resin system is used both within the ply and in the interlayer region, the laminate is able to maintain adequate hot–wet properties, while the impact and damage tolerance improve because of the interleaf created by the modifier particles.

To fully understand the concept of the layered matrix structure in advanced composites, a model system has been formulated. This part of the work examined the processing and toughness improvements of the model heterogeneous interleaved-composite system over conventional thermoset-type systems. The effects of processing on the preimpregnated (prepreg) characteristics and performance as well as the final toughened composite structure and properties were examined.

## Process Modeling

There are two basic techniques for processing a multilayer composite structure that utilizes a heterogeneous interleaf. The first technique involves sprinkling modifier particles on prepreg surfaces and then processing the prepreg in an autoclave to consolidate the final laminate. The second technique involves premixing modifier particles with the base-resin matrix to create a multilayer prepreg structure, which is then processed with typical autoclave curing procedures. This second technique, which utilizes two impregnation steps, is referred to as double-pass impregnation.

In double-pass impregnation, the first impregnation step is used to impregnate the collimated fibers with only the base-resin matrix. The second impregnation step is used to apply a thin layer of particle-modified resin to the top and bottom prepreg surfaces. The resulting multilayer prepreg structure contains a matrix-resin-impregnated fiber bed, with thin layers of particle-modified matrix resin on the top and bottom surfaces. For this toughened prepreg structure to have the same resin content as conventional prepreg systems and at the same time be fully impregnated, the fiber spacing must be closer than that of conventional systems, so that some of the matrix resin may remain on the prepreg surfaces, as shown schematically in Figure 2.

## Conventional Prepreg Structure

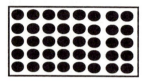

● - carbon fiber
◎ - modifier particle

## Multilayer Prepreg Structure

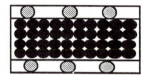

*Figure 2. Comparison of a conventional prepreg structure to a multilayer prepreg structure.*

During processing, only a fraction of the desired total resin content may be applied during the first impregnation step, because more resin will be applied during the second impregnation. The amount of resin that can be applied during the first impregnation is a function of the modifier-particle diameter. Ideally, the thickness of the resin film applied during the second impregnation will be equal to the modifier-particle diameter, thereby obtaining an even monolayer distribution of particles. The amount of resin that may be applied during the first impregnation can be calculated if the diameter of the modifier particle, final desired resin content, and fiber areal weight are known:

$$w_{rl} = \frac{\left(A_f/(1 - w_r) - A_f\right) - 2D_p\rho_r}{A_f + \left(A_f/(1 - w_r) - A_f\right) - 2D_p\rho_r} \tag{1}$$

where $w_{rl}$ is the first-pass weight fraction of resin, $w_r$ is the final desired weight fraction of resin, $A_f$ is the areal weight of the fibers (g/m$^2$), $D_p$ is the average diameter of the modifier particles (m), and $\rho_r$ is the density of the resin (g/m$^3$).

The effect of modifier-particle diameter on the first-pass resin content is shown in Figure 3 for a prepreg containing a fiber areal weight of 145 g/m$^2$, final resin content of 35 wt %, average modifier-particle diameter of 20 μm, and a resin density of 1.2 g/cm$^3$.

After processing, the resin distribution in both the final prepreg and the laminate is extremely important in the multilayer structure. The prepreg resin distribution can be viewed schematically as shown in Figure 4, where A represents the amount of resin on the prepreg surface, B represents the amount of impregnated resin, and C represents the amount of resin waste (*15*). In conventional prepreg processing, resin left on the prepreg surface is undesirable. However, in processing the multilayer prepreg structure, the amount of resin on the prepreg surfaces becomes an important design parameter. Prepreg tack and porosity characteristics are directly affected by the prepreg resin distribution (*15–18*).

Examination of the resin distribution in the final laminate structure is important to understanding the processing of the multilayer structure. The processing sequence for a multilayer composite morphology is schematically illustrated in Figure 5. Initially the prepreg contains a thin layer of particle-modified resin on the top and bottom surfaces. This resin layer thickness is equal to the modifier diameter. When the prepreg is layed-up for the

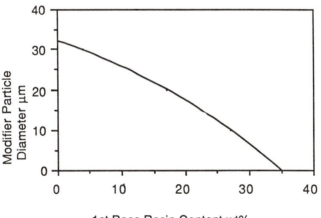

*Figure 3. Effect of modifier-particle diameter on first-pass resin content.*

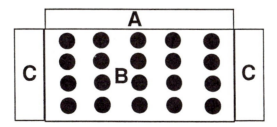

*Figure 4. Schematic of prepreg resin distribution.*

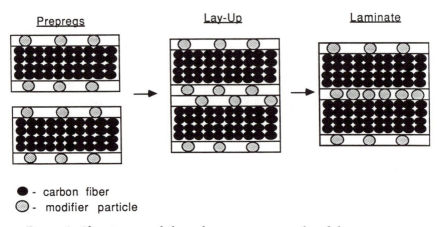

**Prepregs**            **Lay-Up**            **Laminate**

● - carbon fiber
◎ - modifier particle

*Figure 5. Changing morphology during processing of multilayer structure laminate.*

autoclave curing process, the fiber bed of each prepreg layer is separated by a particle-modified resin-layer thickness that is equal to two modifier-particle diameters. During consolidation, the resin is able to flow, which creates a heterogeneous interleaf thickness equal to one modifier-particle diameter that contains twice the modifier-particle concentration that was initially on the prepreg surfaces.

Limitations of the maximum modifier-particle diameter that can be used in processing can be calculated from volumetric and geometric arguments. If the thickness of the interlaminar region is assumed to be equal to the modifier-particle diameter, the maximum particle diameter then can be calculated and will be a function of the fiber areal weight, final resin content, and maximum fiber-packing arrangement. The calculation is based on the concept that in a void-free composite, there must be enough matrix resin to fill all the void areas surrounding the fibers and all the void areas surrounding the particles between plies. As the fibers become closer packed, smaller amounts of resin are required within each ply and more resin is available for the interlaminar region. For the assumption that the maximum fiber-packing arrangement is between a square array and a hexagonal packed structure, as shown in Figure 6, we can calculate the maximum modifier-particle diameter:

$$D_{P\max} = \frac{A_f\big((1 + (V_r/V_f))\pi - 2(1 + \cos\theta)\big)}{\rho_f \pi} \tag{2}$$

where $A_f$ is the fiber areal weight (g/m$^2$), $D_{P\max}$ is the maximum modifier diameter (m), $V_r$ is the volume fraction of resin, $V_f$ is the volume fraction of fibers, $\rho_f$ is the fiber density (g/m$^3$), and $\theta$ is the shift angle as defined in Figure 6 (for maximum effect $\theta = 30°$).

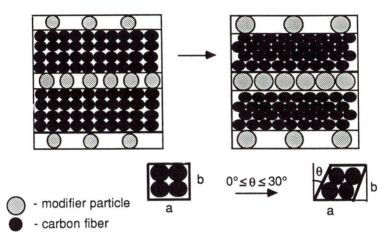

*Figure 6. Maximum fiber packing arrangements.*

An interesting relationship observed from this calculation is that the maximum modifier diameter is a linear function of the fiber areal weight. Therefore, the modifier diameter may be increased by increasing the prepreg fiber areal weight.

To examine the effects of the prepreg process on the prepreg resin distribution and performance, a previously developed prepreg methodology was extended (15). This methodology is based on the combination of the important operating parameters of temperature, pressure, and production speed into a dimensionless number known as the prepreg flow number (PFN):

$$\text{PFN} = \frac{KP_e}{\mu V Y_f} \tag{3}$$

where $K$ is the collimated fiber bed permeability $(\text{m}^2)$, $P_e$ is the effective pressure (Pa), $\mu$ is the viscosity of the impregnating resin (Pa · s), $V$ is the production line speed (m/s), $Y_f$ is the fiber bed thickness (m). The PFN represents a ratio of the impregnation residence time to the time required for full impregnation. Therefore, impregnation may be considered complete if the PFN is greater than or equal to 1. Thus analyzing the different passes with the aid of the PFN, the prepreg resin distribution may be described as a function of the prepreg operating conditions.

## Experimental

The model resin system used for this study was epoxy-based with the following composition: 60% tetraglycidyldiaminodiphenylmethane (TGDDM;

Ciba-Geigy), 40% diglycidyl ether of bisphenol A (DGEBA; Epon 828; Shell Chemical Company) 10-parts per hundred parts resin (phr) polyethersulfone (PES; Imperial Chemical Industries), and 42-phr diaminodiphenylsulfone (DDS) curing agent (Ciba-Geigy). The particles used to modify the base-resin system were semicrystalline nylon 6 (1002 D NAT; Atochem Corporation). The nylon particles were produced through a precipitation process and were approximately 20 μm in diameter. Modifier particles were mixed with the base-resin system before the second-pass impregnation. The concentration of the nylon particles was measured as phr in the base-resin mixture.

All composite samples were processed using a laboratory-scale hot-melt prepreg machine that has been described previously in detail (19). The reinforcing carbon fibers were 12K T-800 (Toray Industries). The prepreg that was produced had a fiber areal weight of 255 g/m$^2$ and a final resin content of 35 wt %. Although a fiber areal weight of 255 g/m$^2$ is larger than conventional areal weights of 145 or 190 g/m$^2$, it was used because of thin-film production limitations.

Fracture-toughness testing was done on a mechanical testing apparatus (4505 Instron). Mode-I interlaminar fracture-toughness testing was performed using double cantilever beam samples. Mode-II interlaminar fracture-toughness testing was performed using end-notch-flexure specimen configurations (20).

## Results and Discussion

In double-pass impregnation, the objective of the first-pass impregnation is to produce a low base-resin content prepreg. The amount of resin that can be applied is controlled by the modifier-particle diameter. In this work, the 20-μm nylon 6 particles applied to both top and bottom prepreg surfaces only allowed a first-pass resin content of 25 wt % resin. The lower resin content increased the difficulty of creating a void-free prepreg structure because more fiber compaction must be obtained than for conventional 35 wt % resin content. Because of the low resin content, large void areas were observed at the center of the prepreg under all processing conditions, but were smallest as the PFN exceeded 1. The voids were concentrated at the center of each fiber tow.

The objective of the second-pass impregnation is to apply a thin layer of particle-modified base resin on the top and bottom prepreg surfaces. The optimum second-pass processing condition represented a trade-off between prepreg porosity and prepreg tack. At a second-pass PFN of 0.001, only a small fraction of the applied resin impregnated the fiber bed, which created a prepreg with a porous structure but a high level of tack. As the second-pass PFN exceeded 1, the prepreg was fully impregnated but had a low level of tack. A decrease in the tack of the prepreg surface was because only thermoplastic particles were present on the prepreg surface and all the thermosetting resin had impregnated the fiber bed.

After the autoclave processing, the final laminate structure contained the multilayered structure shown in Figure 7. All laminates were void-free and appeared to be independent of prepreg morphology. The thickness of the interlayer resin varied slightly across the laminate width but averaged 20–30 μm. To obtain a constant particle-bed thickness across the laminate width, an even monolayer of particles must be achieved on the prepreg surfaces. Therefore, the important prepreg characteristic of laminate morphology is the distribution of particles on the prepreg surfaces.

To evaluate the structure–property relationships, the interlaminar fracture toughnesses of multilayer composite systems were measured in modes I and II at concentrations of 0-, 10-, 20-, and 30-phr nylon 6 particles in the second-pass impregnation resin. A sample with no nylon particles was processed to represent a conventional composite structure with no interleaf. Results of the mode-I fracture-toughness testing are shown in Figure 8. The results indicate no improvement in the mode-I interlaminar fracture toughness by creating a multilayer structure. All mode-I values were approximately 300 J/m², including the unmodified sample. This observation can be explained by examining the fracture surface for the location of crack growth. In all samples, the crack propagated at the interface between the interleaving resin and the fiber bed or within the surface fibers of the fiber bed. Therefore, increasing the modifier concentration would not increase the fracture toughness because the crack propagation remained at the interface.

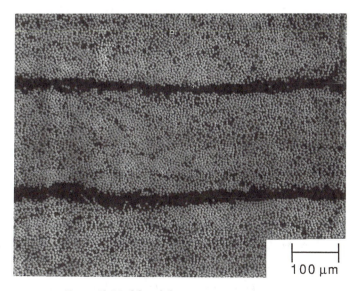

*Figure 7. Model multilayer composite structure.*

Creation of the multilayer structure had a definite improvement in the mode-II fracture toughness as shown in Figure 9. The unmodified mode-II fracture toughness had a value of 530 J/m² and a maximum mode-II fracture toughness occurred in the 20-phr sample at 1135 J/m². Above a 20-phr particle concentration, a decrease in the mode-II fracture toughness was

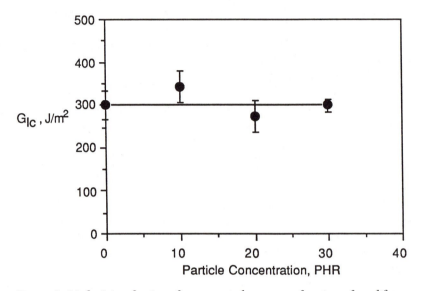

*Figure 8. Mode I interlaminar fracture toughness as a function of modifier-particle concentration in second-pass impregnation resin.*

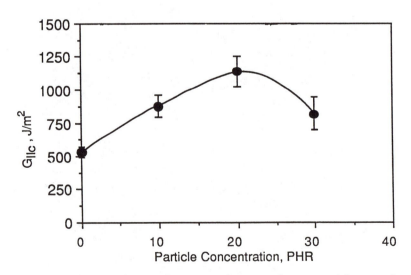

*Figure 9. Mode II interlaminar fracture toughness as a function modifier-particle concentration in second-pass impregnation resin.*

observed. This decrease in fracture toughness can also be explained by examining the fracture surfaces. In the 10- and 20-phr samples the majority of crack propagation occurred in the interlaminar resin-rich region, whereas in the 30-phr sample intralaminar crack propagation occurred. Consequently, the decrease in fracture toughness of the 30-phr sample is because the crack leaves the interlaminar region and jumps into the fiber bed of intralaminar region.

## Summary

A model multilayer prepreg structure has been formulated and analyzed in this work. The importance of resin distribution and fiber packing was identified and utilized to create a realistic model system. The prepreg resin distribution was found to influence porosity as well as final prepreg tack. The model prepreg was successfully produced in a double-pass impregnation, and the operating conditions during each impregnation step were used to control the resin distribution and produce a final prepreg with controlled levels of porosity and tack. Obtaining an even monolayer of modifier particles on the prepreg surfaces was found to be of importance. The particle distribution controlled the uniformity of the interleaf thickness.

Creation of the multilayer composite structure exhibited no improvement in the mode-I interlaminar fracture toughness, but showed definite improvements in the mode-II interlaminar fracture toughness. A maximum in the mode-II fracture toughness occurred at a nylon particle concentration of 20 phr in the second-pass impregnation resin. In larger nylon concentrations, the crack propagation occurred not in the interlaminar region, but within the fiber bed, thus reducing the measured fracture toughness.

Overall, this work demonstrated that a model multilayer toughened thermoset-based system can be successfully developed and used for detailed study of manufacturing as well as performance-related issues in advanced structural composites.

## Acknowledgment

Financial assistance for this work was provided by Boeing Commercial Airplane Group through project support to the Polymeric Composites Laboratory at the University of Washington. The assistance of Dr. Keith Riew of BFGoodrich in identifying particles of different compositions and makeup for this project, as well as the help of R. S. Schaffnit of Boeing Commercial Airplanes for establishing performance characteristics for the model systems are gratefully acknowledged. The carbon fibers provided by Toray Industries and the mechanical testing capabilities of the Instron Corporation through their participation in the Polymeric Composites consortium are also gratefully acknowledged.

# References

1. Hunston, D. L.; Moulton, R. J.; Johnston, N. J.; Bascom, W. D. *Toughened Composites*; ASTM Standard Technical Publication 937; American Society for Testing and Materials: Philadelphia, PA, 1987; pp 74–94.
2. Krieger, R. B. *29th International SAMPE Symposium*; Society for the Advancement of Material and Process Engineering: Covina, CA, 1984; p 1570.
3. Masters, J. E.; Courter, J. L.; Evans, R. E *31st International SAMPE Symposium*; Society for the Advancement of Material and Process Engineering: Covina, CA, 1986; p 844.
4. Hirschbuehler, K. R. *SAMPE Quaterly*, **1985**, *17*, 46–49.
5. Masters, J. E. *Sixth International Conference on Composite Materials*; Elsevier Applied Science Publishers, Ltd.: London, 1987; vol. 3, p 146.
6. Masters, J. E. *34th International SAMPE Symposium*; Society for the Advancement of Material and Process Engineering: Covina, CA, 1989, p 1792.
7. Hirschbuehler, K. R.; Stern, B. A. U.S. Patent 4,539,253, 1985.
8. Evans, R. E; Hirschbuehler, K. R. U.S. Patent 4,604,319, 1986.
9. Jabloner, H.; Swetlin, B. J.; Chu, S. G. U.S. Patent 4,656,207, 1987.
10. Chu, S. G.; Jabloner, H.; Swetlin, B. J. U.S. Patent 4,656,208, 1987.
11. Odagiri, N.; Suzue, S.; Kishi, H.; Nakae, T.; Matasuzaki, A. European Patent 0,274,899, 1988.
12. Odagiri, N.; Muraki, T.; Tobukuro, K. *33rd International SAMPE Symposium*; Society for the Advancement of Material and Process Engineering: Covina, CA, 1988; p 272.
13. Hecht D. H.; Gardner, H.; Qureshi, S.; Manders, P. European Patent 0,351,026, 1989.
14. Hecht, D. H.; Gardner, H.; Qureshi, S.; Manders, P. European Patent 0,351,025, 1989.
15. Ahn, K. J.; Seferis, J. C. *34th International SAMPE Symposium*; Society for the Advancement of Material and Process Engineering: Covina, CA, 1989; p 63.
16. Ahn, K. J.; Seferis, J. C.; Price, J. O.; Berg, A. J. *35th International SAMPE Symposium*; Society for the Advancement of Material and Process Engineering: Covina, CA, 1990; p 2260.
17. Hoisington, M. A.; Seferis, J. C.; Schaffnit, R. S. *Proc. ACS Div. PMSE* **1990**, *63*, 797.
18. Hoisington, M. A.; Seferis, J. C.; Schaffnit, R. S. *22nd International SAMPE Technical Conference*; Society for the Advancement of Material and Process Engineering: Covina, CA, 1990.
19. Lee, W. L.; Seferis, J. C.; Bonner, D. C. *SAMPE Quarterly*, **1986**, *17*, 58.
20. Carlsson, L. A.; Pipes, R. B. *Experimental Characterization of Advanced Composite Materials*; Prentice-Hall: Englewood Cliffs, NJ, 1987; pp 160–170.

RECEIVED for review March 6, 1991. ACCEPTED revised manuscript June 16, 1992.

# Multiphase Matrix Polymers for Carbon–Fiber Composites

Willard D. Bascom[1,‡], S-Y. Gweon[1], and D. Grande[2]

[1]Materials Science and Engineering Department, University of Utah, Salt Lake City, UT 84112
[2]Boeing Materials Technology, Boeing Commercial Airplanes, Seattle, WA 98124

*The effect of different matrix polymers on the impact resistance of carbon–fiber-reinforced polymers is reviewed. These polymers include unmodified, tetrafunctional epoxies, elastomer-modified epoxies, epoxies plasticized with low levels of thermoplastics, thermoplastics, and thermoplastic-modified thermosets (TMT). The results of some recent work on the impact resistance of TMT matrix composites is discussed. The TMTs were of two types: thermoplastic particles interlayered between plies and co-continuous interpenetrating network polymers. Suppression of interlaminar longitudinal cracking was found to correlate with improved resistance to impact damage.*

$\mathbf{S}$TRUCTURAL APPLICATIONS OF CARBON–FIBER-REINFORCED POLYMERS (CFRP) in aircraft, space vehicles, and automobiles can be seriously limited by low "damage resistance." This term generally refers to impact damage inflicted by low-velocity impacts ($\sim 100$ m/s) such as runway debris striking an aircraft during takeoff or landing or a careless mechanic dropping a tool on a composite wing. This type of impact damage is especially serious because there is extensive internal damage but very little surface damage, thus making inspection difficult, if not impossible. Significant reduction of structural performance—"low damage tolerance"—can result. Work at the NASA Langley Research Center (1) during the late 1970s and early 1980s showed that low-velocity impact damage to CFRP could cause a loss of up to

‡ Deceased.

0065–2393/93/0233–0519$06.00/0

50% in compression strength before any surface damage was detectable (Figure 1).

Low damage tolerance compromises the potential use of CFRP as weight-reducing metal substitutes on airframe structures. Consequently, considerable research has been and continues to be devoted to understanding the failure mechanisms associated with low-velocity impact and finding means of improving impact resistance. Early investigations demonstrated that impact damage primarily consisted of delamination between the carbon–fiber plies and transverse cracking through the plies. Furthermore, it was determined that this damage occurred because of the low fracture energy of the highly cross-linked epoxy matrix polymers. Figure 2 shows a section cut through the impact-damaged region of a low-fracture-energy matrix resin laminate (2). The extensive longitudinal and transverse cracking is clearly evident.

Replacement of these epoxy polymers with higher fracture energy "tough" polymers significantly reduced the extent of impact damage with a corresponding improvement in postimpact compression strength. However, the tough matrix polymers available at that time were unsatisfactory for structural CFRP primarily because of their low compressive strength, especially under hot and wet conditions.

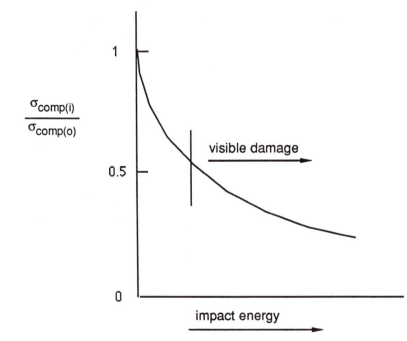

*Figure 1. Schematic of the effect of low-level impact on the compressive strength of carbon-fiber composites with low-toughness matrix resins.*

Approximate impact centre

1.0 mm

*Figure 2. Photomicrograph of polished section through the impact-damage region of a CFRP having a brittle matrix. (Reproduced with permission from reference 2. Copyright 1986 Chapman and Hall.)*

## Modified Epoxies and Thermoplastics

One route to increasing the fracture energy of epoxy and other thermosetting polymers is to incorporate a dispersed elastomeric phase. This technology has been successfully used to formulate high-peel-strength adhesives (3). Liquid elastomers, notably the carboxyl-terminated polybutadiene–acrylonitrile (CTBN), are added to the liquid epoxy. During cure an elastomer–epoxy copolymer phase separates to form micrometer-size inclusions. Numerous investigations of CTBN-modified epoxies (3–6) revealed that the dispersed phase increased the epoxy fracture energy from $\sim 100 \text{ J/m}^2$ to as much as $3500 \text{ J/m}^2$ (6). These investigations found that the toughening mechanisms

**Table I. Effect of Elastomer Modifiers on the Neat-Epoxy Fracture Energy and Composite Interlaminar Fracture Energy**

| Material | Fracture Energy $(kJ/m^2)$ |
|---|---|
| Unmodified epoxy neat resin (Hexcel 205) | 0.27 |
| Elastomer modified neat resin (Hexcel F-185) | 5.1 |
| 205/glass cloth | 1.0 |
| F-185/glass cloth | 4.4 |
| 205/graphite cloth | 0.60 |
| F-185/graphite cloth | 4.6 |

involved dilation of the inclusions and shear yielding of the epoxy matrix. The effectiveness of a CTBN elastomer on the fracture energy of an unreinforced difunctional epoxy and on the interlaminar fracture energy[3] of glass- and carbon–fiber-reinforced composites is shown in Table I. The neat-resin toughness was increased by nearly a factor of 20 ×. The interlaminar fracture energies of the glass- and carbon–fiber-reinforced composites were increased by ~4 × and ~8 ×, respectively. The reduced elastomer effectiveness in the composites compared to neat resin was attributed to the fibers constraining the development of the crack-tip damage zone (7).

Yee (8) demonstrated that for highly cross-linked epoxies the toughening action of the elastomers is considerably reduced because the shear-yield strength of the epoxies is too high to allow the dilation and shear-yielding deformations required for high toughness. In structural adhesives, the epoxies are difunctional with relatively low shear-yield strengths. However, for CFRP the matrix is typically a highly cross-linked tetrafunctional epoxy in which elastomer inclusions have very little effect on toughness. Consequently, elastomer-modified epoxies are not extensively applied in high-performance structural composites where a highly cross-linked and high-modulus matrix is critical.

There have been efforts to modify the base epoxies by the addition of thermoplastic modifiers such as polyether sulfone, polycarbonate, and polyphenylene oxide. Because these thermoplastic additives have high glass-transition temperatures and high moduli, the high modulus of the tetrafunctional epoxy is not sacrificed and the toughness and, thus, the delamination resistance are enhanced. The effectiveness of this approach was marginal because of the low solubility (1–5%) of the thermoplastics in epoxy formulations. The increase in interlaminar fracture energy in CFRP is at best a factor of 2, which does not translate into an acceptable increase in impact resistance. In

[3] In the past mode-I interlaminar fracture energy was used widely as a measure of impact-damage resistance. There is increasing evidence that mode-II interlaminar fracture energy is a better impact-resistance index. Nonetheless, mode-I interlaminar fracture energies and mode-I fracture energies of neat resins have proved useful in the development of damage tolerant composites.

some instances it was possible to induce phase separation of the thermoplastic (9), but the resulting high viscosity and the slow thermoplastic dissolution and dispersion created serious practical processing problems.

Another approach to improved impact-damage resistance is to replace thermosetting matrix materials with thermoplastics that are inherently tough. Previous experience with thermoplastics such as polysulfone were discouraging because of their tendency to creep under low stresses and their poor resistance to attack by organic solvents. However, currently available engineering thermoplastics, notably, polyetheretherketone (PEEK), the polyamide–imides (Torlon), and polyphenylene sulfide (PPS), have low creep characteristics and are solvent resistant. Composites of these thermoplastic matrix polymers reinforced with carbon–fiber offer considerably higher damage resistance than the thermoplastic-modified epoxies (Table II). However, the high-temperature and high-pressure processing conditions required for thermoplastic matrix composites are totally different from the processing conditions for thermosetting polymers, and this difference would necessitate a major investment in new equipment. In addition, some of the thermoplastic polymers develop a degree of crystallinity during processing that affects their mechanical properties and is very dependent on heating and cooling rates. Consequently, for large structures (e.g., an airplane wing) the properties in a thin section may not be the same as in a thicker section because of different thermal histories. Capital investment and difficult processing conditions have deterred the use of thermoplastics as matrices in CFRP, at least for large structures. Thermoplastics well may be the matrix of choice for the production of small components at high production rates using conventional and relatively inexpensive compression-molding techniques.

## Multiphase Thermosetting Polymers

A recently developed class of matrix polymers offers high toughness and can be processed using conventional methods already in place for thermosetting

**Table II. Polymer and Interlaminar Fracture Energies of a High-Modulus Epoxy and Typical Thermoplastics**

| Polymer | Fracture Energy (Unreinforced Polymer) $(kJ/m^2)$ | Interlaminar Fracture Energy $(kJ/m^2)$ |
|---|---|---|
| Tetrafunctional epoxy[a] | 0.159 | 0.262 |
| Polysulfone[b] | 2.50 | 1.200 |
| Polyetherimide[b] | 3.30 | 0.935 |
| Polyetheretherketone[c] | 4.80 | 2.10–2.70 |

[a] W. D. Bascom, unpublished data.
[b] Reference 7.
[c] Reference 17.

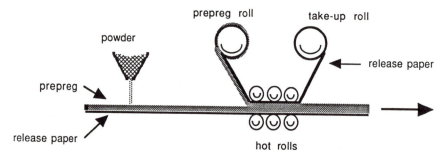

*Figure 3. Schematic of interleaving thermoplastic powder between preimpregnated plies. (Adapted from Toray Industries Inc., European Patent Application 87311364.1, December 23, 1987.)*

polymers. Similar to the elastomer-modified epoxy resins, enhanced toughness in the new polymers is achieved by incorporating a dispersed phase within the epoxy matrix. Also, like the CTBN-modified epoxies, the dispersed phase does not significantly affect the high modulus of the epoxy matrix or its heat distortion temperature, which are critical to the performance of structural CFRP.

Thermoplastic-modified thermosetting (TMT) polymers involve the incorporation of thermoplastic domains within the thermosetting matrix. This morphology can be accomplished by distributing thermoplastic polymer particles between layers of epoxy-impregnated carbon–fiber. Particle diameters range in size from a few micrometers to submicrometers. One method of "interleaving" polymer particles is shown in Figure 3.

A second way to develop a TMT polymer is by chemically dispersing the thermoplastic into the epoxy to form a co-continuous interpenetrating network morphology. The detailed chemistry of commercial TMTs is proprietary information. Further details regarding the TMT resins can be found in reviews by Recker et al. (*10*) and Kubel (*11*).

## Damage Resistance of TMT Matrix Composites

The impact resistance of three commercial TMT-matrix CFRP materials was studied using a repetitive impact technique in which an 8 × 5-cm laminate panel is end-supported as shown in Figure 4 and repetitively impacted with increasing impact energy. The panel stiffness was measured at each impact event. After various levels of cumulative impact energy, the panels were examined using ultrasonic C-scan to locate the damaged region. The laminate was cut around the damaged zone and then potted in a clear epoxy. Sections, 1.9 mm thick, were sequentially cut and polished. Each section was examined using light microscopy to determine the extent and type of damage. Maps were then generated to show the extent of damage through the laminate. Details of this technique are given in reference 12.

The change in stiffness for a 32-ply 0°–90° fiber orientation panel of IM6/3501-6 is shown in Figure 5. The 3501-6 matrix is a brittle, amine-cured tetrafunctional epoxy. From C-scans and microscopy of the sections through the damaged zone it was determined that the first sharp decline in stiffness at

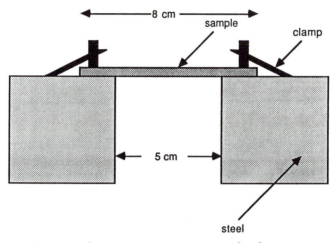

*Figure 4. Fixture used to support composite samples for repetitive impact testing.*

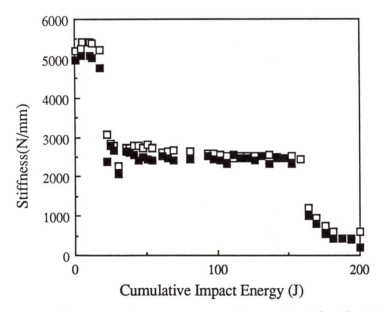

*Figure 5. Stiffness vs. cumulative impact energy for IM6/3501-6, $(0–90)_{8s}$. The open and filled points correspond to two different methods of calculating laminate stiffness (see reference 12).*

~ 25-J cumulative impact energy corresponds to the damage extending through the laminate thickness. The second drop in stiffness at ~ 150 J corresponds to the damage extending to the edges of the laminate. A damage map taken after 110-J cumulative impact energy is shown in Figure 6.

Similar impact tests were done on 32-ply 0°–90° fiber orientation coupons of AS4/APC-2 (APC-2 is a product of ICI Ltd, United Kingdom, that is based on PEEK). The change in stiffness with increasing cumulative impact energy is shown in Figure 7. Unlike the IM6/3501-6, there is a gradual loss in stiffness with continued impact loading and the retention of stiffness of the PEEK matrix laminate was higher than for the 3501-6 matrix laminate. A damage map for the AS4/APC-2 material after a cumulative impact energy of 115 J is presented in Figure 8. Note that the extent of damage is much less than for the 3501-6 material (Figure 6) after a similar impact loading. These differences are consistent with the higher damage resistance of PEEK composites compared to 3501-6 composites.

Repetitive impact tests were conducted on panels of T800/3900 (Hexcel Corp., Dublin CA), IM7/977-1, and IM7/977-2 (ICI Fiberite, Tempe, AZ).

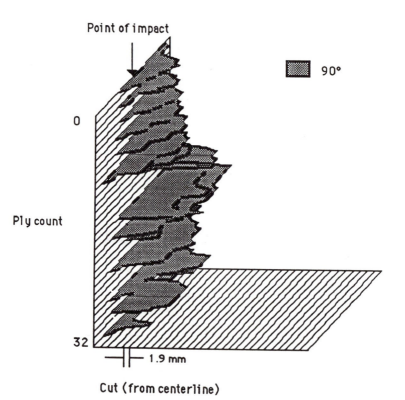

*Figure 6. Damage map of IM6/3501-6, (0–90)$_{8s}$, after a cumulative impact energy of 110 J.*

The 3900 matrix is a particulate interleaf TMT and the 977-1 and 977-2 matrices are co-continuous interpenetrating-network TMTs. In all cases the panels were 32-ply 0°–90° fiber orientation and were cured in an autoclave using the manufacturers' suggested cure cycle.

The changes in stiffness with increasing cumulative impact energy for the laminate materials are presented in Figures 9–11. Damage maps are presented in Figures 12–14.

The T800/3900 exhibited only a slight loss in stiffness out to a cumulative impact energy of nearly 250 J. The loss in stiffness by the IM7/977-1 was similar to that of the AS4/APC-2 (Figure 7). The behavior of the IM7/977-2 was similar to that of the IM7/3501-6 except that the second drop-off in stiffness occurred after 200-J cumulative impact energy for the 977-2 matrix compared to 150 J for the 3501-6. Based on earlier work (*12*), this difference suggests the 977-2 has a higher resistance to longitudinal delamination than the 3501-6. At this point in time, we do not have any explanation for the initial increase in panel stiffness, which is especially noticeable for the IM7/977-1 (Figure 10).

The damage map for the T800/3900 (Figure 12) was taken after a cumulative impact energy of 225 J. The damage has extended through the laminate thickness. However, this extent of damage is usually realized at much lower impact energies; at about 50 J for IM7/3501-6 and 150 J for

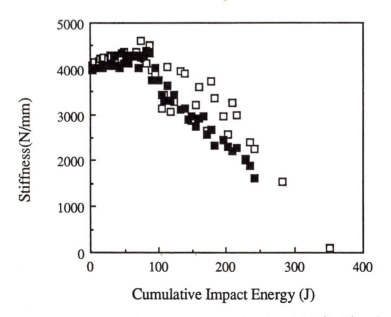

*Figure 7. Stiffness vs. cumulative impact energy for AS4/APC-2, $(0–90)_{8s}$. The open and filled points correspond to two different methods of calculating laminate stiffness (see reference 12).*

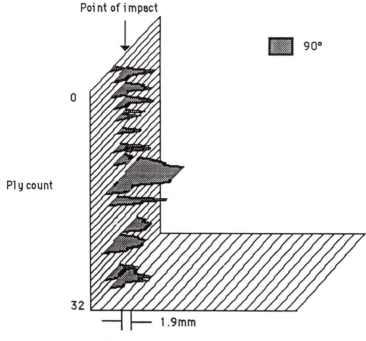

*Figure 8. Damage map of AS4/APC-2, (0–90)₈ₛ, after a cumulative impact energy of 115 J.*

AS4/APC-2. Damage maps for 977-1 and 977-2 are presented in Figures 13 and 14. After 150-J cumulative impact energy, the 977-1 showed only limited damage, whereas the damage of the 977-2 extended through the laminate thickness. Comparison of the damage maps for the 977-1 with that of 3501-6 (Figure 6) indicates much less lateral damage for the 977-1 material, which confirms the earlier suggestion that the 977-1 laminate is more resistant to longitudinal delamination than the 3501-6 laminate.

Photographs of polished sections through the damaged region of a T800/3900 coupon are shown in Figure 15. Because of the differential polishing rate of the continuous matrix phase compared to the interleaf particles, the particles can be seen in the interplay regions. Note that transverse cracks are blunted by the particles, thereby preventing the development of interlaminar, longitudinal cracking. This blunting mechanism is shown schematically in Figure 16.

Efforts to observe the dispersed phase in the 977-1 and 977-2 laminate have not been successful either by simple polishing or by acid–permanganate etching.

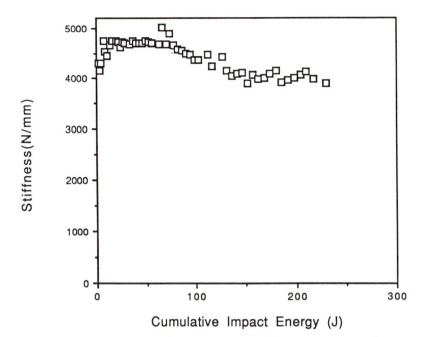

Figure 9. *Stiffness vs. cumulative impact energy for T800/3900, $(0–90)_{8s}$.*

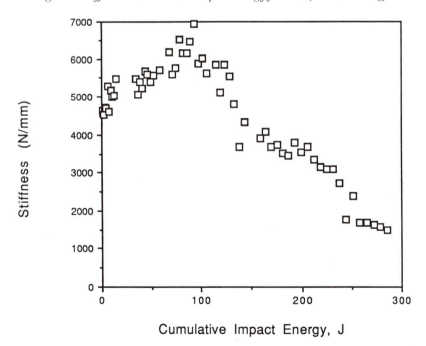

Figure 10. *Stiffness vs. cumulative impact energy for IM7/977-1, $(0–90)_{8s}$.*

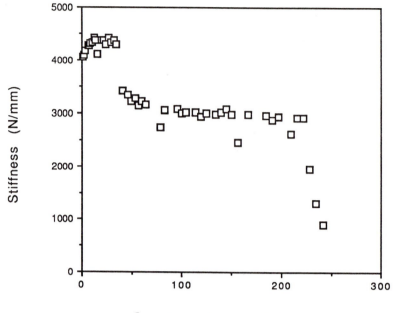

Figure 11. Stiffness vs. cumulative impact energy for IM7/977-2, $(0-90)_{8s}$.

Based on these results, the damage resistance of the three TMT composites in the repetitive impact test rank as follows: T800/3900 > IM7/977-1 > IM7/977-2. This ranking[4] appears to be related to the relative amount of transverse vs. longitudinal cracking in the damage zone. Polished sections of the damage regions were examined to determine the number of transverse cracks per millimeter of longitudinal cracking and the results are shown in Table III. It would appear that suppression of longitudinal cracking correlates with the impact-damage resistance ranking.

## Test Methods for Damage Tolerance

Both government (13) and industry (14) have standard tests for the impact tolerance of CFRP coupons. These tests measure the compression (strength) after impact (CAI) and the procedures specify the laminate configuration (ply lay-up and size), the test fixtures, the impacting procedures, and the postimpact loading procedures. Stated simply, the coupon is impacted and then tested for residual compression strength.

---

[4] This ranking relates only to impact-damage resistance. The relative merit of CFRP materials depends on a number of other factors that could outweigh differences in impact resistance.

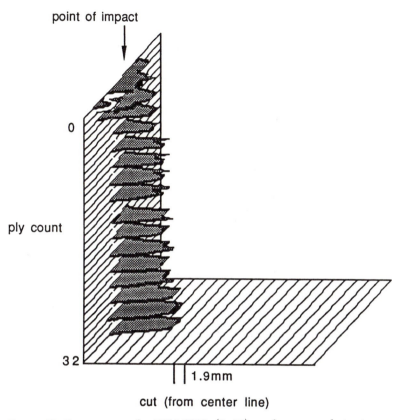

*Figure 12. Damage map for T800/3900, $(0–90)_{8s}$, after a cumulative impact energy of 225 J.*

The impact-damage resistance of the material influences the extent of damage that results from the impact event. Impact-damage tolerance of the coupon is determined by the material's compressive strength, the size and nature of the damage, and the structural response of the damaged panel during the compressive loading. Although the compression test portion of the typical CAI test measures the damage tolerance of the test coupon, the performance of materials in these tests is largely determined by their damage resistance during the impact portion of the test. Thus, the results obtained from CAI coupon tests reflect a combined result of impact-damage resistance and impact-damage tolerance. It is often helpful in practice to consider impact-damage resistance and impact-damage tolerance as two distinct topics.

The CAI tests are costly in terms of time and materials. Consequently, simpler tests are frequently used to evaluate a matrix resin for improved CFRP impact resistance. The most common tests are for the fracture energy

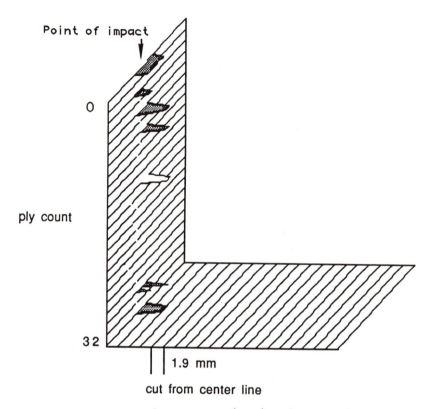

*Figure 13. Damage map for IM7/977-1, $(0–90)_{8s}$, after a cumulative impact energy of 150 J.*

of the neat resin or for the mode-I interlaminar fracture energy. Hunston (7) showed that there is no linear correlation between resin and interlaminar fracture energies, and Kam and Walker (15) found that mode-I interlaminar fracture energies do not correlate with CAI test data. There is, however, increasing evidence that CAI data do correlate with mode-II[5] interlaminar fracture energy. For example, the collection of data for a variety of matrix resins that is presented in Figure 17 indicates a strong correlation between CAI and mode-II delamination fracture energy ($G_{IIc}$). To determine if this correlation between CAI and $G_{IIc}$ is universal and, if so, its physical meaning requires additional study.

The repetitive impact with increasing impact energy test reported here is not a replacement for CAI testing. Nonetheless, it conserves materials,

---

[5] In mode-I testing, the crack propagates in an opening or cleavage mode. In mode-II testing, the crack propagates by in-plane shear.

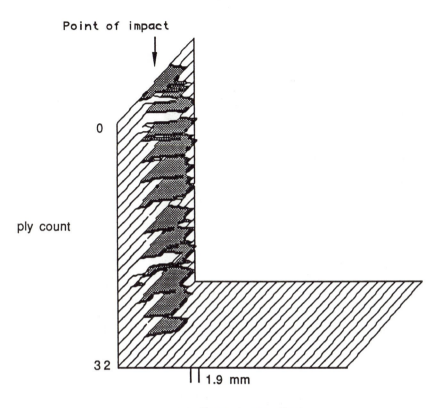

Figure 14. Damage map for IM7/977-2, $(0–90)_{8s}$, after a cumulative impact energy of 100 J.

measures the onset of specific damage mechanisms, and allows relative ranking of impact-damage resistance.

## Summary

The relative merit of matrix polymers for damage tolerance as measured via CAI coupon tests is summarized in Table IV. Despite the low mode-I interlaminar fracture energies, the TMT matrix resins have CAI values in the same range as the thermoplastics. The TMT-based composite currently can be fabricated using existing conventional thermoset processing methods, which grants a temporary cost advantage over the thermoplastics. In the future, this cost differential is likely to be reduced as novel cost-effective methods for processing thermoplastics are developed (16). All other factors

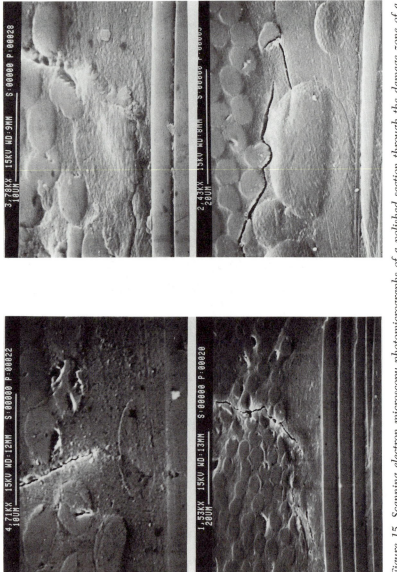

*Figure 15. Scanning electron microscopy photomicrographs of a polished section through the damage zone of a T800/3900 laminate. Note the blunting of transverse cracks by the interply particles.*

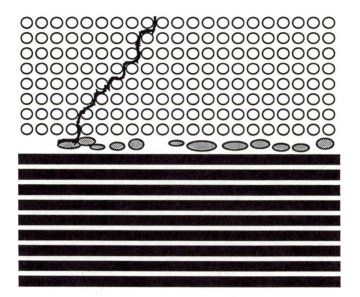

*Figure 16. Schematic of transverse crack blunting by interply particles.*

### Table III. Ratio of Transverse Cracks to Longitudinal Cracks for Three TMT-Matrix Composites

| Material | Transverse Cracks per Longitudinal Cracks (mm) |
|---|---|
| T800/3900 | 5.1 |
| IM7/977-1 | 1.1 |
| IM7/977-2 | 0.55 |

### Table IV. Typical Fracture Energy and Compression after Impact Values

| Polymer Type | Fracture Energy $G_{Ic}$, (kJ/m²) | | CAI (MPa) |
| | Bulk | Interlaminar | |
|---|---|---|---|
| Unmodified epoxy | 0.10–0.20 | 0.15–0.25 | 140–170 |
| Modified epoxy | 0.2–0.5 | 0.20–0.50 | 170–200 |
| CTBN–epoxy | 2.0–6.0 | 0.50–2.5 | |
| Thermoplastics | 1.0–3.0 | 0.75–1.25 | 200–400 |
| TMT | | 0.2–0.5 | 200–400 |

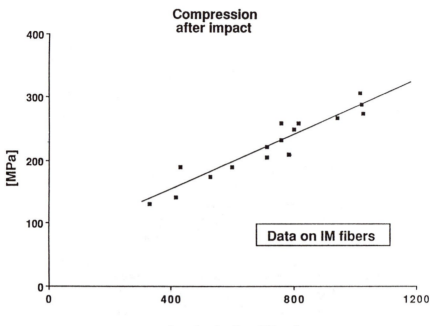

*Figure 17. Correlation between compression after impact (CAI) and mode-II interlaminar fracture energy (G$_{IIc}$). (Supplied by H. G. Recker, BASF Structural Materials, Inc.)*

being equal, the cost of start-up equipment or equipment replacement for thermoset processing may give the edge to thermoplastics.

## Acknowledgments

Partial funding for this work was provided by National Aeronautics and Space Administration contract NAS–18889 under the direction of J. G. Davis and W. T. Freeman (NASA Langley Research Center) and from NASA grant NAG1–705 directed by J. Crews (NASA Langley Research Center). We thank Dr. H. G. Recker of BASF Structural Materials, Inc. for providing an original copy of Figure 17.

## References

1. Rhodes, M. D.; Williams, J. G.; Starnes, J. H. *Effect of Low-Velocity Impact Damage on the Compressive Strength of Graphite–Epoxy Hat-Stiffened Panels;* National Aeronautics and Space Administration: Washington, DC, 1977; Technical Note D–8411.

2. Bascom, W. D.; Boll, D. J.; Weidner, J. C.; Murri, W. J. *J. Mater. Sci.* **1986**, *21*, 2667.
3. Bolger, J. C. In *Treatise on Adhesion and Adhesives*; Patrick, R. L., Ed.; Dekker: New York, 1973; Vol. 2, p 1.
4. Sultan, J. N.; Laible, R. C.; McGarry, F. J. *Appl. Polym. Symp.* **1971**, *16*, 127.
5. Bascom, W. D.; Hunston, D. L. In *Treatise on Adhesion and Adhesives*; Patrick, R. L., Ed.; Dekker: New York, 1989; Vol. 6; p 123.
6. Bascom, W. D.; Cottington, R. L.; Jones, R. L.; Peyser, P. *J. Appl. Polym. Sci.* **1975**, *19*, 2545.
7. Hunston, D. L. *Composites Tech. Rev.* **1984**, *4*, 176.
8. Yee, A. F.; Pearson, R. A. *Toughening Mechanism in Elastomer-Modified Epoxy Resins*; National Aeronautics and Space Administration: Washington, DC, 1983 and 1984; Contractor Reports 3718 and 3852.
9. Bucknall, C. B. In *Advanced Composites*; Partridge, I. K., Ed.; Elsevier Applied Science Publishers: London, 1989; p 145.
10. Recker, H. G.; Alstadt, V.; Eberle, W.; Folda, T.; Gerth, D.; Heckmann, W.; Ittemann, P.; Tesch, H.; Wever, T. *SAMPE J.* **1990**, *26*, 73.
11. Kubel, E. J., Jr., *Adv. Mater. Process.* **1989**, *136*(2), 23.
12. Bascom, W. D. *Fractography of Composite Delamination*; National Aeronautics and Space Administration: Washington, DC, July 1990; Contractor Report 181965.
13. *Standard Tests for Toughened Resin Composite*; National Aeronautics and Space Administration: Washington, DC. 1983; NASA Reference Publication 1092; rev. ed.
14. *Specification BSS 7260 Type I*; Boeing Commercial Airplane Co.: Seattle, WA.
15. Kam, C. Y.; Walker, J. V. *Toughened Composites*; Johnston, N. J., Ed.; American Society for Testing and Materials: Philadelphia, PA, 1985; ASTM Standard Technical Publication 937, p 9.
16. Silverman, E. M.; Forbes, W. C. *SAMPE J.* **1990**, *26*, 9.
17. Leach, D. C. In *Advanced Composites*; Partridge, I. K., Ed.; Elsevier Applied Science Publishers: London, 1989; p 43.

RECEIVED for review March 6, 1991 ACCEPTED revised manuscript August 2, 1991.

# Thermal Characterization of the Cure Kinetics of Advanced Matrices for High-Performance Composites

**J. M. Kenny, A. Trivisano, and L. Nicolais**

**Department of Materials and Production Engineering, University of Naples, P. Tecchio, 80125 Naples, Italy**

*The reaction kinetics of commercial high-performance matrices has been characterized by differential scanning calorimetry. Standard tetraglycidyldiaminodiphenylmethane–diaminodiphenylsulfone epoxy, toughened epoxy, and bismaleimide matrix pre-impregnated plies have been studied. The effect of diffusion control phenomena on the reac -tion kinetics, associated to the evolution of the glass-transition temperature as a function of the degree of polymerization, has been considered in the formulation of a modified nth order kinetic model. Isothermal and dynamic tests have been used to calculate and verify the model parameters. The model is able to describe incomplete reactions in isothermal tests and heating rate dependence of dynamic test results. When the model is integrated into a master model, it can be used for the description of the processing of high-performance composites.*

$\mathbf{N}$EW CHEMICAL SYSTEMS HAVE BEEN INCORPORATED in the last 5 years into the family of high-performance matrices for structural composites. Classical methane–diaminodiphenylsulfone (TGDDM–DDS) epoxy systems that dominated the field for many years have been modified with rubber inclusions or with a thermoplastic second phase to improve impact behavior. Also bismaleimide resins are now commercially available for higher service temperatures. The processing behavior of these thermosetting matrices is strongly affected by the kinetics of their polymerization reaction.

0065–2393/93/0233–0539$06.00/0

High-performance thermoset-based composites are generally produced by the autoclave lamination process. During the process, the consolidation of the preimpregnated plies (prepregs) is accompanied by polymerization reactions (cure process) and rheological changes of the matrix that strongly influence the final properties and the quality of the laminate. The cure process is coupled with marked heat generation as a result of the exothermic nature of the thermosetting reactions. The relative rates of heat generation and transfer determine the values of the temperature and, therefore, the values of the advancement of the reaction and the viscosity through the thickness of the composite. The processing conditions in the autoclave should be designed as a function of the chemorheological properties of the matrix and should consider the heat transfer characteristics of the composite–tool–environment system.

Uncontrolled polymerization causes undesired and excessive thermal and rheological variations that induce microscopic and macroscopic defects in the composite part. Therefore, processing of polymeric composites based on thermoset matrices requires optimization of the cure cycle parameters as well as adequate formulation of the reacting system. The first step for the development of an intelligent system leading to the design, optimization, and control of high-performance composite processing is given by the availability of adequate information about the reaction kinetics of the thermosetting matrix (1).

In previous papers (1, 2), a general model for the thermo–chemorheological behavior of classical epoxy and polyester matrices during the processing of thermoset-based composites has been proposed.

Kinetic and rheological models that correlate the thermal and the chemorheological behavior of different composite matrices to the molecular and chemical characteristics of the reactive systems were integrated into a heat-transfer model. The master model is able to describe the behavior of the main variables during the composite processing and can be used for the simulation of the process under different processing conditions. When suitable sensors are adopted, the master model can be used for optimization and control purposes.

Isothermal and dynamic experiments conducted by differential scanning calorimetry (DSC) have been widely used (1–5) for the indirect determination of cure advancement in a thermosetting system. Cure-advancement information can be processed to construct the kinetic submodel. A recently developed approach (5) made it possible to account for diffusion control effects in the formulation of the kinetic model by considering the effect of the reacting system glass-transition temperature evolution as a function of the degree of reaction. This chapter concentrates attention on the calorimetric characterization and on the development of the kinetic submodel for different high-performance thermoset matrices: standard epoxy (TGDDM–DDS), toughened epoxy (also based on TGDDM–DDS systems), and bismaleimide

(BMI) matrices. The developed kinetic models are verified by comparison with experimental results obtained in isothermal and dynamic tests.

## Kinetic Behavior

Although the reactions occuring during the processing of thermoset-based composites are very complex, empirical kinetic equations have been success-fully used to describe the general behavior of these systems. Considerable research activity has been reported in the field of epoxy resin and matrix cure. Complete reviews on the kinetic characterization of epoxy systems by differential scanning calorimetry (DSC) have been presented by Prime (3) and Barton (4). The fundamentals of the kinetic characterization of the cure reaction of thermosetting matrices are discussed in this section.

Isothermal and dynamic experiments conducted by DSC are used widely for indirect determination of cure advancement in a thermosetting system that assumes heat evolution during the polymerization reaction is propor-tional to the extent of reaction. DSC results are used also for the formulation and verification of theoretical and empirical kinetic models and for the calculation of the related parameters. To interpret experimental DSC data, the degree of reaction ($w$) is defined as (3)

$$w = H(t)/H_T \tag{1}$$

where $H(t)$ is the heat developed during the reaction between the starting point and a given time $t$ and $H_T$ is the total heat developed during the cure and is calculated by integrating the total area under the DSC curve. The DSC thermogram gives the instantaneous generation of heat by the reactive system. The reaction rate is given by the expression

$$\frac{dw}{dt} = \frac{1}{H_T} \frac{dH}{dt} \tag{2}$$

This information can be processed to construct a kinetic model that describes the degree of reaction as a function of time and temperature. When an overall kinetic process characterized by a generic degree of reaction ($w$) is assumed, the empirical kinetic equations for the cure process of thermoset-ting matrices are normally expressed in the form

$$dw/dt = Kf(w) \tag{3}$$

where $K$ is the temperature-dependent rate constant and $f(w)$ is a function of the extent of reaction to be determined by best fitting the experimental results.

The kinetic behavior of TGDDM–DDS systems has been reported in the literature ($1$, $5$–$7$). Stark et al. ($6$) analyzed the reaction kinetics of high-performance commercial TGDDM–DDS-based matrices by applying an $n$th-order kinetic model:

$$dw/dt = K(1 - w)^n \qquad (4)$$

No direct determination of the kinetic parameters was attempted because of the complexity of the reaction mechanisms. The influence of the reactive kinetics of TGDDM–DDS matrices, described by a single $n$th-order model (equation 4), on the chemorheological behavior of different commercial prepregs in the autoclave process was reported by Kenny et al. ($1$). Mijovic et al. ($7$) proposed a slightly more complex kinetic model to interpret isothermal DSC tests on TGDDM–DDS formulations:

$$dw/dt = (K_1 + K_2\, w^m)(1 - w)^n \qquad (5)$$

where $K_1$ and $K_2$ are the reaction constants and $m$ and $n$ are the reaction orders. The inclusion of a second kinetic constant allows consideration of the autocatalytic behavior observed in the polymerization reaction of TGDDM–DDS systems. Equation 5 also may be considered a general empirical model for the description of the kinetic behavior of reactive matrices. Equation 5 can be modified to account for the particular behavior of specific chemical systems. If the maximum reaction rate occurs at the starting point (time $= 0$), then $K_2 = 0$ and the kinetic model corresponds to an n-order equation. Diffusion-control effects can be included in the kinetic model. In isothermal processes at low temperatures, the polymerization reactions are not completed. When the increasing glass-transition temperature ($T_g$) of the reactive system approaches the isothermal cure temperature, the molecular mobility is strongly reduced and the reaction becomes diffusion-controlled and eventually stops ($8$). A method to develop a kinetic model that includes the dependence between the final level of polymerization and the test temperature was described ($5$) recently. The final expression of the $n$th-order model in this case is

$$dw/dt = K(w_m - w)^n \qquad (6)$$

where $w_m$ is the maximum degree of reaction reached by the reactive system at a given temperature. Equation 6 predicts the expected behavior: The reaction rate during an isothermal process will be zero when the degree of reaction becomes equal to $w_m$. The dependence of $w_m$ on $T$ can be obtained experimentally from isothermal DSC tests.

Equation 6 can be applied to interpret results of dynamic DSC tests performed at a constant heating rate because $w_m$ is continuously growing with temperature and the rate constant ($K$) is a function of temperature. The value of the total heat reaction ($H_T$) developed in the dynamic test is used as a reference value for the computation of $w$.

Despite considerable research activity reported in the field of kinetic characterization of thermosets, available information on the correlation between isothermal and dynamic DSC experimental results is sparse. Moreover, data on the influence of a second phase on the reactive behavior of toughened TGDDM–DDS systems and the kinetic characterization of commercial BMI matrices is not available currently. In this study, a new method to characterize the kinetic behavior of different high-performance matrices through isothermal DSC experiments is proposed. The kinetic model, which accounts for diffusion-control effects in the later part of the cure process, is verified through isothermal and dynamic experiments.

## Experimental

Three commercial prepregs are studied: unidirectional Fiberite HY-E/HMF1034K prepreg with standard TGDDM–DDS epoxy matrix, unidirectional Fiberite HY-E1377-2T prepreg with toughened TGDDM–DDS matrix, and Fiberite 986 bismaleimide (BMI) resin.

The calorimetric characterization was carried out in a differential scanning calorimeter (DSC; Mettler TA 3000) operating in the range of temperatures between $-50$ and 450 °C, in a nitrogen atmosphere and was equipped with a liquid nitrogen cooling system. The tests were performed on samples of 40–50 mg of resin or prepreg.

## Results and Discussion

**Dynamic Tests**. A complete calorimetric characterization was performed on the three prepregs to develop the kinetic polymerization reaction model and to calculate the model parameters. Figure 1 shows the dynamic thermograms obtained on the three prepreg materials using a heating rate of 10 °C/min. Despite the complex curing reactions associated to the processing of these matrices (9–11), a single peak signal was obtained in the three cases. For modeling purposes, a single empirical model describes the overall kinetic behavior and varies the model with different experimental results.

The total heat of reaction values developed during dynamic tests, performed at different heating rates and referred to the mass fraction of resin in the prepregs, are presented in Table I. The data variability can be attributed to the indeterminate resin content in the different samples. No correlation can be established between the heat of reaction and the heating rate. The

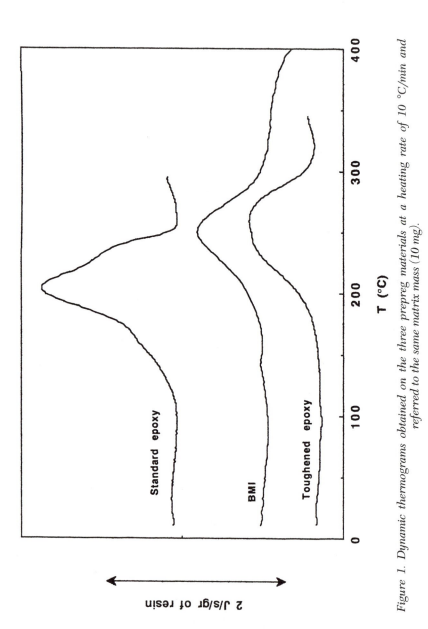

Figure 1. Dynamic thermograms obtained on the three prepreg materials at a heating rate of 10 °C/min and referred to the same matrix mass (10 mg).

**Table I. Total Heat of Reaction Associated to the Peak
of Dynamic Thermograms**

| Run No. | Standard Epoxy | | Toughened Epoxy | | BMI | |
|---|---|---|---|---|---|---|
| | HR | $H_T$ | HR | $H_T$ | HR | $H_T$ |
| 1 | 3 | 467.8 | 2 | 521.4 | 3 | 419.6 |
| 2 | 5 | 456.4 | 5 | 492.3 | 5 | 432.3 |
| 3 | 10 | 455.3 | 10 | 409.3 | 10 | 386.7 |
| 4 | 15 | 446.1 | 10 | 463.0 | 10 | 413.4 |

NOTE: Heat values were obtained at different heating rates and referred to the mass fraction of resin in the prepreg. HR = heating rate (°C/min); $H_T$ = heat of reaction (J/g).

average values of the data reported in Table I were used as a reference value of the total heat of reaction for modeling purposes. Figure 1 results suggest that the toughened epoxy system is characterized by a reactivity lower than the standard prepreg. The thermogram peak is shifted to higher temperatures and the heat of reaction is slightly lower. The thermogram of the BMI system reveals reactive characteristics similar to the toughened matrix, suggesting that the main network structure can be developed under the same processing conditions as the TGDDM–DDS systems.

**Isothermal Tests.** The dynamic test information is not sufficient to describe the complete behavior of the polymerization reaction. For example, autocatalytic and diffusion-controlled effects are only detected in isothermal experiments. To complete the calorimetric characterization, isothermal tests are performed at different temperatures on the three prepregs studied. Thermograms obtained at temperatures in the range of normal processing conditions are reported in Figures 2–4. The results are expressed in terms of reaction rate (1/s) as a function of time (min) and have been corrected, using a previously described procedure (5), to allow for the inaccuracy of the first portion of the thermogram data as a consequence of the stability of the test temperature. The shape of the corrected thermograms indicates that a single rate constant model (equation 4) can be used to fit the data.

Several isothermal tests at different temperatures were performed on the same prepregs and the same initial behavior was observed. The values of the total heat of reaction ($H_c$), calculated after correction of the experimental signal for the inaccuracy of the initial zone, are reported in Table II. Note that the calculated values of the total heat developed during isothermal tests are significantly lower than the heat developed during dynamic tests. During isothermal tests at low temperatures, the polymerization reactions are not completed and the system reaches a final extent-of-reaction value that is an increasing function of the test temperature. The incomplete reaction obtained during isothermal processes has been explained in terms of

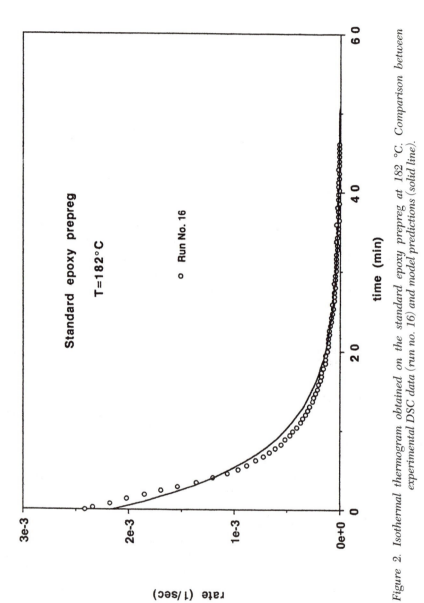

*Figure 2. Isothermal thermogram obtained on the standard epoxy prepreg at 182 °C. Comparison between experimental DSC data (run no. 16) and model predictions (solid line).*

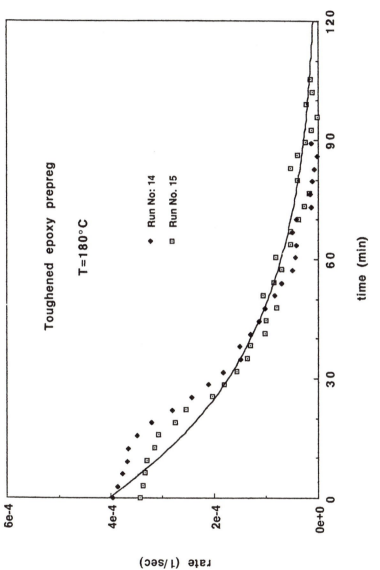

Figure 3. *Isothermal thermogram obtained on the toughened epoxy prepreg at 180 °C. Comparison between experimental DSC data (run nos. 14 and 15) and model predictions (solid line).*

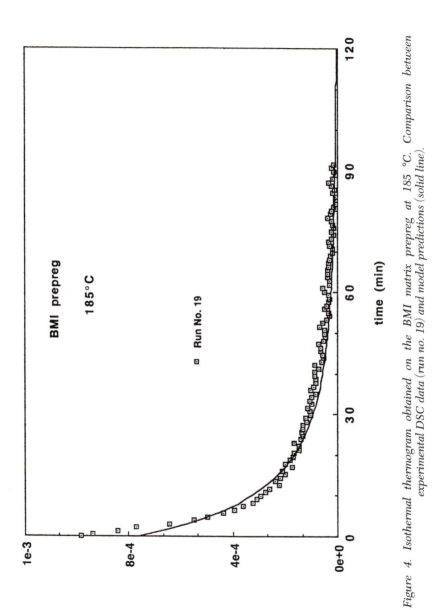

*Figure 4. Isothermal thermogram obtained on the BMI matrix prepreg at 185 °C. Comparison between experimental DSC data (run no. 19) and model predictions (solid line).*

**Table II. Heat of Reaction and Maximum Degree of Reaction Developed
in Isothermal Tests**

| Run No. | T (°C) | Standard Epoxy $H_c$ (J/g) | $w_m$ | Toughened Epoxy $H_c$ (J/g) | $w_m$ | BMI $H_c$ (J/g) | $w_m$ |
|---------|--------|------------|-------|------------|-------|------------|-------|
| 5  | 150 | 312.2 | 0.68 |       |      |       |      |
| 6  | 150 | 329.1 | 0.72 |       |      |       |      |
| 7  | 158 | 350.5 | 0.77 |       |      |       |      |
| 8  | 158 | 368.3 | 0.81 |       |      |       |      |
| 9  | 165 | 377.9 | 0.83 |       |      |       |      |
| 10 | 165 | 377.0 | 0.83 |       |      |       |      |
| 11 | 170 | —     | —    | —     | —    | 211.6 | 0.51 |
| 12 | 175 | 406.6 | 0.89 | —     | —    | 223.2 | 0.54 |
| 13 | 175 | 401.2 | 0.88 |       |      |       |      |
| 14 | 180 | —     | —    | 397.5 | 0.84 | 228.5 | 0.55 |
| 15 | 180 | —     | —    | 380.4 | 0.81 |       |      |
| 16 | 182 | 407.1 | 0.89 |       |      |       |      |
| 17 | 182 | 443.2 | 0.97 |       |      |       |      |
| 18 | 185 | —     | —    | 416.2 | 0.88 | 314.1 | 0.76 |
| 19 | 185 | —     | —    | 406.3 | 0.86 |       |      |
| 20 | 190 | 451.4 | 0.99 | 434.6 | 0.92 | 299.6 | 0.73 |
| 21 | 190 | 456.0 | 1.00 | 412.9 | 0.88 |       |      |
| 22 | 195 | —     | —    | 443.1 | 0.94 | 362.1 | 0.88 |
| 23 | 195 | —     | —    | 445.9 | 0.95 |       |      |
| 24 | 200 | 460.4 | 1.00 | 455.4 | 0.97 | 402.1 | 0.97 |
| 25 | 200 | —     | —    | 446.4 | 0.95 |       |      |

diffusion-controlled effects that are a consequence of the loss of mobility of the reacting molecules in the developed network (8). The structural changes produced by the polymerization reactions are associated with an increase of the glass-transition temperature $(T_g)$ of the reactive polymer. When the increasing $T_g$ approaches the isothermal cure temperature, the molecular mobility is reduced strongly and the reaction becomes diffusion-controlled and eventually stops. In consideration of this assumption, the original kinetic model given by equation 4 has been modified as shown in equation 6 to interpret isothermal results. The model in equation 6 predicts that the reaction rate becomes zero when the degree of reaction becomes equal to $w_m$.

Taking the average value of the total heat developed during dynamic tests as a reference allows the determination of the final degree of reaction achieved during isothermal tests ($w_m = H_c/H_T$). Values of $w_m$ computed from all isothermal tests and for the three systems studied are also shown in Table II. For modeling purposes it is convenient to determine the behavior of $w_m$ as a function of the isothermal test temperature. This behavior is reported in Figure 5 where a linear dependence between $w_m$ and $T$ is observed in the

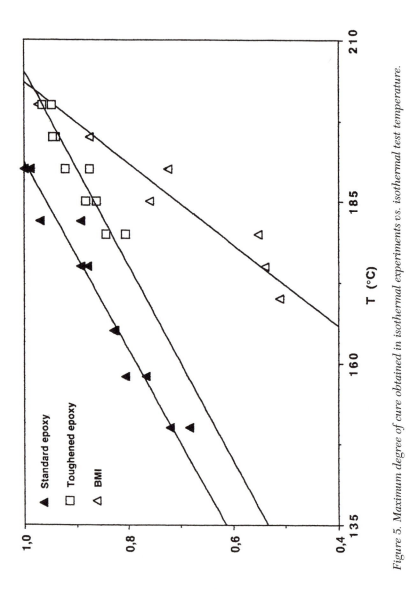

*Figure 5. Maximum degree of cure obtained in isothermal experiments vs. isothermal test temperature.*

three cases. The toughened epoxy matrix data are represented by a straight line parallel to the line representing the standard TGDDM–DDS system, but shifted to lower final conversions. This behavior confirms the observed lower reactivity of the toughened matrix in dynamic tests, which can be attributed to the effect of the second thermoplastic phase. The BMI system is characterized by a higher dependence of $w_m$ with $T$, which suggests a more critical definition of processing conditions.

The correlation between $w_m$ and $T$ is similar to the dependence between the glass-transition temperature and the degree of reaction for a reactive polymer. As previously discussed, it can be assumed that the $T_g$ value reached by the polymeric matrix during the isothermal test is similar to the test temperature value. The dependence of the $T_g$ with degree of reaction has been correlated by the Di Benedetto equation (*12*). Considering that the diffusion control phenomenon is certainly governed by vitrification but cannot exactly correspond to the Di Benedetto analysis, a simple linear dependence is used to express the empirical dependence of $w_m$ with $T$:

$$w_m = p\,T + q \tag{7}$$

where $p$ and $q$ are constants determined from the linear behavior shown in Figure 5 and reported in Table III.

**Model Development.**   To determine the value of the kinetic parameters, a regression analysis method was applied. Preliminary graphically computed values are used as initial values for a computer regression program to find the best values of constants $n$ and $K$ (equation 6). The values of the kinetic constant $K$, computed as a function of $1/T$ in semilogarithmic plot, are shown in Figure 6. The data are well fitted by a straight line, which indicates an activated behavior represented by an Arrhenius type equation:

$$K = K_0 \exp\left(-E_a/RT\right) \tag{8}$$

**Table III. Model Parameters Obtained from the Thermal Characterization of the Systems Studied**

|  | $H_T$ (kJ/g) | $E_a$ (kJ/mol) | $\ln K_0$ (1/s) | $n$ | $p$ | $q \times 10^3$ (1/K) |
|---|---|---|---|---|---|---|
| Standard Epoxy | 456.4 | 62.4 | 10.4 | 1.07 | −2.20 | 6.90 |
| Toughened Epoxy | 471.4 | 69.5 | 10.8 | 0.94 | −2.14 | 6.65 |
| BMI | 413.3 | 48.3 | 6.1 | 1.40 | −6.62 | 16.0 |

where $K_0$ is a pre-exponential constant (frequency factor), $E_a$ is the activation energy, $R$ is the universal gas constant, and $T$ is the absolute temperature. The values of the parameters of the general model represented by equations 6–8 are listed in Table III for the three systems studied.

**Model Verification.** The validity of the proposed model has been verified by comparison between theoretical predictions and experimental data. The results of the model simulation of the isothermal tests performed at the normal processing temperatures on the three prepregs are shown in Figures 2–4. The points correspond to experimental results and the lines are the predictions of the developed kinetic model. Good agreement was obtained in the three cases, which confirms the model formulation procedure .

The form and the parameters of the proposed model (equations 6–8) have been obtained by computing isothermal test data, but the complete kinetic model should also describe the dependence of the reaction rate on the temperature during a dynamic test. Therefore, the complete model was also verified by comparison with experimental dynamic thermograms. Both results compared well in Figures 7–9 where reaction rate measured with different heating rates is shown as a function of time for the three systems studied. Dotted lines correspond to DSC experimental results and full lines correspond to model predictions computed using Table III parameters. The results are satisfactory for all the heating rates analyzed, which additionally confirms the ability of the developed model to reproduce the kinetic behavior of the studied systems.

## Summary

The kinetic behavior of high-performance matrices for carbon fiber composites has been characterized by differential scanning calorimetry. A new approach including diffusion control effects has been adopted for the formulation of the kinetic model represented by a modified nth-order equation. The model reproduces isothermal and dynamic test results. The incomplete reactions in isothermal tests and the heating rate dependence of dynamic test thermograms are well described by the developed model. Some difference in reactivity was observed, but the calorimetric and kinetic behavior of standard epoxy, toughened epoxy, and bismaleimide prepregs is essentially similar and no significant modifications of the respective processing conditions are expected.

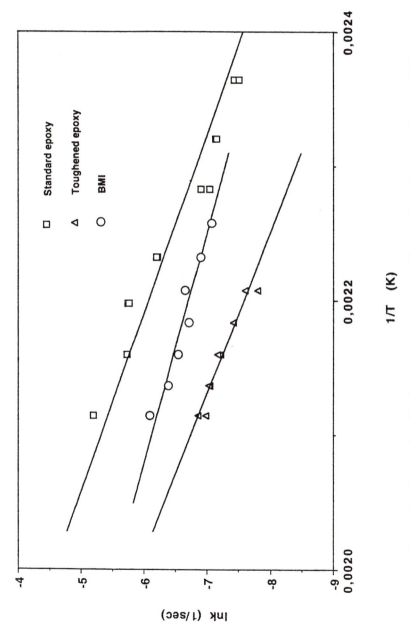

*Figure 6. Arrhenius plot of the kinetic constants computed from isothermal tests as a function of 1/T.*

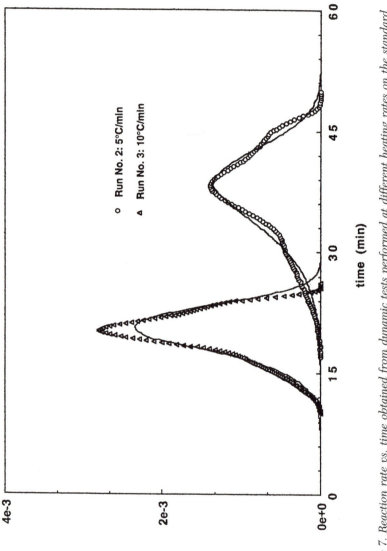

Figure 7. Reaction rate vs. time obtained from dynamic tests performed at different heating rates on the standard epoxy matrix prepreg. Comparison between experimental DSC data (run nos. 2 and 3) and model predictions (solid lines).

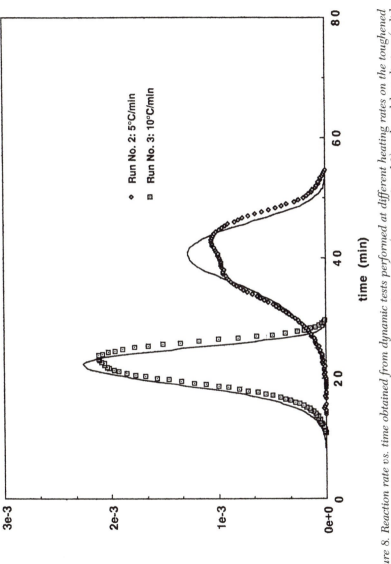

*Figure 8. Reaction rate vs. time obtained from dynamic tests performed at different heating rates on the toughened epoxy matrix prepreg. Comparison between experimental DSC data (run nos. 2 and 3) and model predictions (solid lines).*

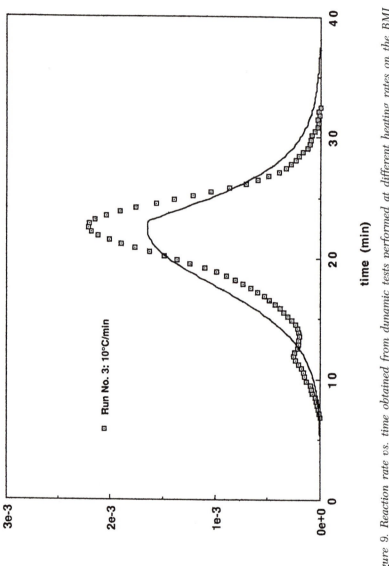

*Figure 9. Reaction rate vs. time obtained from dynamic tests performed at different heating rates on the BMI matrix prepreg. Comparison between experimental DSC data (run no. 3) and model predictions (solid lines).*

## Acknowledgments

The financial support from Aeritalia Saipa for the research concerning this work and from Istituto Mobiliare Italiano for the fellowship to Trivisano are gratefully acknowledged.

## References

1. Kenny, J. M.; Apicella, A.; Nicolais, L. *Polym. Eng. Sci.* **1989**, 29, 972.
2. Kenny, J. M.; Maffezzoli, A.; Nicolais, L. *Compos. Sci. Tech.* **1990**, 38, 339.
3. Prime, R. B. In *Thermal Characterization of Polymeric Materials*; Turi, E. A., Ed.; Academic Press: New York, 1981; Chap. 5.
4. Barton, J. M. In *Epoxy Resins and Composites I*; Dusek, K., Ed.; Advances in Polymer Science 72; Springer-Verlag: Berlin, 1985.
5. Trivisano, A.; Kenny, J. M. *Polym. Eng. Sci.* **1992**, in press.
6. Stark, E. B.; Seferis, J.; Apicella, A.; Nicolais, L. *Thermochim. Acta* **1983**, 77, 19.
7. Mijovic, J.; Kim, J.; Slaby, J. *J. Appl. Polym. Sci.* **1984**, 29, 1449.
8. Enns, J. B.; Gillham, J. K. *J. Appl. Polym. Sci.* **1983**, 28, 2567.
9. Morgan, R. J.; Mones, E. T. *J. Appl. Polym. Sci.* **1987**, 33, 999.
10. Gupta, A.; Cizmecioglu, M.; Coulter, D.; Liang, R. H.; Yavrouian, A.; Tsay, F. D.; Moacanin, J. *J. Appl. Polym. Sci.* **1983**, 28, 1011.
11. Apicella, A.; Nicolais, L.; Iannone, M.; Passerini, P. *J. Appl. Polym. Sci.* **1984**, 29, 2083.
12. Di Benedetto, A. T. *J. Appl. Polym. Sci.* **1987**, 25, 1949.

RECEIVED for review March 6, 1991. ACCEPTED revised manuscript August 1, 1992.

# INDEXES

# Author Index

# Affiliation Index

561

# Subject Index

## A

*Copy editing: Cheryl L. Kranz*
*Indexing: Deborah H. Steiner*
*Production: Margaret J. Brown*
*Acquisition: A. Maureen Rouhi*
*Cover design: Rebecca Lepkowski*

*Typeset by Technical Typesetting Inc., Baltimore, MD*
*Printed and bound by Maple Press, York, PA*

# Highlights from ACS Books

# Bestsellers from ACS Books

*The ACS Style Guide: A Manual for Authors and Editors*
Edited by Janet S. Dodd
264 pp; clothbound ISBN 0–8412–0917–0; paperback ISBN 0–8412–0943–X

*The Basics of Technical Communicating*
By B. Edward Cain
ACS Professional Reference Book; 198 pp;
clothbound ISBN 0–8412–1451–4; paperback ISBN 0–8412–1452–2

*Chemical Activities* (student and teacher editions)
By Christie L. Borgford and Lee R. Summerlin
330 pp; spiralbound ISBN 0–8412–1417–4; teacher ed. ISBN 0–8412–1416–6

*Chemical Demonstrations: A Sourcebook for Teachers,*
*Volumes 1 and 2,* Second Edition
Volume 1 by Lee R. Summerlin and James L. Ealy, Jr.;
Vol. 1, 198 pp; spiralbound ISBN 0–8412–1481–6;
Volume 2 by Lee R. Summerlin, Christie L. Borgford, and Julie B. Ealy
Vol. 2, 234 pp; spiralbound ISBN 0–8412–1535–9

*Chemistry and Crime: From Sherlock Holmes to Today's Courtroom*
Edited by Samuel M. Gerber
135 pp; clothbound ISBN 0–8412–0784–4; paperback ISBN 0–8412–0785–2

*Writing the Laboratory Notebook*
By Howard M. Kanare
145 pp; clothbound ISBN 0–8412–0906–5; paperback ISBN 0–8412–0933–2

*Developing a Chemical Hygiene Plan*
By Jay A. Young, Warren K. Kingsley, and George H. Wahl, Jr.
paperback ISBN 0–8412–1876–5

*Introduction to Microwave Sample Preparation: Theory and Practice*
Edited by H. M. Kingston and Lois B. Jassie
263 pp; clothbound ISBN 0–8412–1450–6

*Principles of Environmental Sampling*
Edited by Lawrence H. Keith
ACS Professional Reference Book; 458 pp;
clothbound ISBN 0–8412–1173–6; paperback ISBN 0–8412–1437–9

*Biotechnology and Materials Science: Chemistry for the Future*
Edited by Mary L. Good (Jacqueline K. Barton, Associate Editor)
135 pp; clothbound ISBN 0–8412–1472–7; paperback ISBN 0–8412–1473–5

---

For further information and a free catalog of ACS books, contact:
American Chemical Society
Distribution Office, Department 225
1155 16th Street, NW, Washington, DC 20036
Telephone 800–227–5558